ADVANCED EMBRYOLOGY

First Published - 2004

ISBN 978-81-7141-774-2

DISCOVERY PUBLISHING HOUSE PVT. LTD.
4383/4B, Ansari Road, Darya Ganj
New Delhi-110 002 (India)
Phone: +91-11-23279245, 23253475, 43596065

web: www.discoverypublishinggroup.com

Contents

1

Hormones in Reproduction

The effect of a hormone is a property of both the hormone and its target tissue(s). The relative similarity of a hormone among different classes of organism does not necessarily imply a commonality of responses. The broadly occurring pituitary hormone prolactin, for example, has a wide variety of responses, including mammogenesis and lactogenesis in mammals, broad patch formation in birds, dermal pigmentation in ectotherms, and water balance and parental care in fish. Many hormones not only stimulate or suppress the activity of the target organ but additionally have some rather general effects. There is a general increase in motor activity with arise in gonadal steroids, and in laboratory rats, for example, exploratory activity increases during estrus. High levels of androgens seem to account for behavioural activity that is sexual in only an ancillary way. In both birds and mammals an increase in persistence toward a goal is positively correlated with serum androgen. In chicks, injections of testosterone elevate the persistence with which they seek a preferred food. Vasopressin has the general effect of elevating blood pressure. In target tissues are cells that contain surface receptors, or binding sites.

Presumably the sensitivity of given tissue is proportional to the abundance of binding sites. Movement of a steroid through the lipophilic cell membrane poses no problem, and concentrations of steroids rapidly accumulate within cells of target tissues. Protein hormones, such as a gonadotropin, contact a receptor on the outer surface of a cell membrane. This may be followed by an adenylyl system, which in turn induces formation of cAMP. Cyclic AMP appears to mediate activities such as ovarian steroid synthesis.

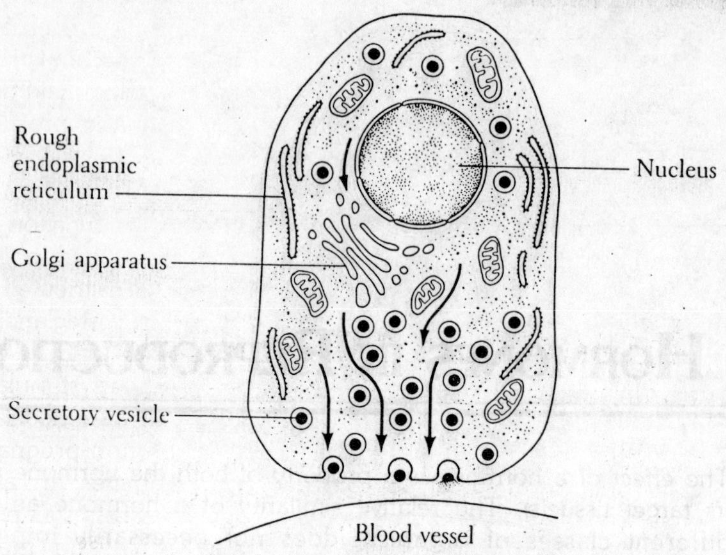

Fig. 1.1. An idealized endocrine cell producing peptide hormones. The pathway from production to packaging into vesicles and release by exocytosis is shown by the arrows.

Internalization of the protein-hormone receptor complex occurs for many hormones, presumably for intracellular metabolic inactivation. It has been suggested that for some protein hormones, internalization may lead to further stimulation of the target cell by mechanisms that remain to be defined but may involve binding to intracellular organelles and/or the nucleus. Within the cell the steroid may form a steroid-receptor complex within the cytoplasm or the nucleus.

Following attachment to an acceptor site on chromatin, the complex may initiate a specific target organ response in addition to accelerating RNA synthesis and cellular metabolism. Also, the presence of estrogen within the cells of the uterus, for example, is followed by an increase in both estrogen and progesterone receptors, which are maximal prior to ovulation. After ovulation and the postovulatory increase in progesterone, receptors of both types decrease. Steroid hormones closely resemble one another and are produced primarily by the gonads and by the adrenal cortex. Some leave their source as prehormones to be transformed to the active state elsewhere, usually at the target organ. Prostaglandins comprise a large number of hormone like lipids that are produced in many tissues. Because they are quickly metabolized in the lungs, they are short lived. They are usually synthesized from prostanoic acid near their target site, and their effects tend to be local. They were first found in seminal fluid and were erroneously believed to emanate solely from the prostate gland, and hence they were called prostaglandins.

Prostaglandins have a broad spectrum of functions, and some are important in reproduction. Some prostaglandins function in parturition and also in the production of proteins within the placenta. More than 10 different prostaglandins are distributed in human semen, but their significance is not known. Prostaglandins also occur in the maternal segment of the placenta and in the uterus, oviduct, and mature follicle. Generally they stimulate smooth muscle contractility and blood flow. They may also modulate steroidogenesis and shorten luteal life span in non-pregnant mammals. A special group of hormones, catecholamines, are also neurohormones, which transmit nervous impulses across synapses or between a nerve and an effector organ. The main catecholamines are epinephrine, norepinephrine, and dopamine. They influence reproductive structures either through the nervous system or by altering the metabolic rate of certain tissues and/or blood flow.

Smooth muscle of tubular parts of the reproductive system of both sexes is innervated by the sympathetic nervous system. There are synchronous fluctuations of norepinephrine and gonadal steroids in the human oviduct. And their local occurrence (histochemically determined) parallels that of adrenergic nerves. The role(s) of catecholamines in normal reproduction is unclear and remains controversial.

The Hypothalamus and its Hormones

The gonads do not operate in isolation from conditions in the rest of the body but are controlled, in part at least, by hormonal messages from both the hypothalamus and the anterior pituitary. Gonadotropins from the anterior pituitary may stimulate metabolism, growth, and vascularization of the gonads in addition to stimulating steroidogenesis. In turn, the output of gonadotropins is affected by varying levels of steroids released by the gonads. In addition, hormones from the hypothalamus not only stimulate the synthesis and release of gonadotropins but may also inhibit secretions from the anterior pituitary. In mammals, and to a lesser

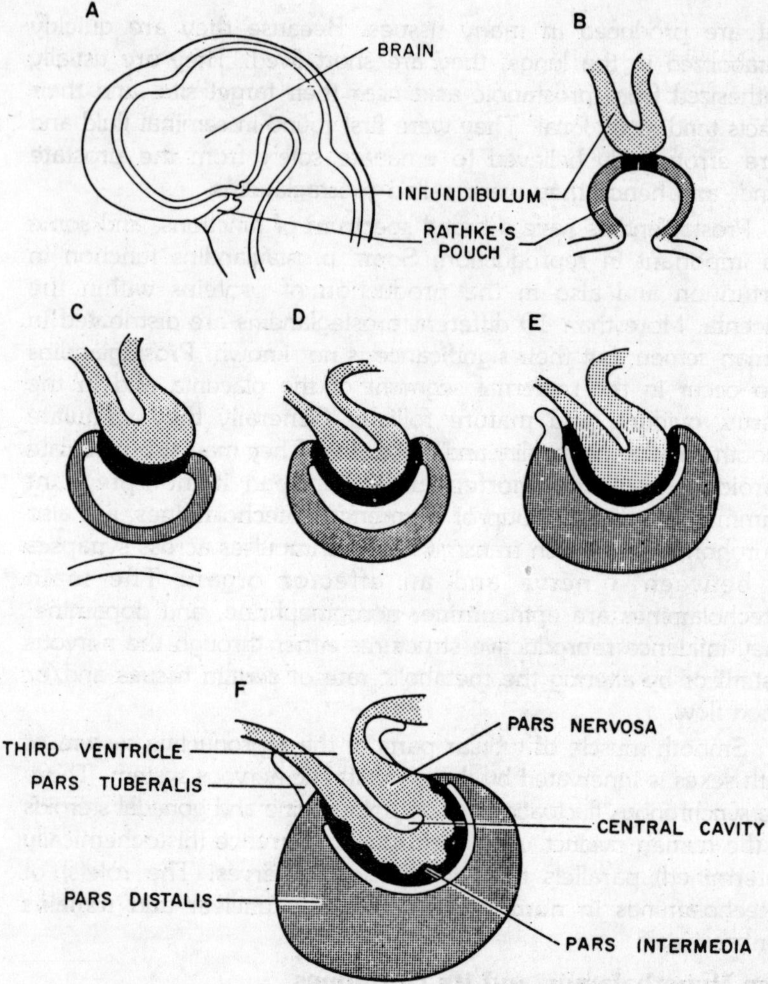

Fig. 1.2. Diagrams showing progressive stages in the embryonic development of the pituitary gland.

degree in some other vertebrates, the hypothalamic-pituitary-gonadal axis is influenced by the pineal gland. This chapter outlines the functions and interrelationships of the hormones controlling gonadal activity.

Structure of the Hypothalamus

The hypothalamus is a loosely defined area in the ventral region of the diencephalon. A posteroventral projection is usually

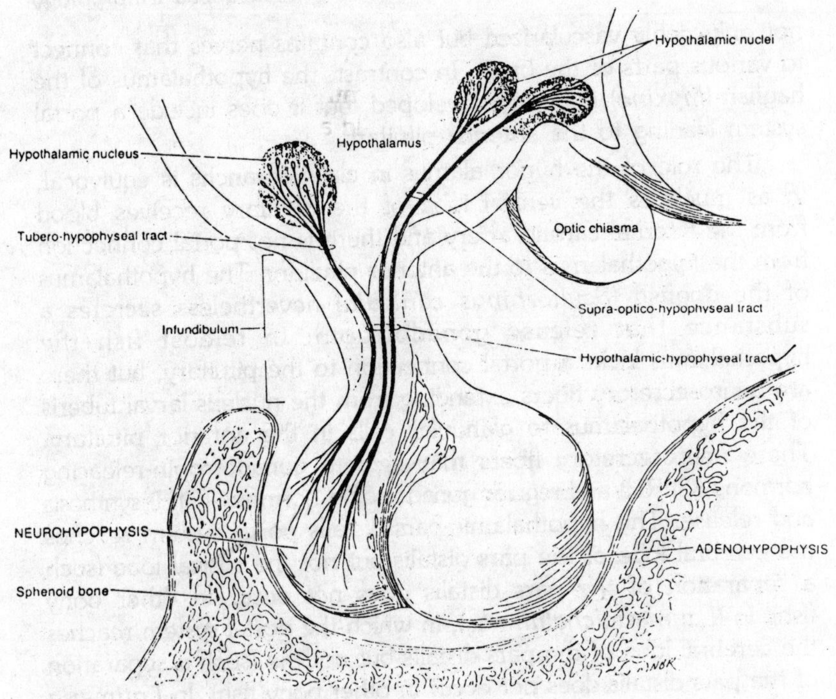

Fig. 1.3. Hypothalamic-hypophyseal tract.

referred to as the posterior pituitary. Two hypothalamic regions, the supraoptic and the paraventricular nuclei contain large secretory neurons that store their products (oxytocin and vasopressin in axon terminals in the adjacent posterior pituitary, from which they are released into the vascular system. The hypothalamus of tetrapod vertebrates has a portal system, a plexus of capillaries that carries hormones to the anterior pituitary.

The median eminence of the avian hypothalamus contains functionally distinct anterior and posterior parts, which are provided with separate portal connections to rostral and caudal lobes of the anterior pituitary. Within the hypothalamus are numerous centers that affect such diverse activities as water balance thermoregulation, appetite, and reproductive activity. These various discrete functions are mediated by different areas and some paired ganglia or nuclei Their connections provide a pathway for responses to both exogenous and endogenous stimuli, including those factors that modify or control reproduction. The hypothalamus of mammals not only richly vascularized but also contains nerves that connect to various parts of the brain. In contrast, the hypothalamus of the hagfish (*Myxine*) is poorly developed, but it does include a portal system leading to the anterior pituitary.

The role of the hypothalamus in elasmobranchs is equivocal, in as much as the ventral lobe of the pituitary receives blood from the internal carotid artery and there is not portal connection from the hypothalamus to the anterior pituitary. The hypothalamus of the dogfish (*Scyliorhinus canicula*) nevertheless secretes a substance that release gonadotropin. In teleost fish the hypothalamus lacks a portal connection to the pituitary, but there are neurosecretory fibers extending from the nucleus larval tuberis of the hypothalamus to glandular cells in the anterior pituitary. These neurosecretory fibers may secrete gonadotropin-releasing hormone (GnRH) and regular gonadotropin hormone (GtH) synthesis and release. The hypothalamic-pars distalis portal system reaches the cerebral lobe of the pars distalis but not the rostral lobe (such a separation of the pars distalis does not occur in other bony fish). In (*Latimeria chalumnae*), in which the portal system reaches the cerebral lobe of the pars distalis but not the (such a separation of the pars distalis does not occur in other bony fish). In *Latimeria* the rostral lobe receives a branch of the internal carotid, resembling the condition found in elasmobranchs.

Hormones of the Hypothalamus

A major role of the hypothalamus is the synthesis of GnRH, which induces gonadotropin production and release by the anterior pituitary. In bony fish, however, gonadotropin release is stimulated by a different hypothalamic secretion. GnRH causes a tonic and low-level release of both LH and follicle-stimulating hormone (FSH). Fluctuating sensitivity of the anterior pituitary and pulsatile GnRH release may account for temporal variations in amounts of LH and FSH release.

Preovulatory increased secretion of estrogen from the ovary is followed by a substantial inhibitory influences on the gonads. GnRH, or a GnRH –like peptide made in the ovary, also affects the ovary directly, depressing the activity of both granulosa and theca cells. It blocks estradiol secretion by granulosa cells and also androgen synthesis by theca cells. In cultured granulosa cells, GnRH suppresses production of cAMP, which is necessary for steroidogenesis. In teleost fish the anterior preoptic area of the brain secretes a gonadotropin-release –inhibiting factor (GRF), which may suppress the release of gonadotropins (Stacey, 1984). Studies on the goldfish and the rainbow trout indicate dopamine as an inhibitor of gonadotropin release from the teleost pituitary. Norepinephrine (or noradrenaline) effects the release of LH and FSH and can thus stimulate ovulation. Norepinephrine in the anterior hypothalamus fluctuates during the estrous cycle, in synchrony with the phasic secretions of gonadotropins. Serotonin, on the other hand, blocks the discharge of GnRH, inhibiting the release of LH. Dopamine, the third catecholamine from this region, blocks the release of prolactin and sometimes LH. A GnRH-associated peptide from the hypothalamus is an inhibitor of prolactin secretion and also a stimulator of the release of gonadotropins in rat pituitary tissue in vivo.

The Anterior Pituitary

The pituitary gland, or hypophysis is a composite organ consisting of an adenohypophysis (formed from Rathke's pouch, a dorsal outpocketing from tissue that also gives rise to the roof of the mouth) and a neurohypophysis (derived from the infundibular process of the diencephalon of the brain floor). The neurohypophysis includes a neural lobe, which joints to the caudalmost part of the adenohypophysis; the neural lobe constitutes the posterior lobe (or pars nervosa) of the pituitary gland. In most vertebrates the apex of the adenohypophysis becomes differentiated as the anterior lobe (or pars distalis). The remainder of the adenohypophysis, which joins the posterior lobe, becomes the intermediate lobe (or pars intermedia). In birds there are rostral and caudal lobes of the adenohypophysis, and both regions secrete gonadotropins.

As pointed out previously, the separate lobes of the avian adenohypophysis are jointed to the hypothalamus by separate portal systems, and this separation may have functional significance. The intermediate lobe is absent in birds and some mammals (e.g., the beaver, armadillo, cetaceans, and probably others). In species lacking the intermediate lobe, a thin layer of nonglandular tissue separates the posterior lobe from the anterior lobe. The adenohypophysis of the hagfish is not discretely divided and is distinct from the neurohypophysis.

The pituitary has a disputed role in gonadal activity in these cyclostomes, but most students find some reduction of gonadal size after hypophysectomy. The pituitary of elasmobranchs consists of a united pars intermedia and pars nervosa and together they constitute the *neurointermediate lobe*, whereas the pars distalis is divided into three lobes (rostral, median and ventral). The ventral lobe of elasmobranchs appears to be the most important in reproduction. It develops a narrow stalk and produces a glycoprotein gonadotropin. Removal of the ventral lobe is followed by gonadal regression.

Hormones of the Anterior Pituitary

Seven protein hormones are synthesized in the anterior pituitary of both sexes. Those of mammals are the best known, and most nonmammals have similar, though perhaps not identical, hormones. Most directly concerned with reproduction are the two gonadotropins –FSH and LH. Prolactin, in addition to having a broad spectrum of functions, affects several important aspects of reproduction, which vary with the taxa concerned. Two other hormones, thyroid-stimulating hormone (TSH) and growth hormone occur widely among the vertebrate classes. The anterior pituitary synthesizes a peptide hormone, adrenocorticotropic hormone (ACTH), which increases the production of steroid hormones from the adrenal cortex. This portion of the pituitary also produces β-lipotropin, from which β-endorphin is derived. Except for prolactin, these hormones are released on specific hormonal signals (releasing hormones) from the hypothalamus. The release of prolactin is suppressed by dopamine from the hypothalamus.

Gonadotropins

Because the roles of gonadotropins have been most carefully studied in mammals. Their functions in these vertebrates will be treated first. Two gonadotropins, FSH and LH control gonadal activity. They are produced by both sexes and exert distinctive effects on the gonads of each. In the female, FSH stimulates the growth and development of the follicle in the early preovulatory phase and initiates steroidogenesis. LH increases vascularization and continues the follicular growth and the synthesis of ovarian steroid hormones. The increased ovarian blood flow facilitates the future passage of the ovarian hormones, estrogen and progesterone, into the general circulation. LH also promotes the conversion of cholesterol into progesterone and estrogen. Macropod marsupials posses both LH and FSH, and their anterior pituitary is regulated by a negative feedback from steroid hormones similar to the negative feedback of eutherians. Follicular growth and ovulation in the tammar wallaby (*Macropus eugenii*) are dependent upon pituitary secretions, and both processes cease after hypophysectomy. Not all gonadotropins come from the anterior pituitary.

The placenta of some mammals produce a glycoprotein called chorionic gonadotropin, which may be similar in its role to both LH and FSH. One of these placental gonadotropin, pregnant mare serum gonadotropin (PMSG), which is often designated as equine chorionic gonadotropin, is commonly used in experimental studies. Human chorionic gonadotropin (hCG), a glycoprotein, has primarily LH activity: it is unable to induce ovulation but does maintain the corpus luteum. The pituitary exerts only a minor influence on gametogenesis in lampreys, and hypophysectomy merely slows the rate of gonadal metabolism.

Moreover, the hypothalamus of lampreys apparently does not control the slight gonadotropic influence of their pituitary. Following the removal of the pituitary in *Lampetra fluviatilis*, however, there may be some follicular growth but no ovulation. Apparently there is only incipient pituitary control of reproductive activity in these primitive vertebrates. The four distinct and separate lobes of the pituitary of elasmobranchs all produce gonadotropins. Gonadotropins from the ventral lobe of the dogfish (*Squalus acanthias*) seem to control steroidogenesis, and maximal secretion of gonadotropins is recorded during the breeding season. It is not known if the ventral lobe of the dogfish produces either FSH, LH, or perhaps a different gonadotropin, but the dogfish (*Scyliorhinus canicula*) produces a gonadotropin (GtH) that has properties of LH.

Teleostean fish are presumed to have two gonadotropins, together designated as GtH, but initially there was thought to be only one hormone. The gonadotropins have distinct effects; one is vitellogenic and the other maturational. There are two types of secretory cells in the anterior pituitary of fish, and the current consensus is that GtH exists in two forms. In fish hypophysectomy is followed by gonadal regression. The release of GtH in teleosts may result not only from the stimulus of GnRH but also from the withdrawal of GRIF. In amphibians, reptiles, and birds two gonadotropins occur, but they may not be identical to mammalian gonadotropins.

In birds LH stimulates the production of estrogen, progesterone, and testosterone, and FSH promotes the growth of the follicle. Prolactin, acting synergistically with FSH and LH, may also stimulate follicular growth. As the follicle increase in size, estrogen synthesis and release increase, and the largest follicles, those nearing ovulation, begin to release more progesterone and less estrogen. Testosterone and progesterone together stimulate. Activity of the avian oviduct. Among the different classes of vertebrates, the testis shows some variation in its responses to LH and FSH.

In amphibians and mammals LH is of paramount importance in stimulating the production of androgens, although FSH and prolactin have a synergistic stimulatory effect. In vitro preovulatory and postovulatory tissue of the painted turtle (*Chrysemys picta*) is steroidogenic when exposed to either mammalian or avian LH. Estrogen predominates from preovulatory tissue and progesterone from postovulatory incubates. The reptilian testis is far more responsive to mammalian FSH, as measured by androgen output, and then it is to mammalian LH, when each gonadotropin is administered separately. Although the potency of gonadotropins in reptiles (squamates, crocodilians, and chelonians) varies with the source of the hormone, reptilian and avian pituitaries stimulate both steroidogenesis and spermatogenesis in the lizard *Anolis*.

Regarding the specificity of testicular response to gonadotropins, reptiles differ more from amphibians, birds, and mammals than these three classes differ among each other. Chronic exposure to ovine FSH stimulates testicular growth and spermatogenesis in the anole, and similar, but less pronounced effects are produced by LH. In vivo FSH tends to have a longer half-life than does LH in lizards, perhaps according for the difference in the observed effects in these reptiles. Gonadotropins from all ectotherms (fish, amphibians and reptiles) resemble, but are not identical to, mammalian LH and FSH, and gonadal response varies from one species to another. In all these vertebrates gonadal response to gonadotropins is greatly influenced by temperature. Two gonadotropins from the amphibian anterior pituitary are comparable of FSH and LH of both reptiles and mammals. Experimental administration mammalian LH and FSH to hypophysectomized frogs indicates that steroidogenesis and spermiogenesis are stimulated by LH and that spermatogenesis is effected by FSH. This pattern seems to hold for most amphibians.

Prolactin

Prolactin and growth hormone are secreted predominately from different parts of the anterior pituitary. Prolactin occurs in all jawed vertebrates, and the prolactin-producing cell has been found in all vertebrates examined except agnaths. Prolactin, growth hormone, and placental lactogen, have structural similarities, suggesting a common evolutionary history. Human prolactin and growth hormone have some 40% of their amino acid sequences in common. Prolactin has numerous known and postulated functions; more than are known for any other hormone, but it similarity to growth hormone may account, in part at least, for the versatility of prolactin. Binding sites for prolactin are known for several diverse tissues.

Prolactin seems to have a variety of functions in teleost fish; it is well-known to enhance osmoregulatory ability and promote lipogenesis. In some groups (Cichlidae) it promotes parental care. In salamanders (Salamandridae) prolactin is reported to promote the movement to water prior to spawning. In birds prolactin not only promotes parental care (e.g., broodiness) but in pigeons and doves (Columbidae) it also stimulates cells lining the crop to synthesize and release "pigeon milk". In mammals prolactin is important not only in the growth of mammary tissue but also in the production of milk. It acts synergistically with both estrogen and progesterone from the ovary as well as with ACTH and growth hormone from the anterior pituitary (see Lactatation).

In lactating rabbits, for example, lactogenesis is interrupted by hypophysectomy but is restored by injections of either sheep prolactin of human growth hormone. Prolactin is under inhibitory control, and its release is prevented by dopamine, the primary prolactin-inhibiting factor, produced in the hypothalamus. Prolactin is released only when dopamine is prevented from entering the portal system to the anterior pituitary. The destruction of his portal system is followed by a rise in secretion of prolactin.

At least some natural release of prolactin may result from the action of a prolactin-releasing factor, which suppresses the action of dopamine. Estrogen is known to inhibit the action of dopamine on prolactin release and thus stimulate the secretion of prolactin. Apart from endogenous influences on prolactin secretion, levels of this hormone may affected by the photocycle. In some examples, these changes are clearly correlated with gonadal changes, which also follow day length. A number of species of teleost fish have circadian rhythms of circulating prolactin. Long days are known to stimulate the release of prolactin in several kinds of birds, for example, the migratory, or Japanese, quail (Coturnix coturnix), and peaks of serum prolactin coincide with gonadal activity. Cattle also have a phase-sensitive period, and prolactin release in increased by both an increase in day length (from 8L: 16D to 16L: 8D) and exposure to skeletal day lengths 6L: 8D: 21:8D (TUCKER, 1981). Prolactin levels also exhibit marked seasonal changes in domestic cattle under a natural photocycle. Prolactin release in sheep and also in the golden hamster are similarly under control of the photocycle.

Prolactin and Gonadal Activity

Prolactin has varying roles in gonadal activity. In the domestic rabbit and in some other domestic and laboratory mammals, prolactin promotes the maintenance of the corpus luteum. Prolactin may inhibit ovulation in the postpartum human female, either via the hypothalamus or possibly by direct effect on the ovary. It is well-established that prolactin, when elevated by suckling young, suppresses ovarian activity and that this suppression reflects the frequency and duration of suckling.

Prolactin may also be released as a reflex from nervous stimulation of coitus, with the accumulation of prolactin in follicular fluid, where it enhances the secretion of progesterone from granulosa cells. It is also released after coition in the male, but with no known significance, although it does induce LH receptors

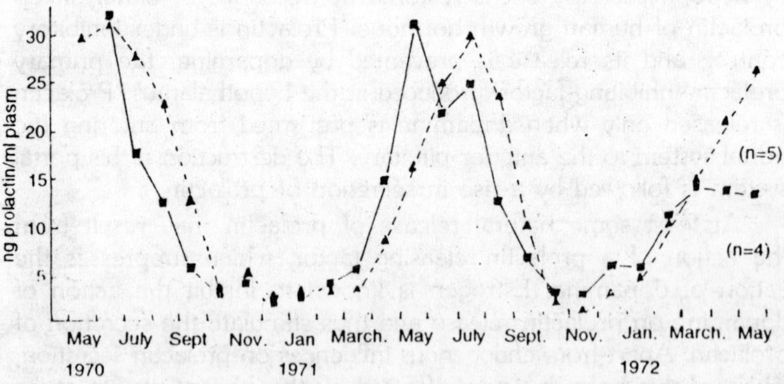

Fig. 1.4. Seasonal variations is basal prolactin levels in nonlactating male and female cattle, from birth to puberty.

in a number of mammals. Experimentally, prolactin has an antigonadal effect in many birds and reptiles, apparently depressing LH and FSH. But whether prolactin is stimulatory or suppressive depends upon the time of day it is administered. There is a circadian rhythm in the sensitivity to exogenous prolactin, in addition to its natural rhythmical release. Prolactin is progonadal in some rodents and some birds.

The Posterior Pituitary

As previously pointed out, the hypothalamus contains large secretory neurons that project into the posterior pituitary, where they release their hormones. The hormone oxytocin plays a major role in uterine contractions and milk letdown of milk ejection, as well as in maternal behaviour (at least in some mammals). The release of hormones from the posterior pituitary is under neural control.

Gonadal Hormones

The steroidogenic cells of the vertebrate ovary are rather similar among the several classes of vertebrates, which is also true for the testis. Gonadal organization may differ from one class to another or within a class, but the cells responsible for steroidal synthesis remain the same. Most gonadal hormones are steroids, ultimately derived from cholesterol, and they are chemically similar.

Progesterone, evolved from cholesterol, can be converted to androgens and estrogens and androgens in both sexes can be converted estrogens. In some mammals the placenta synthesizes progesterone, but in others the bulk of progesterone is produced by the corpus luteum. Progesterone, androgens, and estrogens are also synthesized in the adrenal cortex. These hormones not only have clear-cut physiological effects but are also partly responsible for behavioural patterns in both sexes. In young (sexually immature) vertebrates the adrenal cortex is presumably the major source of sexual steroids.

Estrogens

In most vertebrates estrogen is formed and secreted by granulosa cells of the mature follicle, but it may be formed, in lesser amounts, by the corpus luteum. It has been suggested that estrogen synthesis results from the co-operation of theca interna cells and granulosa cells. The placenta becomes a major source of estrogen in some mammals, and the estrogen appears in bound form in the urine of pregnant mammals of many taxa. Estrogens may also be synthesized by the blastocyst and, in small amounts, by the mammary gland. The source of ovarian steroids in bony fish is not completely agreed upon.

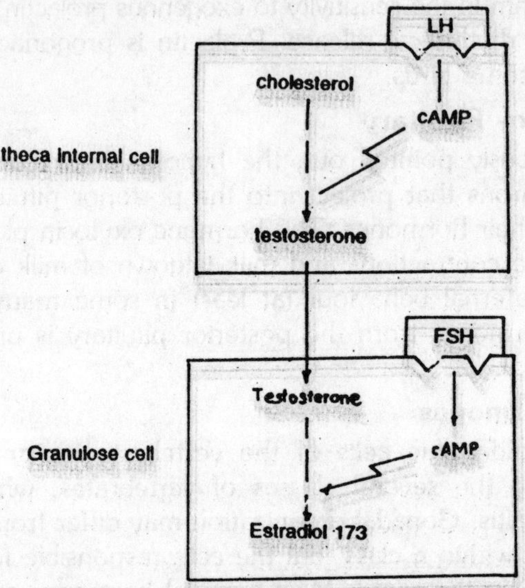

Fig. 1.5. A hypothetical pathway of LH and FSH action in the regulation of the synthesis of estrogen in follicles.

In the guppy (*Poecilia reticulata*) steroid synthesis appears to occurs in the granulosa cells. In the goldfish (*Carassius auratus*) and several species of Pacific salmon (*Oncorhynchus kisutch* and *O. gorbuscha*) theca cells appear to synthesize steroids. Both theca and granulosa cells appear to produce steroids in the preovulatory follicle of the live-bearing bufonid toad (*Nectophrynoides occidentalis*), and after ovulation corpora lutea continue to shows steroidogenic activity. Target organs of estrogen include mammary tissue and the uterus, oviduct and hypothalamus.

A major function of estrogen is the rapid development of such structures as uterine epithelium and mammary tissue through the stimulation of protein synthesis. In humans estrogen promotes fat deposition on the hips and growth of underarm and pubic hair. In these processes estrogen usually acts together with one or more other hormones. Estrogen has immediate effects on target organs, where it may persist for up to two hours (in ovariectomized

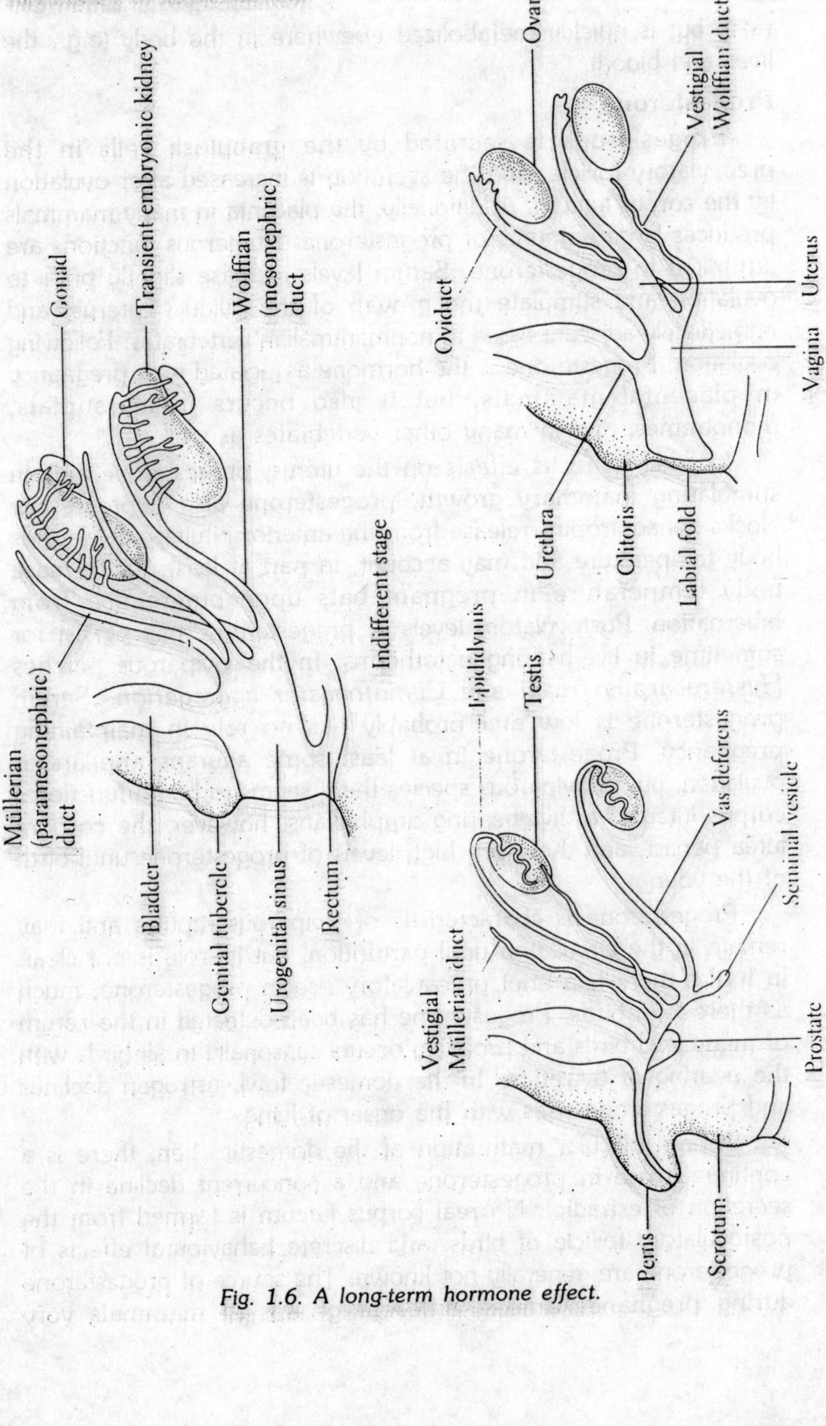

Fig. 1.6. A long-term hormone effect.

rats), but is quickly metabolized elsewhere in the body (e.g., the liver and blood).

Progesterone

Progesterone is secreted by the granulosa cells in the preovulatory follicle, and the secretion is increased after ovulation by the corpus luteum. Additionally, the placenta in many mammals produces large amounts of progesterone. Numerous functions are attributed to progesterone. Serum levels increase slightly prior to ovulation and stimulate the growth of the oviduct, uterus, and immediately adjacent tissue in nonmammalian vertebrates. Following ovulation. Progesterone is the hormone associated with pregnancy in placental mammals, but it also occurs in marsupials, monotremes, and in many other vertebrates as well.

In addition to its effects on the uterus, progesterone aids in stimulating mammary growth, progesterone also depresses or blocks gonadotropins release from the anterior pituitary. It elevates body temperature and may account, in part at least, for increase body temperature in pregnant bats upon emergence from hibernation. Postovulatory levels of progesterone may persist for sometime in live-bearing ectotherms. In the viviparous perches *Hysterocarpus traski* and *Cymatogaster aggregation*. Serum progesterone is low and probably has no role in maintaining pregnancy. Progesterone in at least some anurans appears at ovulation, but in oviparous species there seems to be no functional corpus luteum. In live-bearing amphibians, however, the corpora lutea persist, and there are high levels of progesterone until birth of the young.

Progesterone is characteristic of viviparous reptiles and may remain in the circulation until parturition, but its role is not clear. In turtles there is a brief preovulatory rise in progesterone, much as there is in birds. Progesterone has been detected in the serum of many wild birds and probably occurs seasonally in all birds with the nearing of ovulation. In the domestic fowl, estrogen declines and progesterone rises with the onset of lying.

During follicular maturation of the domestic, hen, there is a continuous rise in progesterone and a concurrent decline in the secretion of estradiol. No real corpus luteum is formed from the postovulatory follicle of birds, and discrete behavioural effects of progesterone are generally not known. The source of progesterone during pregnancy among different groups of mammals very significantly. In the domestic rabbit, the goat, and the pig, preservation of the placenta is completely dependent upon progesterone from the corpus luteum. On the other hand, in some species the placenta gradually assumes the production of progesterone. This is true of the horse, in which the endometrial cups (of embryonic origin) of the placenta synthesize progesterone. In some bats the corpus luteum may regress in early pregnancy (e.g., the African *Nycteris luteola*; or in later pregnancy (e.g., the Australian *Eptesicus regulus*.

Androgens

Androgens are synthesized by the testes of all, and the ovaries of quite a few, if not most vertebrates. Androgens are, also the major product of testicular steroidogenesis and are synthesized by Leydig cells of the testicular interstitial tissue. In the ovary, androgens are produced by theca cells in response to LH stimulation in laboratory rats, and such androgens stimulate both endometrial and myometrial hypertrophy. Androgens may be converted to estrogens in target organs. There are discrete cycles of testosterone in some female reptiles, and peaks may occur with or proceeding peaks of estradiol. The significance of appreciable amounts of testosterone in females may relate to the conversion of testosterone to estradiol. Androgens in female birds account for the slight development of male traits, such as the comb in a hen.

In some species, such as phalaropes (Phalaropodidae), normally high androgen levels in females produce male behavioural traits, such as courtship, and also brightly coloured plumage, which is usually characteristic of only males. Androgens may also function synergistically with other ovarian hormones. Small amounts of androgens are circulating in both sexes and can be detected in the very young. In puppies, for example, behaviourally ineffective levels of testosterone occur in circulation, but in amounts far below those of a sexually mature male. In adult males androgens (most testosterone) induce activity of the Sertoli cells and the seminiferous tubules. Androgens also account for male behaviour, including territoriality and courtship, and for the development of secondary sexual characteristics.

Androgen implants increased calling and aggressiveness in territorial cock red grouse (*Lagopus lagopus scoticus*), and estrogen implants in territorial cocks resulted in a decline in song, loss of territory, and desertion of mates. Both male and female anoles (*Anoles carolinensis*) exhibit male sexual behaviour when treated with dihydrotestosterone. Injections or silastic implants of testosterone induce receptivity in females, presumably being converted (aromatized) to estrogen in the brain.

Relaxin

Relaxin has long been known as a hormone that softens cartilaginous connections in the pelvic girdle. It is a polypeptide produced in the ovaries and especially in the corpus luteum. It is found not only in mammals but also in the sand shark (*Odontaspis taurus*), a viviparous shark that continues to ovulate throughout pregnancy. It is also produced by the placenta of at least some mammals (rabbits and horses). It is most abundant in the luteal phase of the sow, increasing during pregnancy and declining after parturition. By softening the birth canal and relaxing the cervix, relaxin facilitates parturition. Additionally, relaxin inhibits spontaneous uterine contractions, perhaps preserving the placenta

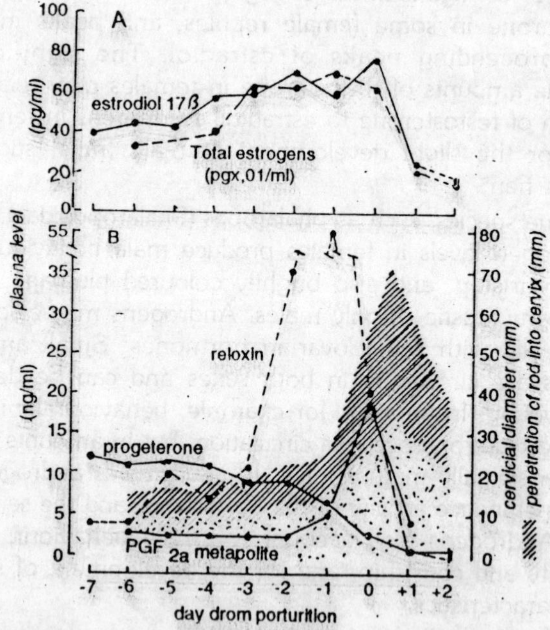

Fig. 1.7. The sudden rise in relaxin one to three days before parturition in the sow, together with levels of (a) estrogens, (b) progesterone, PGF^{2a} metabolities, and cervical dilation.

and fetuses after progesterone has declined, until this suppression in overridden by high levels of oxytocin and prostaglandins. Relaxin inhibits the release of oxytocin.

Oxytocin

As previously pointed out, this hormone a peptide, is synthesized by neurosecretory cells in the hypothalamus and released from the posterior pituitary. It is also produced by the ovary and may play an important role in the ovulatory cycle. In addition, it is synthesized by the placenta. Levels of circulating oxytocin and progesterone parallel each other after ovulation, when the corpus luteum is active, and elevated concentrations of oxytocin have been found to occur in the corpus luteum of the cow. In sheep, concentrations of oxytocin are greater in the ovarian vein than in the ovarian arterial circulation, and the secretion of ovarian oxytocin is probably stimulated by $PGF^{2\alpha}$ from the uterus.

The function of ovarian oxytocin is not clearly understood, but it may inhibit the synthesis of progesterone and thereby hasten the demise of the corpus luteum. Possible roles of gonadal oxytocin are reviewed by Wathes (1984). A major function of oxytocin is the stimulation of smooth muscle contraction: oxytocin causes myofibrin contraction in the lactating breast tissue, inducing milk letdown, and also initiates uterine contractions at the time of parturition. The presence of a fetus in the birth canal causes additional oxytocin release, an example of positive feedback. Oxytocin may be released during intercourse, also stimulating milk letdown in a lactating woman. Oxytocin also occurs in males. In the male rat and rabbit, oxytocin increases after ejaculation and is followed by sexual refractoriness.

Inhibin

It has long been postulated that a gonadal hormone limits the production of FSH. A polypeptide had long ago been detected, but its exact nature was elusive. Its presence is apparent when castration is followed by a rise in FSH. Although it is well accepted that inhibin, from seminiferous tubules in testes and granulosa cells in the ovarian follicle, limits pituitary release of FSH, the nature of inhibin has been difficult to define. When cultured in vitro. Sertoli cells of the rat secrete a non-steroidal substance the suppress the synthesis and release of FSH. This action, which describes the role of inhibin, clearly affects the anterior pituitary and perhaps also the hypothalamus. Analysis of inhibin from porcine follicular fluid indicates its occurrence as two subunits, α and β. There are also dimers of the β subunit, and either subunit β_A or β_B can combine with subunit α and inhibit secretion of FSH. The two subunits of β, moreover, may combine with each other and stimulate the production and release of pituitary FSH. This subunit combination is called activin (Ling *et al.*, 1986). Neither inhibin or activin affects LH or prolactin release.

Feedback

Not only is the ovary sensitive of LH and FSH from the anterior pituitary, but both the hypothalamus and the pituitary respond to varying levels of the steroid hormones secreted by the ovary. This feedback relationship accounts for fluctuations in secretory activity of the hypothalamus, the anterior pituitary, and the ovaries and also for the cyclicity of ovulation in polyestrous species. This phenomenon is best known for mammals but probably occurs in

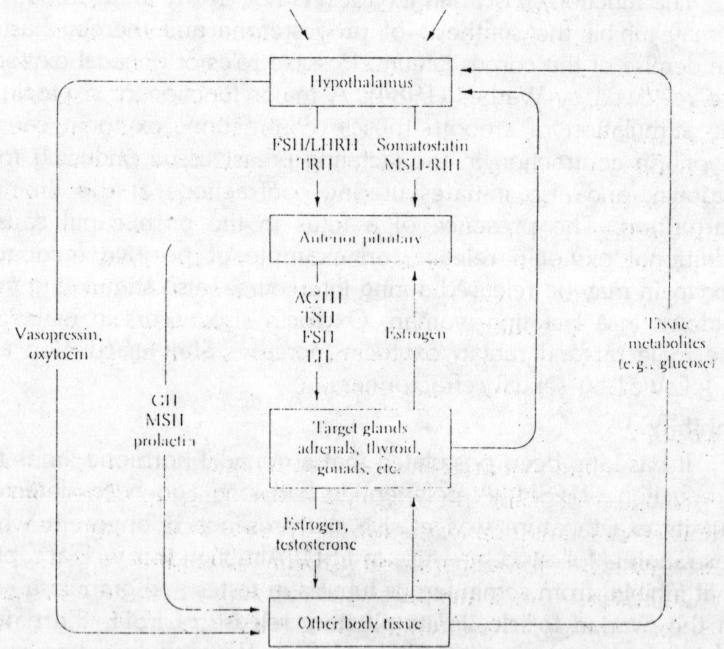

Fig. 1.8. The hierarchial organization of the endocrine system in vertebrates. Note the central roles played by the hypothalamus and anterior pituitary gland and the importance of feedback control loops. The two arrows at the top of the figure indicate the presence of input to the hypothalamus from higher centers of the central nervous system.

all vertebrates. Castration of the rainbow trout (*Salmo gairdnerii*) is followed by a five-fold increase in GtH, which in these individuals is reduced by pituitary implants of 11-ketotestosterone. There may also be a positive feedback in sexually developing teleost fish (e.g., salmonids and eels), which may promote sexual maturation. In juvenile trout testosterone has been shown to directly promote maturation of gonadotrophs. As already pointed out, when circulating levels of estrogens are low, levels of serum LH and FSH are higher. The release of gonadotropins follows stimulation of GnRH from the hypothalamus, through its portal circulation, to the anterior pituitary.

The rise of gonadotropins stimulates the activity in one or more of the larger follicles, causing an increase in plasma estrogens. The rise in circulating levels of estrogens depresses FSH release by increasing the pituitary threshold to GnRH. This is the *negative feedback*. While the circulating estrogen suppresses the release of FSH, there is a continued but depressed release of LH. In as much as the movement of GnRH into the anterior pituitary continues, LH synthesis and accumulation also continues.

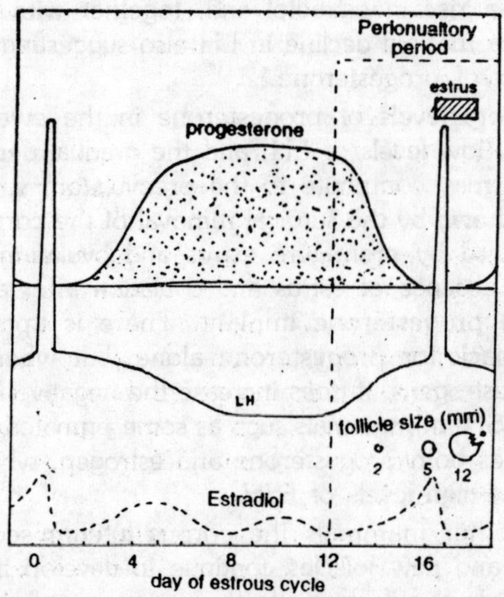

Fig. 1.9. Fluctuations of LH, estradiol, and progesterone in the 16-day ovulatory cycle of the ewe, showing the sharp rise of LH with the disappearance of serum progesterone.

With the gradual increase in the release of estrogens, however, a point is reached at which estrogen from the large follicle(s) reaches a stimulatory level.

This increase of estrogens (especially estradiol), together with the buildup of LH in the anterior pituitary, seems to account for the sudden release of LH that is associated with ovulation. This is a *positive feedback* of estradiol. Thus the preovulatory surge of LH does not require additional GnRH. With the decline of estrogen, as in a sterile ovulatory cycle, the negative feedback disappears. This is followed by a subsequent rise in circulating LH and FSH, indicating a lowering of the threshold of sensitivity of the pituitary of GnRH. Progesterone has a clear negative feedback effect on the release of LH.

In the ovariectomized ewe, Silastic capsules that released levels of progesterone stimulating the midluteal phase produced a significant decline in LH release. But the reduction was not as great as in the midluteal phase of the normal estrous cycle, suggesting that estrogen also plays a role in the reduction of the tonic LH secretion. Prior to the preovulatory surge of LH in the ewe, there is a rise in estradiol and, together with the rise in progesterone, a marked decline in LH also suggesting a negative feedback effect of progesterone.

Postovulatory levels of progesterone in the ewe allow the release of only low levels of LH, and the gradual decline of the corpus luteum may contribute to the preovulatory surge of LH. This effect is shown by the surgical removal of the corpus luteum, which is followed by premature estrus and ovulation. It is also shown by the absence of estrus and ovulation after a lutectomy followed by a progesterone implant. There is apparently no negative feedback for progesterone alone, but when it occurs together with estrogens, it does increase the negative feedback of estrogens on FSH. In mammals such as some primates, the corpus luteum produces both progesterone and estrogen, which in such species prevent high levels of FSH.

In some other mammals the corpus luteum secretes only progesterone, and new follicles continue to develop immediately after ovulation. In the domestic fowl, progesterone has a positive feedback on both the hypothalamus and the anterior pituitary, stimulating the release of LH. This modification in birds enables the rapid development of a full clutch of eggs, with the release of

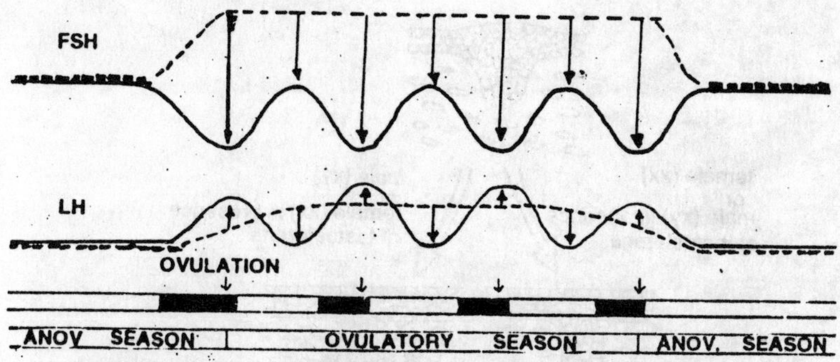

Fig. 1.10. The combined effects of the ovarian and photoperiodic influences on circulating concentrations of LH and FSH in mares.

only a single egg at one time. The negative feedback relationship is also shown by ovariectomy. In the absence of gonadal steroids and inhibin, LH and FSH increase in the general circulation but still reflect the stimulation of long days.

Mammary Gland Development in the Mouse

Hormones are chemical substances which circulate freely throughout the body, affecting the activity—in embryos, the differentiation—of a variety of target tissues which are often widely dispersed and whose responses to the same hormone may be quite different. They can therefore play no primary part in establishing patterns of morphogenesis, but act as switches—on or off—of gene expression in cells whose developmental options have already been narrowly restricted by prior determinative events. Thus, five pairs of mammary rudiments are produced in a specific pattern in all mouse embryos of whatever sex, but in males they regress in response to secretion of the androgenic steroid testosterone by the developing testes.

Using organ culture techniques in 1971, Kratochwil showed that the genetic sex of the mammary gland has no influence on its developmental capacities—glands of male embryos are able to develop in the absence of testosterone, and those of females will regress in the presence of it. The rudiments appear at the 12th day of gestation, consisting of mesenchymal and epithelial components, the epithelium forming buds around which the mesenchyme condenses. In the next two days this forms a

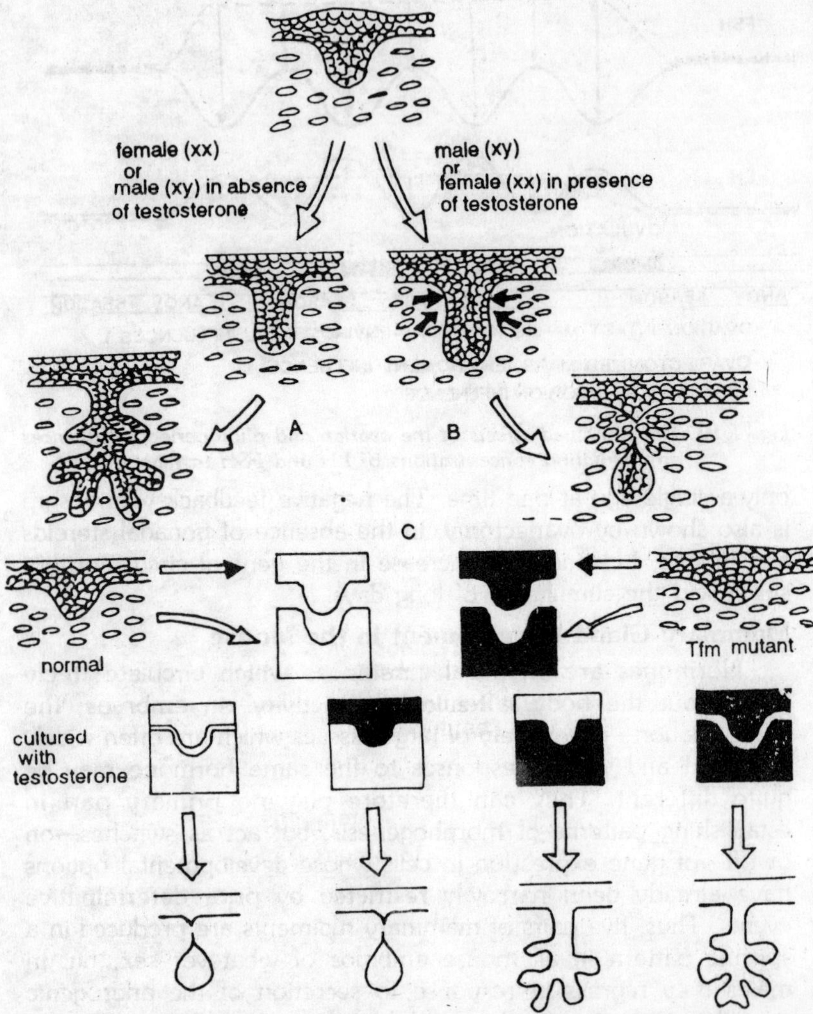

Fig. 1.11. Hormonal control of mammary gland development in the mouse. Development of the mammary gland rudiment in (A) female mice or in male rudiments cultured in the absence of testosterone; (B) male mice or in female rudiments cultured in the presence of testosterone. C. Response to testosterone of cultured recombinations of ectodermal and mesodermal components of rudiments from normal and testicular feminization (Tfm) mutant mice.

branching glandular system in female embryos, but in males the stalk of the bud becomes pinched off by a dense accumulation of mesenchymal cells and the bud itself becomes smaller and necrotic, then is cut off from the stalk, and eventually disappears. In 1976, Kratochwil and Schwartz used a mutant—*testicular feminization* (*Tfm*)- to analyze the hormones—apparently through impaired function of the androgen receptors at the cell surface-so all their secondary sexual characteristics, including the mammary glands, are female in type. In reciprocal recombinations of normal and mutant components in organ culture, regression occurred as a response to testosterone when *Tfm* epithelium was combined with normal mesenchyme, but not when normal epithelium was combined with *Tfm* mesenchyme, indicating that the primary action of the hormone is upon the mesenchymal component, whose response leads in some way to regression and necrosis of the epithelial bud.

Phenotypic Sexual Characteristics

Hormones are prominent in development wherever changes in widely separated tissues have to be mutually adjusted, sometimes over a long period, as in the control of growth or (especially in embryogenesis at critical periods) as general switch mechanisms where several developing systems must be directed into one pathway or another. The commitment to male or female secondary sexual characteristics, including the development of the ducts of the reproductive system, is a clear example of the latter, determined as a response to androgenic or oestrogenic hormones produced by the testis and ovary respectively. It is much less clear to what extent hormones play a part in primary sex determination, which is in many organisms partly controlled by non-genetic factors; the most extreme example is the marine worm *Bonellia*, whose larvae develop into females if reared in isolation, but into minute semiparasitic males if they become attached to the proboscis of a female. Even in some vertebrates the genetic sex-determining mechanism may be very unstable, as in the fish *Lebistes*, where the genetic sex is often over-ruled by environmental influences. In experimental situations the genetic sex of many vertebrates may be altered by hormonal treatment if it is given sufficiently early in gonadal development.

Extensive studies were made by Witschi in the 1930s on the effects of parabiosis (i.e., joining two individuals by grafting at an early stage) in which he showed, e.g., in 1938, that a genetically female toad would be masculinized and produce sperm when grafted to a male, but that there was no effect upon the male partner. Much later, in 1956, Chang and Witschi showed that *Xenopus* males were feminized, to the extent of laying eggs, if oestrogens were added to the water in which they were reared.

Among mammals, a condition sometimes arises in the development of cattle as a result of male and female twins sharing a placental circulation, in which the female is characterized by a partial sex reversal of her ovaries and consequently incomplete differentiation of the reproductive tract, and becomes what cattle breeders call a freemartin. This phenomenon was studied in 1917 by Lillie, who concluded that it arose through the effect of androgens passing over from the developing testes of the bull calf and masculinizing the gonads of the female, just as in Witschi's amphibians. But more recent work, as reported by Short in 1970, suggests this is not the explanation, since it has not proved possible to transform an ovary into a testis or vice versa by hormone treatment in any mammals so far investigated; some circulating substance must be responsible, but its nature and mode of action have not yet been discovered.

The Molecular Basis of Hormone Action

The androgens and oestrogens referred to above are steroids—rather small lipid soluble molecules which probably act by passing through the target cell's plasma membrane and acting directly upon the gene regulatory mechanism; O'Malley and colleagues, working on the effects of oestrogen and progesterone on the developing chick oviduct in 1972, produced evidence that these hormones may act at the level of the genome by altering gene transcription, leading to the formation of new nuclear RNA. The majority of hormones are non-steroids, and most are polypeptides—much larger than steroids and consequently unable to pass through the plasma membrane, so that their mode of action is quite different.

In 1957, working on carbohydrate metabolism in the liver, Sutherland and Rall introduced a concept that has had the most far-reaching influence in research on chemical messengers—that there is a second messenger, cyclic adenosine monophosphate or cAMP (it now appears that another nucleotide, cyclic guanosine monophosphate or cGMP, may be another) which transfers the signal to the metabolic and gene regulatory mechanisms within the cell. Polypeptide hormone molecules are accepted at specific receptor sites which exist only on the target cells, at their cell surface, where cAMP is synthesized through the activity of the enzyme adenylcyclase as a response to the presence of the hormone molecule. Within the cell, cAMP may also be inactivated through the activity of the enzyme phosphodies—terase, related in some way to the hormone molecules at the surface, so the concentration of intracellular cAMP is modulated in both directions. Through its interaction with protein kinases, this has an effect upon the level of many other enzymes, and consequently upon the metabolic state of the cell, which will depend partly upon the cell's previous state; the degree of specificity and flexibility offered by this system is very great, and Bitensky and Gorman in 1972 and McMahon in 1974 have listed a wide range of cellular activities which it has been found to regulate—including morphology, motility and pigmentation of cells, cell division, stability and assembly of microtubules, plasma membrane permeability, gene expression, protein synthesis.

Almost all work so far has been upon adult tissues (the closest to a morphogenetic system is in the slime mould. *Dictyostelium*, where cAMP, secreted outside of the cell, acts as an inducer of cell aggregation) but the cellular activities affected are of the same type as in embryogenesis. We have referred in the last chapter to McMahon's hypothesis that at least some embryonic inductions act through the cAMP system; it certainly the case that it is established as an extremely widespread and possibly universal mechanism for integrating cellular responses with specific external signals, including hormones and neurotransmitters, in adults; it may well turn out to be just as widespread in embryos, and the distinction between induction and hormonal control will in that case sometimes be difficult to maintain.

2

HORMONES IN DEVELOPMENT

The development of a normal full-grown living organism from a single egg cell is an extraordinary achievement. The young individual may start life hidden in the maternal womb or in the eggshell before receiving parental care and help, or it may develop alone, free and exposed. In spite of these differences, for the biologist its development raises questions of morphological and physiological specialization. The physiologist attempts to uncover the mechanisms involved in the chain of events that links genetic information received by the egg cell to its final expression in development. Hormones have been recognized as taking part in these events. They participate in processes of growth, morphogenesis, and differentiation in both plants and invertebrate and vertebrate animals. The most spectacular instance of hormonal intervention is provided by metamorphosis of insects and of anuran larvae. Hormonal control of development can also be demonstrated in higher vertebrates and in man. The first clue for hormonal control of development was derived from clinical observations, and it is therefore rewarding to review the historical groundwork of development endocrinology.

HISTORICAL BACKGROUND

First Clues and Concepts

Studies of the thyroid opened the path. As early as 1850, Curling reported "two cases of absence of thyroid and symmetrical swellings of fat tissue at the side of the neck, connected with defective cerebral development." Fagge, in 1871, confirmed these data and connected cretinism with thyroid deficiency, but the final proof that dwarfism and idiocy depended on thyroid secretion was obtained only during the following twenty years, after the first thyroidectomies in humans and in animals were performed. Schiff, in 1884, avoided the effects of thyroidectomy in dogs with abdominal thyroid grafts. In the early 1890's, myxoedematous patients were treated with injected thyroid extracts or thyroid given orally. At the end of the nineteenth century the need for thyroid hormone to assure normal postnatal growth, stature, and mental maturation was recognized.

The role of the pituitary gland in growth, also first suggested by clinicians, slowly emerged from studies of hypophysectomies performed on dogs by experimenters, culminating in the preparation of growth-promoting extracts in 1921 by Evans and Long. Progress in developmental endocrinology during these early times was also stimulated by embryological, morphological, and biochemical studies of endocrine glands in development animals that paralleled endocrine research on adult organisms. The discovery that the mammalian foetus has active endocrines was not immediately accompanied by any special hypothesis about their role in development. It was supposed that foetal endocrine glands are merely held in readiness for their future postnatal function. Occasionally it was implied that they might exert the same regulatory function in developing animals as in adults.

As early as 1899, Langlois and Rehns extracted a pressor substance from adrenal glands of sheep, rabbit, and guinea pig foetuses. Camus, in 1902, detected secretin in the duodenum of rabbit and guinea pig foetuses. Insulin was extracted from bovine foetal pancreas by Banting and Best in 1922, and growth hormone was obtained from pituitaries of 9- to 11-cm pig foetuses in 1929 by Smith and Dortzbach. Other experimenters concentrated more on the significance of hormones in development.

The concept that hormones acting at a "definite period" of development bring about permanent effects was first presented by Bouin and Ancel (1903), who were studying the testicular interstitial cells as the source of male hormone. They observed the presence of active-looking interstitial cells in sections of the testes of 3-cm pig foetuses— i.e., before sexual differentiation of the genital tract—and suggested that the internal secretion of the fetal testis "imprints on the organism" the essential characters of the male sex, so that castration, even if performed at birth, will no longer be able to suppress sexual differentiation. The correctness of this assumption was later verified by Jost in castration experiments on rabbit foetuses. More recent research extended this concept to the influence of androgens upon the function of the hypothalamus in the rat and upon neural structures involved in sexual behaviour in the guinea pig. Similarly, early treatment of young rats with thyroid hormones results in permanent alterations of thyroid function.

The hormonal theory of sex differentiation was developed and popularized by Lillie's beautiful papers dealing with the bovine freemartin. This work stimulated experimental research on all classes of vertebrates for half a century. The discovery by Gudernatsch that thyroid could provoke metamorphosis in anuran tadpoles is to be regarded as another milestone in developmental endocrinology.

Gudernatsch's experiments—published in two papers under the modest general title of *Feeding Experiments in Tadpoles* and supplemented by other studies in which the effects of thyroidectomies and hypophysectomies on tadpole metamorphosis were investigated—demonstrated unequivocally the requirement of hormones for amphibian development, as well as stimulating interest in pituitary-thyroid relationships. The dramatic morphological and physiological charges produced by thyroxine in tadpoles has no exact equivalent in higher vertebrates. Nevertheless, ideas concerning the role of hormones in development were largely influenced by the demonstration of their role in amphibian metamorphosis.

Maternofoetal Relations in Humans and Other Mammals

The human and, more generally, the mammalian foetus and newborn closely depend upon the mother during intrauterine development and during lactation. A maternal endocrine adjustment to pregnancy and to lactation is a prerequisite for normal foetal development. This extensive and important chapter of the endocrinology of reproduction cannot be considered here. But it should be recalled that uterine accommodation under ovarian hormones is so important for the young that foetal malformations may result from hormonal imbalances in the mother (24-26). Since the foetus produces and needs hormones for its development, and since the mother also produces her own hormones, the

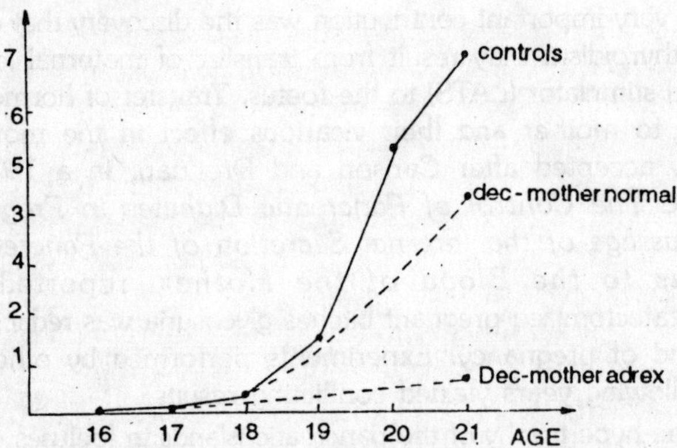

Fig. 2.1. Glycogen contents in the liver of rat foetuses (percent of fresh weight). The heavy line shows the increase of glycogen contents in controls according to age. In foetuses decapitated on day 17 in mother animals which were adrenalectomized (adrex) on day 14, the glycogen is low on day 21. In decapitated foetuses of entire mother animals, the amount of glycogen is two-thirds normal. The interrupted lines arbitrarily join the initial (day of decapitation) and final glycogen values.

possibility of exchanges or of reciprocal help in case of hormonal shortage in either organism was a tempting subject for study.

Kocher, in 1892, was probably the first to suggest that the secretion of the maternal thyroid could act vicariously in the hypothyroid foetus. Such a concept presupposed transfer of thyroid hormone from mother to foetus. It was later contradicted by Dorff's observation in 1934 of prenatal retardation in bone maturation in congenital cretins. The concept of vicarious action of some maternal hormone in the foetal organism supposes at lease that the hormone in question crosses the placenta and that it is necessary for the foetus. One example may be cited: In the rat foetus, glycogen storage in the liver or in the heart is hormonally controlled. If the foetus is deprived of its own adrenocortical hormones by decapitation, heart glycogen remains normal while liver glycogen is reduced by approximately one third. If, in addition to foetal decapitation, the pregnant mother is adrenalectomized, both heart and liver glycogens stores are lowered. Corticosteroids injected directly into the decapitated foetus may replace the maternal hormones. A recent extension of the problem of maternal influences upon the foetus is linked with the finding that some proteins, especially gamma globulins, are transferred from mother to foetus.

A very important contribution was the discovery that congenital hyperthyroidism may result from transfer of maternal long-acting thyroid stimulator (LATS) to the foetus. Transfer of hormones from foetus to mother and their vicarious effect in the mother were largely accepted after Carlson and Drennan, in a 1911 paper entitled *The Control of Pancreatic Diabetes in Pregnancy by the Passage of the Internal Secretion of the Pancreas of the Foetus to the Blood of the Mother*, reported that in pancreatectomized pregnant bitches glycosuria was reduced toward the end of pregnancy. Experiments performed by others during the following years yielded conflicting results.

The hypertrophy of the pancreatic islands in foetuses of diabetic women, noted by Dubreuil and Anderodias in 1920, could lend support to the view of Carlson and Drennan, but it could also result from maternally induced foetal hyperglycemia. Schlossmann insisted that there is a high glucose uptake in large dog foetuses during late pregnancy, and Cuthbert *et al.* noticed that the need for insulin by pancreatectomized bitches is diminished not only during late pregnancy but during lactation as well. The problem of placental transfer of insulin is obviously difficult to solve, and the theory that foetal hormone could act vicariously for the mother is still controversial. The contribution of the foetoplacental unit to steroid production during pregnancy renewed one aspect of this question.

Hormones and Body Growth

There is considerable evidence that indicates three major hormones are concerned in growth- thyroid hormone, growth hormone, and insulin. Only a few pertinent data will be recalled.

Thyroid

In frog larvae, thyroidectomy prevents metamorphosis but does not stop growth; after a prolonged period of development giant tadpoles may be obtained. In mammals, thyroidectomy of young animals stops growth, but it is not certain that prenatal growth depends on thyroid hormones. Athyreotic human babies are of normal size and cease growing more or less completely a few months after birth. At birth they already show signs of hypothyroidism; only limited amounts of maternal thyroid hormones reach the foetus toward the end of pregnancy. Similarly, rabbit foetuses thyroidectomized in utero continue growing despite the reduced level of thyroxine present in foetal blood and other signs of hypothyroidism. It appears likely that either thyroid hormone is not necessary for prenatal growth at all, or if it is needed, minimal amounts suffice to meet the requirements.

Hypophysis

In frog larvae, hypophysectomy impairs metamorphosis but permits continuation of growth at a slow pace. According to Smith in *Rana boylei* retardation of growth becomes very pronounced mainly after the midlarval period and stage of 30 to 32 mm appears as a "critical point" in growth. However, this has not been observed in other species. In hypophysectomized larvae of *Alytes obstetricans*, beef growth hormone or ovine prolactin induces gigantism. In rats and in rabbits, the pituitary gland is not indispensable for growth at early stage. In rats, subnormal growth continues after early hypophysectomy until approximately the end of the fourth week. In young rabbits hypophysectomized on day 52 or 74 subnormal growth continues until approximately 100 days of age. In animals hypophysectomized on day 105, growth stops immediately. Prenatal growth, on the other hand, continues practically normally in decapitated rabbit foetuses. In human infants suffering from hypopituitarism, growth continues for some time after birth. In the guinea pig, the situation is different in that hypophysectomy does not stop growth, although the thyroid and the adrenals are depressed.

Insulin

Babies born to diabetic or prediabetic women are frequently overweight and somewhat oversized. This has been attributed to an overproduction of insulin by the foetal pancreas in response to hyperglycemia. These infants secrete an excess of insulin. Picon succeeded in inducing an excess of growth in rat foetuses given repeated injections of insulin, but this effect appeared only after the stage of 21 days in prolonged pregnancies. Similarly, excess of growth in foetuses of diabetic women becomes conspicuous only after the twenty-eighth week. A complete interpretation of the hormonal control of body growth requires more detailed studies of the various hormonal and metabolic interrelationships. But from the foregoing examples, it would appear that the hormonal requirement for growth varies during different developmental periods.

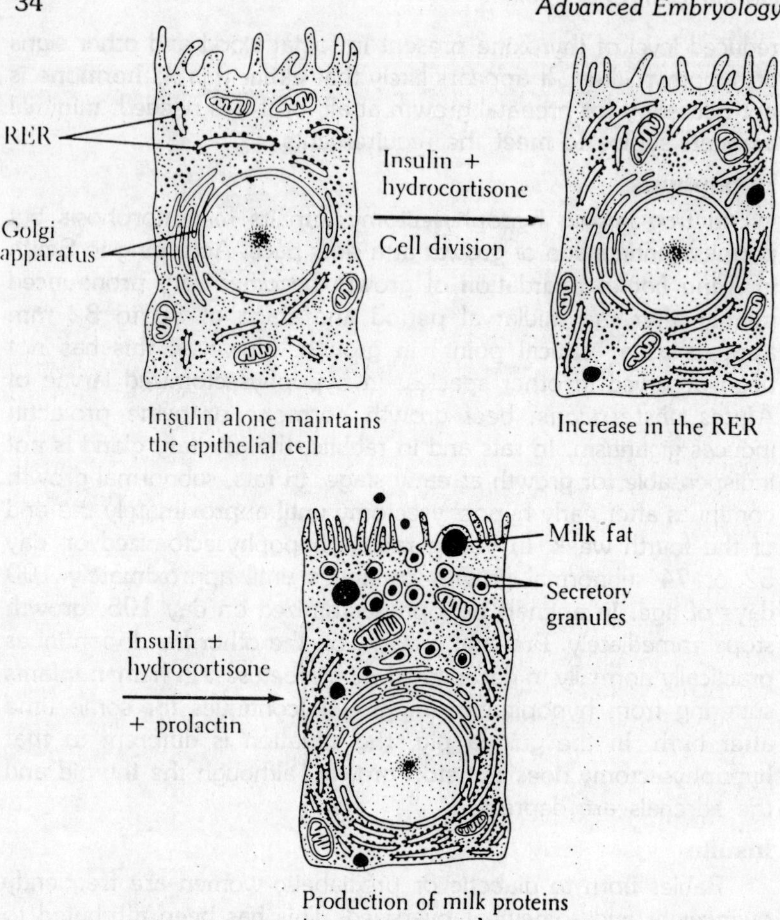

Fig. 2.2. Changes in the structure and function of duct cells in the mammary gland. Insulin and hydrocortisone stimulate the increase of the rough endoplasmic reticulum (RER), and prolactin in combination with these hormones causes the secretion of milk proteins.

The Role of the Foetal Pituitary in Development

The analysis of the role of the foetal pituitary in development affords an example of a rather complex situation concerning correlations between foetal glands, maternofoetal interrelations, and maturational processes. It is well-established that the larval or foetal pituitary gland is necessary for the normal development and function of other endocrine glands, such as the thyroid or adrenal cortex. The modalities of action of the foetal pituitary are still currently under investigation. Only a few selected topics will be considered here.

Significance of the Pituitary Stimulus

The differentiation into specific endocrine glands (thyroid or adrenal cortex) of the early embryonic primordia is independent of the hypophysis, but subsequent growth and functional maturation is controlled by pituitary hormones. The modalities and significance of the pituitary stimulus may be illustrated by examples pertaining to the rat foetus. The effect of pituitary removal by decapitation on the growth of the adrenal glands of rat foetuses varies according to the stage at which decapitation is performed. If the pituitary gland is removed on day 17 or before, the adrenal glands actually continue to grow at a very low rate until birth; their apparent "Atrophy" at birth results from impaired growth and not from decrease in size.

The difference between the growth curves of the adrenals in control and in decapitated foetuses graphically illustrates the role of the pituitary stimulus. If the pituitary is removed on day 18 or 19, after it has already begun to stimulate the adrenals, the adrenals gland maintains the size attained at that stage. The effects of early or delayed decapitation on some other pituitary-controlled systems provide similar curves. Thus, rate of glycogen deposition in the growing liver, expressed as percentage of glycogen in fresh tissue, remains low after early decapitation, it continues at intermediary values if the foetus is decapitated slightly later.

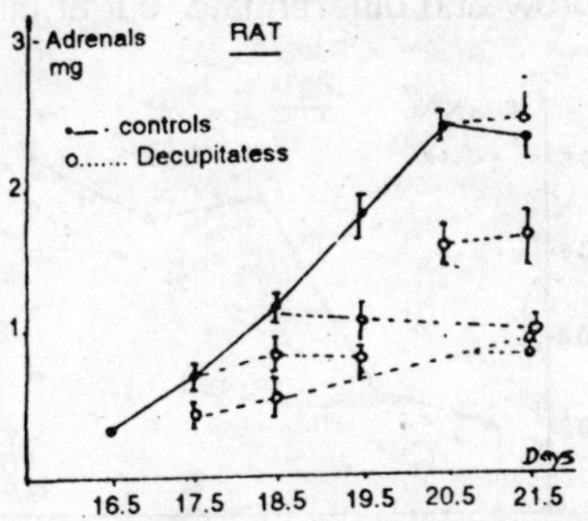

Fig. 2.3. Weight of one pair of adrenal glands from rat foetuses decapitated at various stages, in comparison with controls, confidence interval indicated.

Similar results are obtained in measurements of the rate of methylation of norepinephrine to epinephrine in the adrenal medulla, which in rats and rabbits has been shown to be controlled by the adrenal cortex. A similar curve was obtained by Geloso in a study of the uptake of radioiodine by the thyroid gland of rat foetuses. It appears that the length of time the pituitary gland is allowed to stimulate the target systems determines whether full or partial physiological maturation is attained.

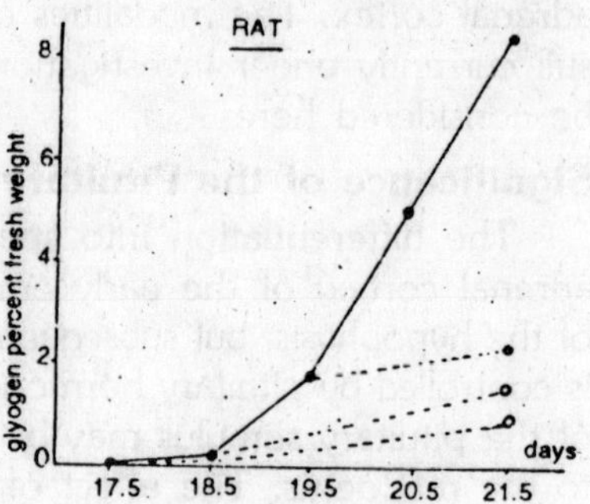

Fig. 2.4. Glycogen contents (percentage of fresh weight) in the liver of rat foetuses from mothe animals adrenalectomized on day 14.

Hypothalamic Control of Pituitary Function

In the rat foetus, the encephalon can be removed easily along with the hypothalamus while the pituitary body remains *in situ* in the sella turcica. In foetuses submitted to such surgery on days 16 to 18, the pituitary gland continues to grow and differentiate, but at birth the adrenals are reduced almost to the same extent as in foetuses deprived of their hypophysis by decapitation. Crude hypothalamic extracts or purified corticotrophin releasing factor (both almost inactive in decapitates correct the size of the adrenals in encephalectomized foetuses. By contrast, the thyroid glands of encephalectomized rat foetuses look normal, and respond to maternally administered propylthiouracil with characteristic hyperplasia. In the absence of the pituitary gland, the antithyroid drug has no effect. In unpublished experiments, Annette Geloso observed that blood thyroxine is somewhat reduced in encephalectomized foetuses studied under conditions of isotopic equilibration.

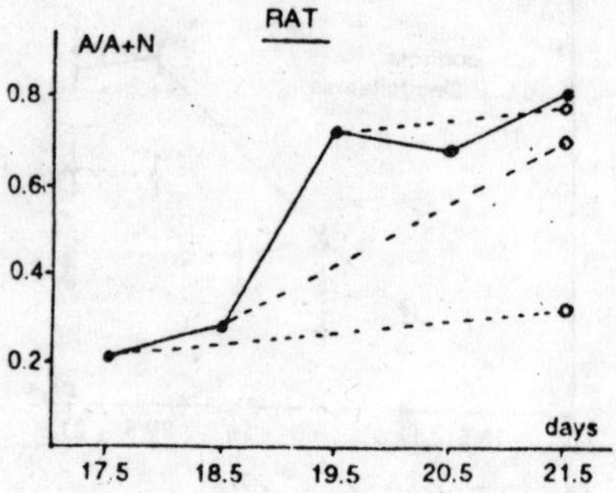

Fig. 2.5. Ratio of epinephrine to epinephrine plus norepinephrine (ratio a/A+N) as measured fluorometrically in one pair of adrenal glands of rat foetuses.

The thyrotropic function of the foetal pituitary thus seems to be much less strictly controlled by the hypothalamus than the adrenocorticotropic function. These results give a clue as to the causes of the condition found in human anencephalia in which (1) the pituitary gland is usually present but has no or abnormal connections with the disturbed hypothalamus; (2) the adrenal cortex is poorly developed: and (3) the thyroid gland is histologically close to normal.

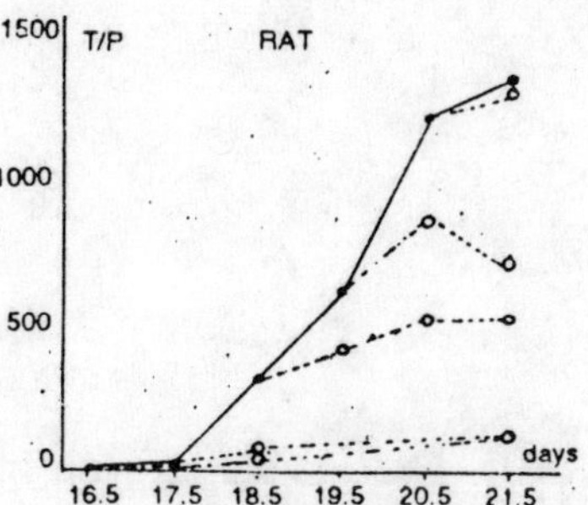

Fig. 2.6. Collection of ^{131}I by thyroid tissue of rat foetuses (t/P=ratio of the radioactivities permilligram of thyroid tissue/blood plasma). Tracer doses of 131I were injected 24 previously

The Human Foetal Adrenal Cortex

The adrenal cortex of the human foetus is exceedingly large, but it shrinks rapidly after birth. The involution concerns mainly the most internal zone (foetal zone) and can be prevented if ACTH is given to the newborn infant. In anencephalic monsters, the foetal zone is absent at birth. It has long been known that the size and histological structure of the adrenal cortex are normal in anencephalic foetuses under 20 to 22 weeks of prenatal age. After this stage, the relative amount and width of the foetal cortex decrease by a process of involution while the permanent cortex becomes somewhat thicker. Data from literature on the weight of the adrenals in control and in anencephalic infants may be helpful in analyzing the process of involution in adrenals of anencephalics.

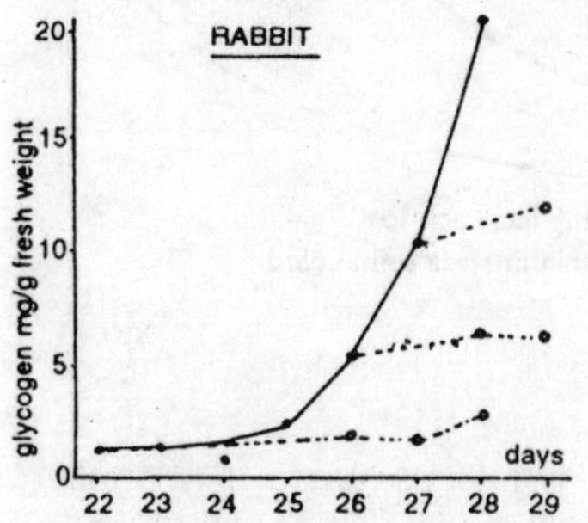

Fig. 2.7. Glycogen contents (milligrams per gram fresh weight) in the liver of rabbit foetuses.

It appears that in anencephalia, growth of the adrenals is stopped after approximately that twentieth week, although after that time some histological reorganization of the cortex still takes place. The curve resembles that of early decapitated rat foetuses, but in anencephalic monsters the primary cause is probably a hypothalamic defect. The question of the identity of the hormone that stimulates the adrenal cortex until birth and which is absent in anencephalia has been often debated. It should first be recalled that in rat and rabbit foetuses, feedback relations between

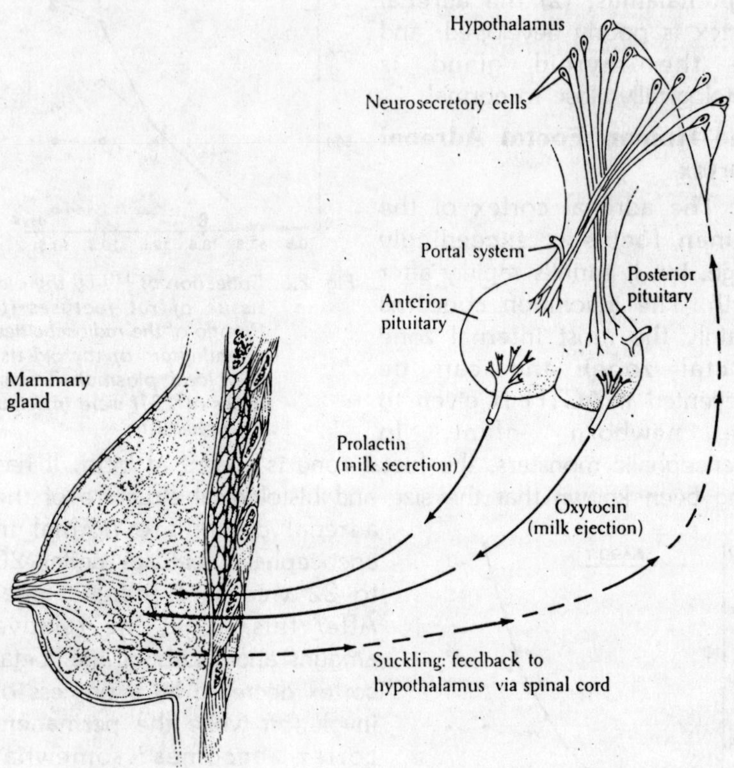

Fig. 2.8. The feedback loops controlling the release of oxytocin and prolactin through the hypothalamus to the pituitary gland.

adrenocorticotropic function of the hypophysis and corticosteroids of maternal or foetal origin are well documented. Similar feedback effects exist in the human foetus.

Depression of the adrenal glands may result either from an excess of corticosteroids given to the mother or from maternal Cushing syndrome. Conversely, impaired production of corticosteroids in adrenogenital syndrome permits an increase in ACTH secretion and adrenal hypertrophy. The intense stimulation of the adrenal cortex in the normal foetus probably results from a feeble feedback inhibition of the foetal pituitary, due to a low level of unbound or free corticosteroids in foetal blood. During human pregnancy, the corticosteroid-binding capacity of the maternal plasma, especially the transcortin globulin, increases steadily under the influence of ovarian hormones.

It is tempting to consider that the high level of maternal corticosteroid-binding proteins is involved in the foetal adrenocorticotropic function. They might cause an outflux of free corticosteroids from foetal to maternal blood, or if appreciable amounts of binding proteins were transferred from mother to foetus, they could directly reduced the concentration of free corticosteroids in the foetal blood. In either case, at birth an increased level of free corticosteroids would depress the hypophyseal activity. This last hypothesis is in accord with the data recently observed in newborn rats by Koch *et al*. In this species, some decrease of the size of the adrenal glands also occurs after birth. The transcortin capacity in the blood of the newborn decreases during the first 10 postnatal days, while free corticosterone increases. After that time the young rat produces its own transcortin, and the percentage of free corticosterone diminishes again.

Corticosteroids and Foetal Hepatocytes

Glycogen deposition in the foetal liver is also under hormonal control. In the rat foetus, it can be prevented if corticosteroids are suppressed by maternal adrenalectomy and foetal decapitation and reinitiated if corticosteroids are given. In the livers of these decapitated foetuses, the activity of several enzymes involved in glycogen synthesis remains low. The foetal hepatocyte thus permits studies of hormonally induced changes. A detailed electron-microscopic cytological study of this cell revealed that in control foetuses, glycogen appears on day 18 as rosettes, or Drochman's alpha particles and increases strikingly during the next three days. These changes are paralleled by enlargement of hepatic cells and development of the reticulum. On day 21, the huge amount of glycogen is composed mainly of small beta particles. In foetuses decapitated on day 17 (in adrenalectomized females), the size of hepatocytes and the reticulum does not increase very much. Glycogen remains scarce and is composed of alpha rosettes.

The cytological aspect of the hepatocytes of 21-day old decapitates is to a large extent similar to that found in an 18-day normal foetus. Corticosteroid deprivation apparently impairs the maturation of the hepatocytes. When cortisol is administered to such decapitates on day 20, two thirds of the normal load of glycogen accumulates in 24 hours. A study employing autoradiography, standard histological procedures, and electron microscopic analysis of hepatocytes revealed that following a single injection of 0.1 mg of cortisol acetate into the decapitates, the sequence of effects can be summarized as follows:

1. The content of glycogen is significantly increased 14 hours after injection. In similar experiments, Plas and Jacquot observed an increase in UDPG-ansglucosylase activity beginning six hours after cortisol administration.
2. Uridine incorporation into the nucleus of hepatocytes is increased six times within one hour following injection.
3. After two hours, pronounced changes in the shape of the nucleoli appear in an increasing number of cells. These nucleoli become irregular and are surrounded with faintly material: they often touch the nuclear membrane. They were considered as "activated." At the same time, granulated basophilic material, most probably RNA, accumulates in the cytoplasm in larger numbers of cells. Moreover, the average diameter and area of the equatorial sections of the nuclei in histological preparations (measured on at least 100 cells) reaches a maximum four hours after the injection.

These observations do not prove that the injected corticosteroid acts directly in the hepatocyte. But they suggest that the stimulation of the foetal hepatocyte involves nuclear changes such as RNA synthesis in the nucleolus, it release into the cytoplasm, permitting enzymatic synthesis, and finally formation of glycogen. This interpretation is supported by preliminary data on RNA fractionation.

3

Hormones in Metamorphosis

The aim of this chapter is to set the scene for the detailed accounts that will be given in the following three chapters. Metamorphosis has long been recognized as a characteristic of insects. Even the earliest systems of classification distinguished between those insects that metamorphosed and those that did not. From paleontological records the oldest insects determined with certainly are winged forms. However, it is fairly firmly established that the origins of the insects lie with the symphylan group of the Myriapoda, and that ancestral primitive insects were wingless and therefore would have lacked a metamorphosis. What exactly do we understand by "metamorphosis"? A dictionary definition may be as follows: "A change in form structure, or function, as a result of development, specifically the physical transformation during postembryonic, as of the larva of an insect to the pupa or the pupa to an adult or the tadpole to the frog."

We speak, within the insects, of the absence of metamorphosis in the groups Collembola, Diplura, and Thysanura, which are described as ametabola since no dramatic changes in appearance occur during development. These insects are wingless and therefore are apterygotes. The rest of the insect orders are primitively winged or pterygote. Of these, the exopterygote insects—the bugs (*Hemiptera*), the may flies (*Ephemeroptera*), the stone flies (*Plecoptera*), the termites (*Isoptera*), the dragonflies (*Odonata*), cockroaches, and locusts (*Orthoptera*), to name a few, have their wings developing externally, hence their name Expoterygota, These are characterized by a series of larval stages or nymphal stages that lack wings and have immature reproductive systems. They show "incomplete" metamorphosis, i.e., a gradual transition to the winged, reproducing adult. Each nymphal stage molts into the next stage, which may have slightly more developed wing buds and reproductive system. However, even here the greatest "jump" or change is generally seen at the last nymphal-adult molt.

At this time the most dramatic growth in wings and reproductive systems occurs. The second group of pterygote insects are endopterygotes with wings developing internally and include the scorpion flies (*Mecoptera*), the alder flies and lace wings (Neuroptera), the caddis flies (*Trichoptera*), beetles (*Coleoptera*), moths and butterflies (*Lepidoptera*), bees and wasp (*Hymenoptera*), and flies (*Diptera*). In these insect orders metamorphosis is described as "complete" i.e., the larval stages completely lack any external evidence of wings and reproductive system, and the transformation into the adult is a two-step process, involving first a most from the larva into a pupa, and then a molt from a pupa into an adult. Often, as in the more specialized forms of the Lepidoptera, Hymenoptera or Diptera, the larva is completely unlike the adult in appearance. Thus, in the Lepidoptera we have the larva or caterpillar metamorphosing into the chrysalis or pupa from which the very different adult moth or butterfly emerges. In this last category there are degrees of change. At the least dramatic, we have cases like *Sialis* the alder fly, or the lacewing or ground beetles, where the larva, pupa, and adult still show strong resemblances, particularly in the abdominal segments and the pupa is capable of movement. Larva and adult differ mainly in the development of compound eyes, wings, and external genitalia.

Significance of the Pupal Stage

Much has been written of the "pupal stage" and its significance in, the evolution of the insects. One of the more recent discussions is that by Hinton (1963), who compares some of the older theories of Berlese and Poyarkoff. Hinton favours the argument that a pupal stage has been necessitated by the development between larva and adult of flight muscles that require a "mold". He homologizes the various nymphal stages of exopterygotes with the larval or nymphal stages of the endopterygotes, the last nymphal stage of exo- with the pupa of endo-, and the adults of both. Berlese was of the opinion that the endopterygote larval stages corresponded to embryonic forms and that all of the exopterygote pupa.

Poyarkoff suggested that the adult exopterygote corresponded to the pupal endopterygote and that the adult endopterygote was an *extra* adult stage. Hinton draws attention to examples of "pupal" forms among the exopterygote orders—Thysanoptera (thrips), Coccidae (mealy bugs), and Aleurodidae (greenhouse white flies). In these, the general structure of the feeding larva or nymph has departed very widely from that of the adult; the structural differences are bridged in the last two nymphal stages. As the differences between the stages have become greater, this has involved a greater degree of structural reorganization to bridge the gap of the internal tissues. As a result, the last two nymphal stages. Instars have become more and more quiescent and have eventually ceased to feed, becoming closely analogous to endopterygote pupae. The structural reorganization required to bridge the gap, in some cases, may be greater than is found in some primitive endopterygotes.

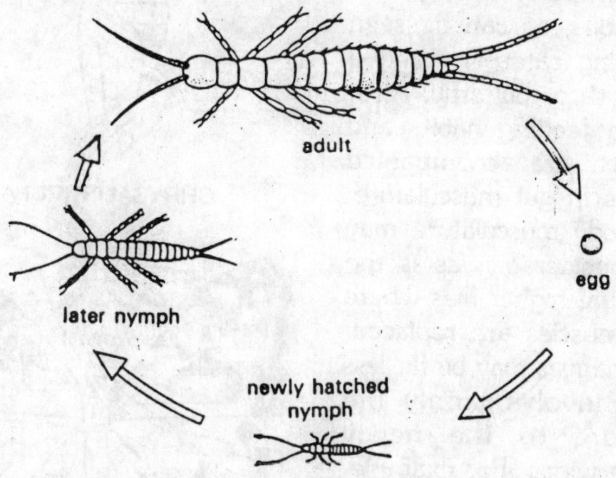

Fig. 3.1. Lepisma (sliver-fish). Stages of life history exhibiting no metamorphosis (ametabolic metamorphosis).

General Changes Involved During Metamorphosis

Besides the development of wings, compound eyes, and of external and internal genitalia, one of the other main changes that occurs in metamorphosing insects may be in their feeding habits. Thus, a caterpillar and a moth or butterfly, or a fly maggot and the adult fly, may have completely different habits, and these are reflected in structural changes in the mouth parts, gut, salivary glands, and gut muscles. In some frogs, a change from the larval herbivorous diet to the adult carnivorous one is accompanied by a shortening of the gut at metamorphosis. In a fly, the gut shortens also, and the larval crop, larval caecae, and larval salivary glands all disappear the very much shorter adult gut develops a crop and an elaborate rectum with characteristic rectal papillae.

Clearly, changes in mouth parts accompany changes in feeding habits, as can be seen in a chewing caterpillar and a sucking moth or butterfly. Also, changes in feeding habits and mouth parts are accompanied by changes in gut musculature. General body musculature may change considerably, as is the case with the higher flies where all larval muscles are replaced, or these changes may be far less extreme, involving only the musculature to the newly developing wings. For example, insects such as the alder fly (Neuroptera), scorpion fly (Mecoptera), or even the generalized members of some of the more specialized families such as Lepidoptera and Diptera, retain essentially larval musculature in the abdomen.

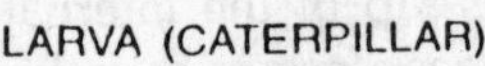

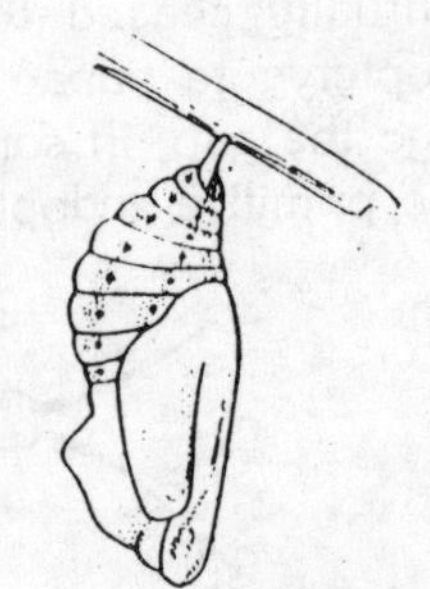

Fig. 3.2. Butterfly (Lepidoptera). Stages of life history exhibiting holometabolic (complete) metamorphosis.

The motile pupa of a mosquito or of a chironomid nonbiting midge is very different from the immobile pupa of a higher fly in which all of the larval muscles are broken down. In the case of an amphibian such as the frog, metamorphosis involves the migration from water to land, and correlated changes in the respiratory system occur. In the pterygote or winged insect there are changes in the respiratory system of the terrestrial larval form during metamorphosis to the flying adult form. The elaborate adult thoracic musculature is heavily tracheated, often with an equally elaborate system of air sacs. A simpler system of air sacs is also found in aquatic forms, where one function would be to provide buoyancy. Insects appear to have evolved from terrestrial ancestors with well developed tracheal systems, paralleled today in the Myriapoda. Aquatic forms therefore are secondary and are found with normal tracheal systems that may have closed spiracles; these become open and functional in the pupal and adult stages. Aquatic larvae may have tracheal gills or a superficial dense tracheation beneath the thin body wall, but these are lost during metamorphosis to the terrestrial flying adult form.

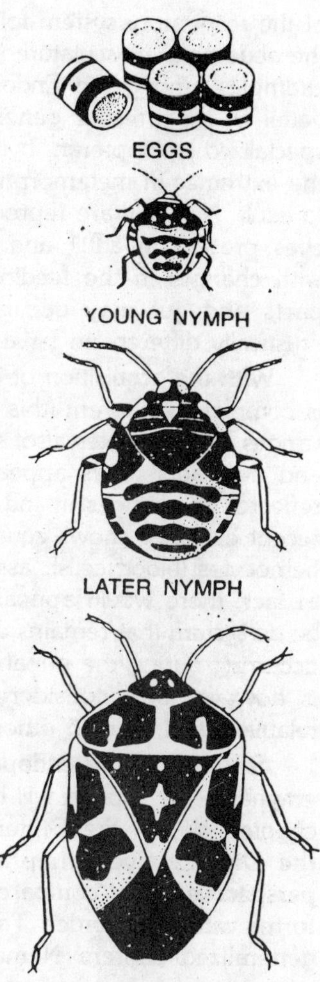

Fig. 3.3. Stink bug (Hemiptera). Stages of life history exhibiting heterometabolic metamorphosis or to say gradual metamorphosis.

The tracheal system is intimately involved in the metamorphic changes undergone by the other organ systems. For instance, it has been found within the Diptera that the degree of breakdown of the respiratory system follows closely the degree of histolysis of the abdominal musculature (Whitten, 1960). The most specialized examples among the Endopterygota will be considered in most detail in the present general survey, for it is here among the specialized lepidopteran, hymenopteran, or dipteran that one finds the extremes in metamorphic change as one proceeds from larva to adult. Not only are reproductive systems, wings, and compound eyes present in adult and absent in younger larval stages, but with changes in the feeding habits extensive changes in mouth parts and gut may occur, and muscle requirements may be drastically different in larva and adult.

With the acquisition of a flying habit the sensory environment is completely different (this occurs to a lesser or greater extent in various exopterygotes also) so that larval sensillae are broken down and new adult ones appear. Thus, drastic internal changes are reflected in changes in the respiratory and nervous systems, and recent evidence shows equally elaborate changes occurring in the hemocytes (blood cells), associated with changes in other tissues. In fact, there would appear, in the specialized endopterygote, to be no system that remains unaffected by the metamorphic changes occurring during the pupal stages. The specialized endopterygote is however best considered first in relation to its generalized relatives, and then to other endopterygotes.

Since all of the endopterygote orders cannot be given equal emphasis, the Diptera will be treated in some detail in the present chapter. Within the Diptera, the specialized problems found in the Diptera Cyclorrhapha (higher flies) are placed in better perspective when compared with those of the more generalized forms within the order. The mosquito is a member of the more generalized Diptera Nematocera and it is significant that the extremes of larval tissue histolysis and reorganization are absent here. Clements (1963) has stated for the mosquito that "metamorphosis starts early in larval life with slow development of imaginal buds, compound eyes and gonads, processes accelerated in the fourth instar, and largely completed in the pupa. It is clear that important changes continue to take place in the first days of adult life that only in the older adult is metamorphosis completed."

Pigmentation of the adult eye ommatidia apparently occurs in *Culex pipiens*, by the second larval instar! The evolution of extremes in metamorphosis is therefore seen in many cases to have occurred within the higher endopterygote orders, just as there has been a general progressive evolution in the insects as a whole form the a—metamorphic apterygotes, through the exopterygotes to endopterygote forms. It should be noted that apterygote a metamorphic insects have a complement of endocrine organs exactly comparable with those of the "higher" pterygotes (Watson 1963). It would appear that the juvenile hormone has the same relative role—that of blocking the factors responsible for differentiation of tissues into adult forms, and a second role, that of stimulating yoke deposition in adult females. In all insects ecdysone has the same function also, that of inducing molting. Changes in hormone balance influence the *progress* of metamorphosis as will be seen in this and succeeding chapters, but the endocrine organs themselves do not *cause* the differences between a non-metamorphosing and a metamorphosing insect.

The question of why the cockroach begins life as a wingless individual or why a fly has evolved into a form that begins life as wingless, legless, eyeless maggot remain unresolved. It is possible that, as Berlese conjectured, the latter represents an embryonic form, hatching prematurely. In the vertebrates the old concept of "ontogeny recapitulating phylogeny," where embryonic features

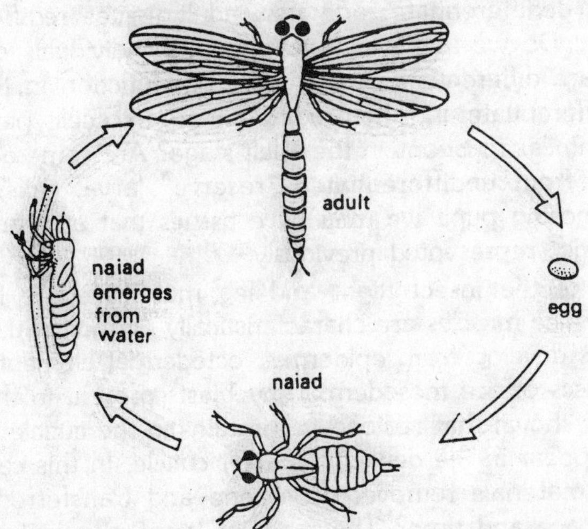

Fig. 3.4. Dragon-fly (Odonata). Stages of life history exhibiting hemimetabolic metamorphosis or to say incomplete metamorphosis.

tend to be retained while the adult form is more susceptible to change, is seen for example in the transient development of gill arches and circulation in the embryos and early stages of development of higher vertebrates.

In an amphibian such as the frog we find the aquatic tadpole, with its fishlike respiratory and corresponding circulatory systems, "metamorphosing" into the adult, terrestrial amphibian with lungs, no gills, and modified gill-arch circulation. If the ancestral insect is a symphylan-diplura type, and the pterygote insect larva is prematurely hatched in the ancestral wingless form, then the acquisition of wings at metamorphosis closely parallels the amphibian condition, with its acquisition of legs at metamorphosis. These are academic questions, the answers highly speculative, but the facts of metamorphosis remain significant to the biologist, with the insect and amphibian remaining striking examples with which to study the process.

Problems Posed By Insect Metamorphosing Tissues

In a metamorphosing pupa one finds various larval tissues degenerating at specific times. What initiates cells death? Can this predetermined event be experimentally altered, delayed, or permanently prevented? How far back in this is the process determined? Also in the developing pupa we have tissues seemingly regress or dedifferentiate, and new adult tissues redifferentiate from them. Do we know what happens to individuals cells? Do cells that are differentiated for a particular function in the larval stage dedifferentiate and then redifferentiate into cells performing like or dissimilar functions in the adult stage? Are some adult cells developed from undifferentiated "reserve" larval cells? In the metamorphosing pupa we may have tissues that appear as new structures not represented previously.

Some of the insect flight and leg muscles come into this category. Since muscles are characteristically mesodermal, can the leg muscles develop from epidermal, ectodermal elements of the imaginal discs or are mesodermal myoblasts present in the discs? It has been shown that radioactive proteins in the cuticle of larval insects reappear in the developing adult cuticle. In this case, how are these materials removed from one and transferred to the other in space and time? The mass of larval muscle in a fly is relatively enormous. At the time of muscle breakdown are the materials lost to the body or is there conservation and use in the build-up of adult tissues? Are the blood cells involved to any great extent in tissue breakdown? Are the cells destined to cell death autocytolyzed or are living cells attacked by phagocytic blood cells? Are there ultracellular components comparable to focal degradation points in other animals, and how general are lysosomes? How does a tracheal system laid down at a previous instar solve the problems of providing respiratory gases, firstly to larval tissues destined to die, secondly to larval tissues to be retained, and thirdly to adult tissues not yet differentiated. all requirements within the same pupal instar? A vital and fundamental problem involves the role of the central nervous system. How much metamorphosis occurs in this system? Are we any further forward now than we were in 1952 when Power, in one of a remarkable series of papers on the *Drosophila* nervous system stated, "The morphogenesis of the insect's nervous system is still largely unexplored in spite of a large bibliography." Among the problems that remained unanswered he named the basic sequence of events in the origin and differentiation of the component parts of the central nervous system, the fate of the peripheral nervous system at the time of metamorphosis, the development of the innervation of the imaginal organs, and the changes that occur in the nervous system correlated with the changes that occur in the various end organs.

An understanding of the events occurring in the nervous system is basic to any understanding of metamorphosis of individual organs, yet work on the central nervous system during metamorphosis is far behind that on the endocrine system. Even here it has been shown possible to change, enhance, or retard metamorphic events by altering the hormonal environment, yet we are still left with the basic problem of control of the endocrine organs themselves and, ultimately, of the brain neurosecretory cells. What, e.g., are the controlling mechanisms behind the circadian rhythms of development in the metamorphosing *Drosophila* pupa, shown so elegantly by Harker (1965a, 1965b)?

The Significance of Metamorphosis to General Biological Problems

Embryology

In the more extreme cases of metamorphic change there is a virtual return of most tissues to an embryonic state with the constituent cells in a undifferentiated form. As metamorphosis proceeds, differentiation of individual cells of the various tissues occurs. Such is the case in a higher fly or bee where the muscular system is entirely broken down and reorganized, and there are correspondingly elaborate changes in the gut, general epidermis and tracheal system, with correlated reorganization of the nervous system. One of the most valuable aspects of insect metamorphosis is as subject material for study of embryological and general developmental problems. Problems of cell and tissue differentiation, of pattern formation, of cell determination and of cell death can all be studied in the metamorphosing insect. Of primary significance also is the fact that insects are amenable to experimental techniques such as those of implantation (e.g., Bodenstein 1950 and Hadron 1951).

Genetics

Studies using the fly *Drosophila*, and to a lesser extent, the hymenopteran *Habrobracon*, have played a large part in our understanding of genetical problems; these are essentially studies of metamorphic events in these insects. The various eye mutants or wing mutants or mutants showing abnormal bristle pattern result from the action of various mutant genes. However, an understanding of the steps which intervene between the formation of the initial protein and the manifestation of the visible effect, is in essence an understanding of the events of differentiation within the metamorphosing pupa, at least where the effects are on the adult phenotype. There may be a large number of developmental sequences between the activity of the mutant gene and the expression of for example, "vestigial wing" in *Drosophila*. Studies on the formation of biological "patterns" are often also metamorphic problems, as in the case of bristle patterns in *Drosophila* (e.g., Sondhi, 1963).

Chromosome Structure and Gene Action

Insects have another tremendous advantage over other organisms in possessing giant chromosomes. It is unfortunate that they are known only from the order Diptera among the insects, but at the same time it makes this order of greater interest. These giant polytene chromosomes arise by chromosomal DNA replication in the absence of corresponding cell division. They are best known from larval salivary glands; in *Drosophila* has been possible to "map" the different bands of the salivary cell chromosomes, so that specific regions represent the positions of particular genes whose action has been determined by genetical means.

Giant chromosomes are characterized by regions of intense "puffing" activity. These regions represent areas of active RNA synthesis. The RNA is thought to be messenger RNA which passes into the cytoplasm, where in association with the ribosomes, it organizes the amino-acids into particular sequences to form specific proteins. In this way the origin of a specific protein could be traced back to the activity of a specific puff at a definite position on the giant chromosomes. Cells other than salivary gland cells may contain polytene chromosomes, although they are relatively smaller. These include cells of Malpighian tubules, male seminal vesicles, nurse cells of the ovary, heart cells, pericardial cells, and even fat body cells. Recently, cells having chromosomes as large as salivary gland chromosomes have been recorded for epidermal cells of the fly foot (Whitten, 1964a). They are responsible for secretion of the various layers of cuticle forming the dorsal surface of the adult foot-pad of the higher flies. These cells are thus intimately involved in general adult "development", whereas salivary gland cells have the disadvantage of undergoing cell death at the beginning of the pupal stage.

Changes in salivary chromosome puffing, brought about by changing the hormonal milieu, have been correlated with effects on "general development". Titles such as "Regulation of gene action in insect development" (Kroeger and Lezzi, 1966), and "Chromosome changes associated with differentiation" (Clever, 1965), indicate the extent to which studies of salivary gland chromosome puffing activity have attempted to relate events at the chromosomal level to events at the level of the whole organism, in this case the metamorphosing pupa. Giant epidermal cells, such as those of the foot-pads in the higher flies, should prove of particular interest in pursuing these problems. Questions of growth by cell division, as opposed to growth by cell size, have also been raised as a result of studies involving metamorphosing insects, particularly the mosquito (e.g., Trager, 1937 and discussed in Clements, 1963).

It has been found that as a rule, cells which increase in size (for example of salivary glands, Malpighian tubules, rectum, abdominal muscles, trichogen cells, larval oenocytes) are incapable of cell division, and expect for Malpighian tubules are incapable of transforming into adult organs, so that these then are the cells which histolyse in the metamorphosing pupa. During growth, replication of chromosomes proceeds without separation and without cell division, resulting in the characteristic polytene chromosome. On the other hand cells that multiply by cell division remain small, and are retained and pass through into the adult; presumably they are undifferentiated and therefore capable of subsequent differentiation.

A seemingly unique condition is found in the various mosquitoes—that of *endopolyploidy* with somatic reduction. Here, in the small intestine, and to lesser extent in other parts of the gut, epidermis, tracheae and neurilemma, cells become polytene but apparently not differentiated. In *Aedes* the cells may be 16 N with 48 instead of the normal 6 chromosomes, with the homologues closely associated in a polytene chromosomes. However, at the onset of population the polytene chromosome splits up and cell division occurs, without chromosome replication. In this way there are produced many small diploid cells which differentiate to reconstitute the small intestine and rectum. An answer to the question of why this should happen is not at present available.

Gross Morphological Changes at Metamorphosis

One of the most striking aspects of insect metamorphosis, particularly to the casual observer, is the dramatic change from larva to pupa at the larval-pupal ecdysis. This is all the more remarkable when one realizes that the pupal "mold" produced at the larval-pupal molt follows the future adult shape, particularly with respect to the large muscle-filled thorax. Yet at the time of its formation none of these adult structures may be developed in some instances, as is the case with the higher flies. The mechanical aspects have been discussed at length recently in a review article by Cottrell (1964): The external shape of the insect is largely limited by the inherent form of the different parts and more particularly by the extent to which the outermost layer of cuticle—the epicuticle—can be unfolded. Then again, the unfolding of the epicuticle and the stretching of the underlying endocuticle is brought about by hydrostatic pressures produced by two activities: swallowing of air or water and the contraction of muscles. The relative importance of these two may differ in different insects. Final form is also attained by the expansion of soft, not yet expanded parts, but not of any prehardened areas. There is a very exact relationship between the two. Kohler has shown that the eversion of the wing buds in *Ephestia*, the clothes moth, is brought about by hydrostatic pressures and by muscles opening the mouths of the peripodial cavities. Fraenkel showed that in *Calliphora* eversion of the pupal head is brought about by peristaltic abdominal contractions. Cottrell discusses other examples also. Personal observations with *Drosophila* and *Sarcophaga* have shown the presence of a series of larval suspensory muscles that seem to play a role in attaining pupal form.

Typically, the cyclorrhaphan nephrocytes are shown as slung from the two lobes of the salivary glands. In fact, their attachment in both *Drosophila* and *Sarcophaga* (and probably other Cyclorrhapha also) is by way of a muscle that runs along the proximal third of each lobe of the salivary glands, with firm attachment between gland, muscles, and nephrocytes being maintained by the continuous connective tissue basement membranes (micrographs in Whitten, 1964). These muscles are attached at one end near the spiracles and at the other on the gut. Two other muscles, with origins in the prothoracic region and insertions asymmetrically on the midgut, pass through the body cavity to their points of attachment on the midgut. These muscles, show rhythmical contractions and resemble the alary muscles, being clearly striated under phase optics. In the fly, a "cryptocephalic pupa" has its head still invaginated, but leg and wing discs are everted though flaccid and unexpanded; the abdomen is larval in shape being disproportionately long, and larval tracheae are still within the body. Simultaneously with eversion of the head, there is extension of the wings and legs and contraction of the abdomen. Contraction of the abdomen must reduce the internal volume and help force blood into the legs and wings and head. A large air bubble present before, and absent after eversion of the head, also probably plays some part.

Observations of internal changes reveal a dramatic overall contraction of the entire system of longitudinal visceral muscles and unpaired suspensory muscles. This simultaneous contraction of the muscles effects the shortening of the larval gut to the dimensions of the much shorter adult gut. At the same time, contraction of the longitudinally running visceral muscle through the midgut caecae reduces these to four small knobs on the gut.

Their cells later histolyze, and these structures are not represented in the adult gut. At the same time, the contraction of the "slinging" or suspensory muscles plays a role in the form of the pupa and future adult. By anchoring the gut and salivary glands they seem to help prevent their being everted into the head along with the brain. This is probably also helped by muscles such as the alary muscles, which also attach to tracheae. At the same time the muscles, whose origins are on the body wall behind the prothoracic spiracles, may help during their contraction to draw in the "neck" region between head and thorax. Nothing is fortuitous, and the establishment of shape in the pupa is a very precise process. Attainment of adult shape is equally intriguing.

Clearly, it is genetically determined that the slightly bilobed saclike pupal leg appendage of the higher fly should develop by epidermal cell retraction, cell division, and differential growth into an adult leg, yet little is known of the factors that determine this shape. There is a suggestion in the development of the fly foot that the ecdysial membrane (c.f. Taylor and Richards, 1965), the *first* secretion of the adult epidermal cells, may play a mechanical role in maintaining shape in the delicate, otherwise naked cytoplasmic extensions of the various epidermal cells that form hairs of various sizes. Once the adult structures are laid down, it can be seen from Cottrell's account that the adult shape is attained by a dual process—by the prehardening of structures that will *not* expand after ecdysis (the various leg bristles and the terminal pretarsal structures shown in the fly foot) and the postecdysial expansion of soft structures that subsequently harden, e.g., the wings and general abdominal cuticle. Evidently, timing is delicate and there are elaborate control mechanisms involved. Again it is with the flies that much work has been done.

Fraenkel and Hsiao (1962) and Cottrell have separately found a "darkening factor", named by former "bursicon". More detailed discussion is given in Cottrell's review. It is sufficient here to note that the factor is released after the flies have ceased "digging" their way through the soil and that darkening can be produced in these "digging" flies by injection of blood from air swallowing, darkening flies. The cessation of digging and the onset of air swallowing and darkening are closely interrelated; "shut off" of digging movements is permanent. It cannot be related to the hardening process and, according to Cottrell, the mechanism of "shut off" is probably nervous. Push through the ground is accompanied by and partially brought about by alternate eversion and retraction of the blood-filled sac at the anterior end of the head—the ptilinum. The muscles producing the movement breakdown within three days of emergence, demonstrating the remarkable co-ordination of events. The nervous stimulation for cessation of digging may well result from signals received (or no longer received) by the tiny tactile hairs dispersed over the surface of the ptilinum (Whitten, 1963c;). Each hair possesses a sensory nerve fiber, and avoiding reactions with soil particles are seen to result from stimuli received by these hairs.

When the fly appears above ground, no resistance is felt by the ptilinum, and signals cease to be received by the sensory nerve fibers. This in itself could *initiate* the chain reaction which involves retraction of the ptilinum, air swallowing, darkening, and finally breakdown of the ptilinal muscles. Particularly puzzling is that if the darkening factor is a hormone whose release is necessary for the final synchronous darkening of all areas to be hardened and darkened following expansion, what is responsible for the progressive predarkening of areas such as the legs before ecdysis? This occurs from thorax forward and backward and proximodistally along the legs. The predarkening process has been used as a very reliable aging device, e.g., for general development aging in *Drosophila* (Harker, 1968a, 1965b) and in the development of the fly leg (Whitten, 1968). Are the preecdysial and the postecdysial hardening and darkening processes really controlled and effected by different agents?

Developmental Rate in Metamorphosing Pupae: Internal Clocks

A very different but equally intriguing aspect of the metamorphic process occurring between larval and adult stages can be followed in two papers by Harker in a study of developmental rates and their controlling mechanisms. Working with the fruit fly *Drosophila melanogaster*, Harker (1965a) first studied the course of development in two strains in an attempt to establish the factors affecting the time taken to complete three stages of: head eversion to yellow eye, yellow eye to wing pigmentation; wing pigmentation to eclosion. Time measurements were taken for pupae entering each stage at each particular hour of the day when insects were kept at 12-hour light; 12-hour darkness; 12-hour bright light: 12-hour dim light; or in continuous darkness. She found that the duration of each stage in both strains is affected by the time of day relative to the light cycle at which the stage is entered. The duration of each stage for pupae kept in continuous darkness is affected by the time of day at which the stage is entered, relative to the light cycle to which they had been exposed as larvae. She further found that the time interval curves for all three stages of any one strain take the same form.

Consequently, because of the very wide range of development rates, dependent on the time within the light cycle at which each stage begins, a population in the larvae all pupate within a 24-hour period will continue to produce adult flies over several days. The eclosion rhythm is a population effect and does not reflect the phasing of individuals at a dawn eclosion; the majority of individuals emerge at dawn because of the summation effect of circadian rhythms of development at earlier stages. In a second series of experiments, Harker (1965b) studied the time taken for *D. melanogaster* to complete the same three stages of development for pupae entering each stage at each particular hour of the day in cycles of 12-hour light; 12-hour darkness; 4-hour light: 20-hour darkness; 18-hour light; 6-hour darkness. Similar experiments were carried out on insects in which the larvae were subjected to the light cycles but the pupae remained in constant darkness.

It was found that the duration of each stage is affected by the time of day, relative to the light cycle, at which the stage is entered. When pupae are in constant darkness, the development rate of each stage is affected by the time of day at which the stage is entered relative to the *particular light cycle* to which the larva had been exposed. It was further shown that the eclosion rhythm, arising as a "summation effect" of rhythms of development at earlier stages, may become bimodal in light cycles with suitable photofractions. It was further found that the rate of development of a pupa entering a stage during the light period is related to the time interval since the light "on" signal; the preceding dark period has no effect. Finally, the development rate of a pupa entering a stage during the dark period is affected by the time interval elapsing since the light "off" signal, but may also be affected by the previous light "on" signal, although there is no simple relationship between them.

Harker concluded that as the development rates are maintained in constant darkness, the rate is affected by factors following a *diurnal rhythm*. The form of the rhythm is determined by both light "on" and light "off" signals, but the *timing* of the rhythm is determined by the two signals acting independently of each other.

Metamorphic Changes in Individual Systems

In order to follow the types of problems presented by individual tissues one needs to look at each system in turn, for each presents its own characteristic problems. Cell differentiation and pattern formation may be best studied in various epidermal cell types, cell death perhaps in muscle systems, along with considerations of blood cell relationships, cell determination in the various imaginal rudiments, and controlling mechanisms in endocrine and nerve - tissue relationships. As general a picture as possible will be given in the space available, but slightly more emphasis will be given, in tracing sequences in individual systems, to members of the Diptera Cyclorrhapha, relating them as far as possible to events occurring in the lower Diptera Nematocera and other insects. For each system the same question will be asked. "In what way does the mosquito (Diptera Nematocera) differ from the fly (Diptera: Cyclorrhapha), and how then do other insects compare with both?".

Metamorphic Changes in the Alimentary Canal

Changes occurring in the alimentary canal vary from one insect type to the next and cannot be covered comprehensively here. The cyclorrhaphan system again will be taken as an extreme example of the type of changes that occur. The feeding apparatus in larva and adult are quit different: It has been mentioned earlier that the larval gut is longer, possesses a larval "crop" and four midgut caecae and is otherwise more or less a long, straight tube. The adult gut lacks the crop and midgut caecae, and is characterized instead by the development of an adult crop and of a rectum containing four "rectal papillae" of characteristic appearance but until recently of little known function. The mouth parts of the larva consist of a pair of strong mouth hooks; the adult mouth appendages—labial, maxillary, and mandibular—are highly specialized and arise from the median portion of the frontal sac of the larva, an imaginal rudiment lying between the brain lobes. Changes in the gut during metamorphosis involve degeneration of the midgut epithelium, disappearance of the midgut caecae, disappearance of the labial "crop" and the development of adult crop and rectum from two "imaginal disc" areas located at the junctions of foregut and midgut and midgut and hindgut, respectively. Also, the larval salivary glands breakdown, and adult salivary glands develop from an "imaginal disc" area located between the base of the larval gland and the duct. From the undifferentiated cells of this area the adult glands arise by cell mitosis, growth and differentiation. Other imaginal areas may include labial and a ring of peri-anal cells. Co-ordinated longitudinal contraction of the gut muscles and those of the midgut caecae bring about shortening of the gut and reduction of the caecae to stumps. This contraction throws the basement membrane into elaborate folds.

Slightly later, the basement membrane breaks down as does that the other tissues in the pups (Whitten, 1962a), and the gut lining cells degenerate, adult epithelium being formed from small interstitial cells in the gut wall. The age of a midpupa can be fairly accurately followed by determining the length and degree of development of the adult crop. In the cyclorrhaphan pupa the Malpighian tubules are retained throughout pupation and evidently function at least in the late pupa, where they are filled with a milky material. This is passed into, the accumulates in, the rectum and later partially contributes to the gut content or "meconium", which is eliminated from the gut at the time of emergence. Virtually nothing is known of muscle reorganization. Do the adult visceral muscles arise from dedifferentiated gut muscle cells or do some or all arise from myoblasts, as has been suggested in older literature and again recently by Crossley (1965) for certain of the abdominal skeletal muscle cells in *Calliphora*, and by Daly (1964) for a large proportion of the thoracic muscles in *Apis*, the bee?

In the case of the fly gut, the muscles apparently disappear, with adult muscles appearing in the late pupa. However, it is very probable that the muscles dedifferentiate, to redifferentiate later in a fashion similar to the cyclical muscle changes show in *Rhodmus* by Wigglesworth (1956). Here the muscle fibers disappear between molts, with only the sheath and nuclei remaining; then at the onset of each molt the muscles are completely differentiated again. Changes in the mosquito gut are essentially similar (Clements, 1963). The phenomenon of endopolyploidy, discussed earlier, is of particular interest.

Metamorphic Changes Undergone by the Heart

The heart during metamorphosis is an organ that has received relatively little attention. It reportedly ceases to function in many insect pupae, and in some it has been described as reversing the direction of heartbeat. In *Sialis*, the alder fly, a very fine study has recently been made, by Selman (1965), on the circulatory system. In this species he notes that, as is the case in other larvae including those of dipteran flies, the larval heart is not innervated, whereas for some at present inexplicable reason the adult heart is innervated. In general, the heart changes relatively little during metamorphosis. For instance, in the mosquito (Clements, 1963) the heart and associated pericardial cells are said to pass unchanged from larva to adult. Personal observations on *Sarcophaga* and preliminary work on *Drosophila* suggest a closely similar pattern of events for these cyclorrhaphans. The heart is seen to pass without dissolution through to the adult stage, but in the midpupa metamorphic changes do occur. Also the larval posterior tracheal plexus is absent.

The heart wall itself becomes somewhat thickened, and this thickening does not extend to the aorta. Correlated with the increase in wall material, the intrinsic heart cells become active and display well-developed polytene chromosomes (Whitten, 1963a). The smaller anterior and the large posterior pericardial cells undergo changes in the midpupal stage, and the larger cells also develop well defined polytene chromosomes. The thickening of the heart but not of the aortal region produces a humped effect at the junction of the two regions, and this comes to lie in the raised abdominal region immediately posterior to the constriction between adult thorax and abdomen. In a midpupa the heart muscles have degenerated and degeneration is also seen in the alary muscles. In these latter, there would appear to be events occurring that are comparable with the cyclical muscle dedifferentiation and redifferentiation in the abdomen of *Rhodnius*. In a fly midpupa, heart pumping ceases and there is virtually no circulation of blood. Developing adult nerves can be followed migrating over the heart surface of a mid to late pupa.

An attempt to review events recorded for other insects will not be made; the scarcity of information concerning metamorphic changes can be seen from the recent review of Jones (1964). The absence of circulation during the midpupal period would seem to be pertinent to questions of hemocyte relationships in the pupal stages, and these will be discussed later in more detail. There are many other peripheral issues in connection with the developing adult circulation during metamorphosis, including the development of pulsatile organs, dealt with in current textbooks on entomology. As with all the other systems, the more one looks into the problem the more one realizes how little, relatively, is known of the subject. Few insects have their circulatory systems as thoroughly investigated as has the alder fly *Sialis* in Selman's recent study.

Metamorphic Changes in the Fat Body, Oenocytes and Nephrocytes

The fate of the fat body varies in different insects. For instance within the Dipters, in a generalized dipteran such as the mosquito, it is said to pass from larva to newly emerged adult unchanged and little if at all diminished in size (Clements, 1963). Changes in the higher flies are however somewhat more complex. It was formerly considered that the larval fat body of the fly broke down in the early pupa and thereafter ceased to function (e.g., Bodenstein, 1950). However, what happens at pupation is that the connective tissue sheath (basement membrane) that binds the cells into a composite organ breaks down and release the component cells (Whitten 1962a). The breakdown, as with all the events in the metamorphosing pupa, is perfectly timed and co-ordinated.

Breakdown of thoracic segments of the fat body occurs in the cryptocephalic pupa while the abdominal portion breaks down later, in the early pupa. Although appearing diffuse, the fat body arises segmentally and functions as such in later life. A branch of a mid-dorsal nerve has been traced to the fat body in *Sarcophaga*, and the basement membranes of the nerve and fat body are confluent. The fat body in this fly would therefore seem to be innervated. The cyclorrhaphan larval fat body cells function right into the adult stage. They have been the subject of much recent research, particularly in *Drosophila*, where they seem to play an important part in intermediary metabolism in the developing pupa. Changes in the fat body are characteristic of certain *Drosophila* mutants.

In many insects including flies and moths a separate adult fat body originates within the pupal stage. Older workers have suggested that adult fat body cells may have originated from hemocytes. The whole subject is by no means clear. Evidence in the Diptera Cyclorrhapha recently has suggested that this is not as unlikely a possibility as might at first be supposed (Whitten, 1964b). Phagocytic hemocytes, which at the time of pupation ingest larval fragments, later undergo a second cycle of activity. They persist right into the adult stage, and at this time appear very like the adult fat body cells. The mosquito is seen not to have a distinct adult fat body, and the larval fat body is said not to histolyse in the early adult (Clements, 1963).

Certainly phagocytic hemocytes would be relatively much less active in the mosquito where tissue histolysis is less extensive. *Oenocytes* are cells which are thought to be involved in intermediary metabolism, in growth and in reproduction (Wigglesworth, 1965). In many insects including the mosquito and the fly there are two generations—larval and adult. The larval oenocytes in the fly are large cells with polytene chromosomes, located segmentally in groups along the side of the body. Adult oenocytes are smaller, appear in the pupal stage and are often associated with the adult fat body cells.

In the mosquito the larval cells grow in size, develop polytene chromosomes, and breakdown in the pups. The many smaller adult oenocytes appear in the abdomen during the fourth larval instar (Clements, 1963). In the fly, *ventral nephrocytes* thought to be excretory in function and related to pericardial cells, are slung as a garland between the lobes of salivary glands [attached by muscles and basement membrane material]. They break down in the early pupa, at the time of disappearance of the salivary glands. Comparable cells are present in the mosquito where in some species they also form a chain between the salivary lobes.

Metamorphosis of the Muscular System

Among the Endopterygota there is a considerable range in the degree to which muscles are retained from larva to adult. In some of the more primitive orders, but also in some of the more primitive members of the more specialized orders including the Diptera, the majority of the larva muscles are carried over into the adult stage. The muscles, particularly of the abdomen, are carried over and changes occur mainly in the muscles of the thorax that are responsible for flight. In the mosquito for example, larval abdominal muscles are used in the pupa for swimming and these degenerate early in adult life. The adult abdominal muscles arise from myoblasts in the pupal stages. However, among the specialized from larva to adult. In the thorax of the honeybee, a very comprehensive survey made recently by Daly (1964) should that no larval muscles pass unchanged into the adult, that 85 percent of larval muscles are associated with the development of adult muscles, and that 15 percent of the larval muscles degenerate and have no relationship to adult muscles. Some 61 percent of adult muscles arise from aggregations of myoblasts, including those of appendages and direct flight muscles.

Crossley (1965), in a recent study on the blowfly *Calliphora erythrocephala*, similarly showed the origin of certain of the abdominal muscles from myoblasts, an observation substantiating the opinions of several of the older workers at the turn of the century (e.g., Perez, 1910). The description by nuclei that subsequently differentiate into adult muscles in the same location as the larval is difficult to conceive but is all the more fascinating if actually the case. In recent years interest has centered largely on the breakdown of muscles and the cause of the breakdown. This remains one of the most intriguing problems of insect metamorphosis.

Of particular interest is the fact that not all larval muscles degenerate at the same time. Some are retained while their neighbours degenerate. Initial work on this subject was in the Lepidoptera (see e.g., Finlayson, 1956). More recently, an extremely interesting series of papers have appeared by Lockshin and Williams (1965), also on the Lepidoptera. Whereas the earlier workers concentrated on muscles that were carried over from the larva into the early pupa and then degenerated. Lockshin and Williams studied muscles that passed into the adult and were involved in emergence, these degenerated three days later under normal conditions. Similar differential breakdown occurs in other insects: e.g., in the cyclorrhaphous Diptera certain larval muscles do not degenerate with the remainder after puparium formation (Cottrell, 1964), but they do so some 48 hours later.

Likewise, there are muscles that degenerate in the early adult, including those of the ptilinum and other muscles expressly involved in the process of eclosion. The ptilinal muscles degenerate following the sequence of events that starts with cessation of digging. Similar controlling mechanisms to those shown by Lockshin and Williams for the lepidopteran may apply in these and other cases. They have concluded that the initiation of cell death in the muscle cells is brought about by the coordination of several factors-humoral, nervous, and an ability of the cell to respond to the factors. Thus, the breakdown of certain intersegmental muscles in the adult of first potentiated by exposure to the molting hormone ecdysone during the first few days of adult development. Three weeks later, when development is completed, actual breakdown is triggered by a neural mechanism.

The muscle cells attain the capacity to respond to this during the three days before emergence. Johnson (1959) has shown that where degeneration of flight muscles normally occurred in adult aphids this was prevented by decapitation. Further, the thoracic muscles of decapitated aphids broke down when they were joined in parabiosis with intact aphids. The complete story may well be similar to that shown for the postadult muscle breakdown in Lepidoptera. Much has been done and much remains to be done on this subject.

It is an important biological problem because the whole question of "programmed cell death" is involved. The problems of muscle dedifferentiation and subsequent redifferentiation are equally interesting. Why should the muscles of *Rhodinus* (Wigglesworth; 1956) dedifferentiate between molts, and shortly before molting redifferentiate contractile muscle fibers? The same problem is posed in the metamorphosing pupa by the larval muscles that the transformed into adult muscles, constituting the 39 percent of the total thoracic musculature in the adult honeybee (Daly, 1964). Reorganization of the alary muscles in the fly pupa presents a similar problem. Another problem is posed by muscles that develop in the pupal stage and degenerate before emergence.

The functions of muscles of this type are discussed at length by Daly for the honeybee, and their possible involvement in determining shape is also considered. Hinton (1961) has described how the disposition of the indirect flight muscles of insects can be brought about at the time of the larval-pupa molt by activity of the epidermal cells. He demonstrates this with *Simulium*, which as a generalized nematoceran member of the Diptera has (unlike the more specialized Cyclorrhapha) its epidermal cells carried over from larva to adult. The indirect flight muscles at the time of the larval-pupal molt are undifferentiated strands, attached to the epidermis but in relative positions quite unlike the relative positions of the corresponding future adult muscle. He concludes that the definitive shape of the adult mesothorax and the final position and orientation of the skeletal muscles are determined by the way the epidermis grows between the time of secretion of pupal cuticle (larval-pupal molt) and the time that the larval cuticle is shed (larval-pupal ecdysis).

The Controlling Centers of Metamorphic Change: Endocrine Glands and the Nervous System

A brief look will be taken of the hormone system that controls the metamorphic changes in the various organs. As, was seen earlier, the endocrine system is essentially similar in all of the groups of insects. The corpus allatum secretes juvenile hormone, the prothoracic (ecdysial) gland cells (and the ventral gland cells, their homologues in the apterygote thysanurans) secrete ecdysone, the molting hormone, while the corpus cardiacum, which has been assigned various functions, contains distinct glandular and storage regions. The cephalic endocrine system in the Thysanura Apterygota is seen to resemble closely that of the generalized pterygote (Watson, 1963). It is currently thought that the brain hormone, secreted by neurosecretory cells of the brain, stimulates the prothoracic glands as the principal target organs.

The secretion of ecdysone in the presence of juvenile hormone in the winged pterygote insects produces a juvenile form; the secretion of ecdysone in the presence of a reduced titer of juvenile hormone initiates metamorphosis, so that at the time of the last larval molt in an endopterygote insect, a pupa is formed instead of a further larval stage. With relatively few exceptions it has been shown that after emergence the corpus allatum again becomes active, and the hormone secreted stimulates the deposition of yolk in the female.

Metamorphosis of the Nervous System

Along with the endocrine system, the nervous system is the controlling center of the body's activities, so that changes in the nervous system are fundamentally concerned with metamorphic changes in other systems. It has already been seen for example, that the nervous system plays a role in the process of differential muscle breakdown. Also, the neurosecretory cells of the brain are virtually the controlling agents of the endocrine system: through regulation of the release of brain hormone molting is controlled; through regulation of the release of juvenile hormone maturation is controlled (Schneiderman and Gilbert, 1964). What then happens to the nervous system itself during metamorphosis? As was noted earlier in reference to Power's work, surprisingly little is known about metamorphosis of the insect nervous system at the individual cellular level.

Neurometamorphosis has been the subject of study recently by several workers, e.g., pipa and Woolever (1965) and Ashhurst and Richard (1964a and b), both for Lepidoptera. Many older workers considered the fate of individual nerves; evidence is extensive, although often contradictory, and Wigglesworth (1965) should be consulted for major references on the subject. For *Drosophila* along (Bodenstein, 1950) there is no detailed account of nervous system changes. What happens to the paired lateral nerves of the larva? Do these disappear in the pupal stage? Are the adult nerves all new developments? How is the relatively longer and narrower ventral thoracico-abdominal nerve mass of the larva converted into the shorter, broader thoracic mass of the adult where the whole abdominal center is reduced to the small posterior region? An answer to some of these questions will be attempted below in discussing metamorphic changes in the cyclorrhaphan nervous system.

Power (1952) drew a graph for the *Drosophila* nervous system development in the pupa, plotting volume increases in the components of the central nerve mass, of cells relative to fibers. He concluded that there was essentially little cell breakdown, but that there was a great increase in neuropile. Clearly, where larval organs break-down their nerve supply is no longer required. Conversely, where new adult organs are formed, a new nerve supply also has to be available. Do the old nerve fibers degenerate? Are the old nerve fibers taken over by other tissues? Do new neurons come into being in the pupa? Do the larval neurons dedifferentiate, their nerve processes degenerating, and later redifferentiate and form new nerve fibers? These are problems touched on but not completely solved for any of the insects in which changes are extreme. The field lies open for investigation at the light and electron microscope levels.

Of course, the problem is far more extreme in a case like the Cyclorrhapha than in, say, the Diptera Nematocera such as mosquitoes or crane flies, where most tissues and organs are carried over more or less unchanged from larva to adult. In the mosquito, it is said that central nervous system grows by cell multiplication in the larva, with very slight increase in cell size. Only slight histolysis occurs in the pupa, and in the adult there is again cell multiplication (Trager, 1937). However, even here we will have the large flight muscles developed. How is the new innervation to these developed? How do nerve fibers grow out of and into (sensory) a central nerve mass that is invested in a thick neural lamella? Breakdown of the connective tissue sheath would seem to be one prerequisite for this (Whitten, 1962a, Ashhurst and Richards, 1964; Pipa and Woolever, 1965).

General Metamorphic Changes in the Cyclorrhaphan Nervous System

Instead of complete breakdown and reorganization of the peripheral nervous system there is actually continuity in the *basic* framework from larva to adult (Whitten, 1965). At the time of puparium formation there are numerous mitoses among the sheath cells of the lateral abdominal nerves. Similar mitoses are present in the recurrent nerve from hypocerebral ganglion to the midgut junction, and the mitoses are paralleled in the string of cells (mesodermal gonad duct, rudiment) arising posteriorly from the testis and ovary. This burst of mitotic activity is followed by the breakdown of the sheaths, thus allowing cell growth and migration to occur. In the ventral nerve mass, growth occurs along the region and the net result is a dramatic sweeping backward of the abdominal nerves, aided by a general contraction in length of the abdominal nerves themselves (brought about by the shortening of individual sheath cells now that the sheath has gone). In this way the sheath cells, be cell division, growth, and migration, lay down the skeleton of the adult nervous system. This system is basically the same as the larval in general plan, but it differs in the details of individual branching.

Differences are most extreme in the thorax where tremendous growth of the sheath cells, from the ventral nerve mass, form the framework of the nerves to and from the flight muscles, legs, wings, and halters. The *external framework* is thus laid down. Internal to this, individual nerve fibers may degenerate and new fibers regenerate. Evidence so far suggests this and also that there may be dedifferentiation and redifferentiation of fibers of individual neurons. Pipa and Woolever, using electron microscopy, observed little evidence for "axoplasmic degradation", but the situation in

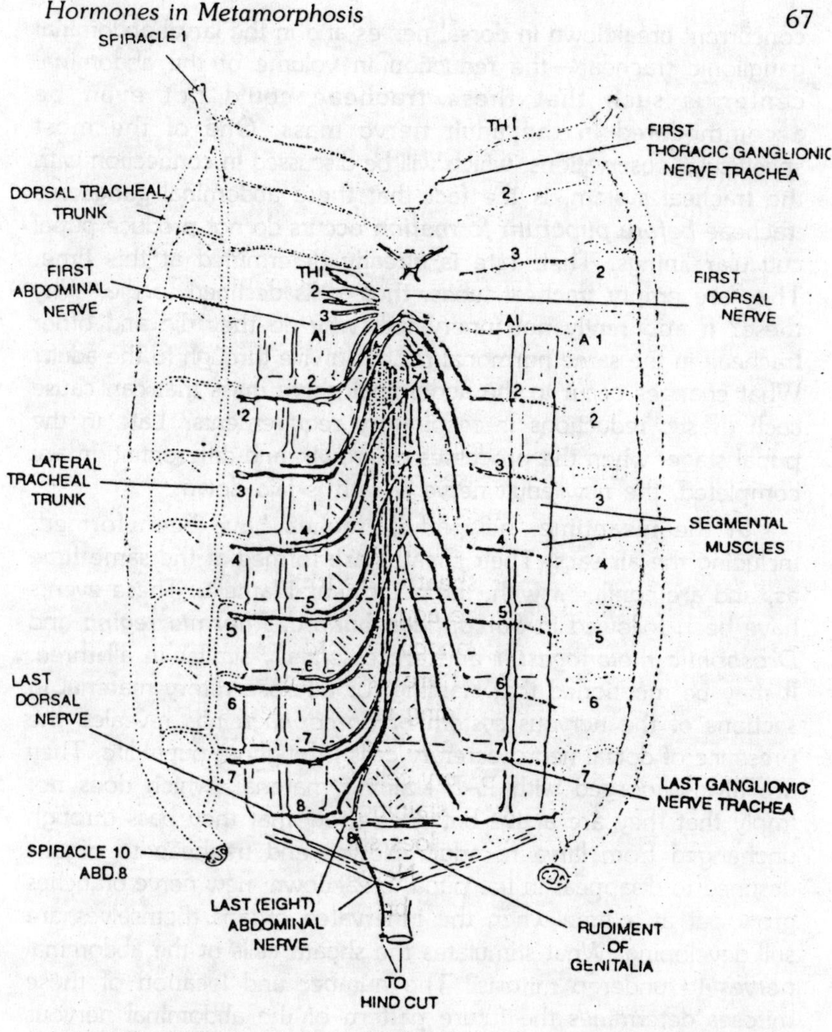

Fig. 3.5. Dorsal view of larval nervous system in relation to the segmental tracheal system, showing the segmental leteral and dorsal nerves, together with some of the larval muscles, the first and second leg imaginal discs, and the genital disc, Sarcophaga and other Cyclorrpha.

the lepidopteran nerve cord may well be less extreme than in the cyclorrhaphan system.

Muscle breakdown and tracheal histolysis are certainly less extreme in the former group of insects. Early in pupation, at the time of breakdown of the larval neural lamella, there is a concurrent breakdown in dorsal nerves and in the larval abdominal ganglionic tracheae—the reduction in volume of the abdominal center is such that these tracheae could not even be accommodated in the adult nerve mass. One of the most spectacular observations, which will be discussed in connection with the tracheal system, is the fact that these abdominal ganglionic tracheae *before puparum formation* occurs do not produce pupal cuticular linings. Their fate is already determined at this time. They are empty tracheal tubes, their cells destined to die. Why these? If apparently not innervated, why do they die and other tracheae in the same hormonal milieu survive through to the adult? What changes occur in the abdominal nerve mass that can cause such drastic reductions in respiratory requirements? Late in the pupal stage, when the processes of growth and differentiation are completed, the new adult nerve sheath is laid down.

In the meantime, new adult trachea have been formed, including the air sacs. Their sheaths are formed at the same time as, and are confluent with, the adult neural lamella. These events have been followed in *Sarcophaga bullata*, *Phormia regina* and *Drosophila melanogaster* and are essentially similar in all three. It may be mentioned that staining for neurosecretory material in sections of the nervous system of *Sarcophaga* has revealed the presence of dorsal neurosecretory cells throughout pupal life. They are always gorged with PAF staining material, which does not imply that they are active but does show that they pass through unchanged from larva to adult. Nerves and tracheae to organs destined to disappear in the pupa, breakdown; new nerve branches grow out at a time when the innervated organs themselves are still developing. What stimulates the sheath cells of the abdominal nerves to undergo mitosis? The number and location of these mitoses determines the future pattern of the abdominal nervous system. How do the nerves at this stage know where future adult structures will develop and require innervation? This brings us into the realm of cell differentiation and pattern formation, problems that will be considered in more detail later. The coordination of the different metamorphic processes is amazing, as is the realization that all these events occur within, in response to, and at speeds determined by light darkness cycles such as those shown by Harker for *Drosophila* (1965a and b).

Hemocyte Activity in Relation to Metamorphosing Tissues

Older works on the hemocytes were largely concerned with the role of blood cells in relation to developing tissues. More recently, work has centered more on hemocyte identification, origins, and their relative numbers at various stages, together with their role in coagulation. The vast amount of work up to 1962 is reviewed by Jones (1962). Since that time renewed interest has arisen in the part that blood cells may play in 1962, described briefly the fate of blood cells during the life history of the alder fly, *Sialis lutaria*. Here, phagocytes were found to ingest all particles freed in the hemolymph as a result of the cytolysis of the cells of the fat body, and in the gills the epidermal cells are ingested by the blood cells also only after cytolysis.

The phagocytes or morula cells, as they are later called, were said to disintegrate immediately before darkening of the cuticle. Ashhurst and Richards (1964b) described adipohemocytes in the vicinity of the nerve cord at the time of breakdown of the sheath in the lepidopteran *Galleria mellonella*. The participation of these cells in the breakdown of the neural lamella was substantiated by autoradiographic studies by Shrivastava and Richards (1965) and by electron microscopic observations by Pipa and Woolever (1965) on the same species. The latter demonstrated the presence within the adipohemocyte of neural sheath fibrils. These blood cells were not found to be involved in formation of the later adult nerve sheath. In the Diptera Cyclorrhapha similar conclusions had been reached by Whitten (1964b) in an investigation of the blood-tissue relationships primarily in *Sarcophaga bullata*, with observations also on *Phormia regina* and *Drosophila melanogaster*.

It was found that phagocytic hemocytes underwent very complex activities at the time of puparium formation. They were responsible for the ingestion of all tissue fragments in the hemocoele; they were also located in such consistent and characteristic positions in the developing pupae as to suggest that they participated in the transfer of materials to the developing tissues, particularly hypodermal cells of the general body surface and internal tracheae and muscle tendons. As with the lepidopteran nerve cord, these blood cells were not found closely associated with the developing adult neural lamella. The blood picture is outlined.

The sequence of events for cell type 5 shows nuclear and not cytoplasmic division to form enormous phagocytic bodies, which then over an extremely short period of time ingest larval fragments, including whole nuclei. These cells correspond to the granular hemocytes of Jones (1962), and in their later gorged form—as "spherules" —they correspond to the "spherule cells" and "spheres of granules" of earlier workers. These cells have recently been followed at light and electron microscope levels (Whitten, 1968) through the pupal to the adult stage. Their continuing close relationship with the developing adult tissues has convincingly shown the importance of the cells not only in removing larval fragments but also in processing and passing on materials to the developing adult tissues.

By the time the adult emerges the blood cells contents are depleted. These cells undoubtedly correspond to the type F hemocytes described by Crossley for *Calliphora* (1964). In a later paper (1965), Crossley described breakdown of type F cells in the midpupa, whereas it will be shown below in a study of co-ordinated development in the fly foot that the corresponding cells in *Sarcophaga* pass through into the adult stage, functioning even at the time of emergence, in the ingestion of selected epidermal cells. We have now reached the stage in studies on insect hemocytes where it is clear that phagocytic blood cells play as important a role in tissue destruction as phagocytic blood cells do in a metamorphosing amphibian.

There is also no doubt what so ever, in the cyclorrhaphan dipteran pupa, that the phagocytic blood cells can and do play an extremely important part in providing "processed larval tissues" as raw materials for the developing adult cells. Blood cells have been variously described as giving rise to adult tissue. Attention may be drawn here, where the possible origin of the cyclorrhaphan adult fat body cells from the phagocytic hemocytes is suggested. This suggestion is not new, having been made by several of the older workers at the turn of the century.

Electron microscopy has substantiated, rather than refuted this suggestion. As with other systems, there is a vast source of information awaiting investigation with modern techniques. There is no evidence to suggest that phagocytic hemocytes attack normal, healthy insect tissues. The beginnings of focal degradation in foot pad tenant hair cells. This occurs before any attack by the hemocytes which only subsequently ingest the pycnotic tenent nuclei and surrounding cytoplasm. Figure parallels closely micrographs of vertebrate tissue in which focal degradation is occurring (e.g., Swift and Hruban, 1964).

Lysosomes

The cyclorrhaphan phagocytic hemocyte provides ideal material for study of lysosome origin and fate and function. DeDuve's article on the lysosome (1963) describes lysosome activity with respect to the tail resorption and tissue breakdown in the metamorphosing frog. A similar story can be given for lysosome activity in the cyclorrhaphan phagocytic hemocyte. Figure 3.10 shows a typical picture of different stages of degradation of *larval* fragments in the vicinity of the nucleus in a day four (at 25°C) pupa. Acid phosphates has been demonstrated in these cells and in their apparent homogoues in *Calliphora crythrocephala* (Crossley, 1964). This is particularly good material for such studies of lysosomes in that within a time span of 12 days at 25°C these blood cells can be followed at all stages of their various phagocytic activities.

The Reproductive System

Since the reproductive system matures in all insects regardless of whether they undergo metamorphosis or not, considerations of general development of the reproductive system are not directly relevant to the present discussion. However, certain aspects bear on the subject. In the apterygote a metamorphic insects, the reproductive system is laid down at the time of hatching from the egg, and maturation occurs later, although even in the Apterygota, genitalia may be undeveloped at the time of hatching and develop in later instars. In insects with complete metamorphosis, the larvae may hatch with the reproductive system in an embryonic state. Gonads are present but the ducts are limited to the mesodermal elements that extend from the gonad distally, to end posteriorly in a coelomic sac or ampulla, located in different segments in male and female.

During metamorphosis, development of the reproductive system is completed when the genital imaginal rudiment develops both the external genitalia and the ectodermal elements of the internal reproductive system, including median ducts, accessory glands and spermathecae. When growth is completed and ectodermal and mesodermal ducts have met and fused the breakdown of cells and continuity of the ducts to the outside is established. In one of the besi-known insects, *Drosophila*, onset of metamorphosis in characterized by immediate and rapid growth of the imaginal discs, including the genital disc. In this same example an anomaly exists in that the ectodermal component is considered to give rise to the entire duct system right to the gonad. This anomaly has recently been doubted in the higher fly *Sarcophaga*, where the mesodermal ducts have been shown to be present in the larva in a typical embryonic state. The ducts themselves represent true examples of "imaginal rudiments" (Whitten, 1965). The onset of papation is characterized by extensive mitoses in this strand of cells.

Subsequent dissolution of the connective tissue "basement membrane" and shortening of the strand (closely similar to events in the peripheral nerves) produces the proximal mesodermal genital duct in both male and female. This meets up and fuses with the ectodermal components derived from the genital imaginal disc. Thus *Sarcophaga* falls into line with other endopterygote insects in which metamorphosis involves passing from the embryonic to the adult condition. It would be surprising if *Drosophila* is really exceptional in this respect. During metamorphosis, extensive changes occur within the reproductive system, the Degree of maturation attained by the time of adult emergence varies from male and female and from species to species. In many cases gametes are fully developed at the time of emergence of the adult insect. Review articles of King (1964) and Telfer (1965) provide interesting and instructive reading on this subject.

The Epidermis in the Metamorphosing Pupa

The epidermis and its derivatives have been left for consideration until last because in many respects this is the most important and interesting cell layer to the insect developmental biologist. Included here are the imaginal discs which are essentially epidermal sacs containing nerve and tracheal endings along with the hemolymph and contained blood cells. Strictly speaking, the ectodermal regions of the reproductive system and of the gut also fall into this category, but they have already been considered briefly along with the rest of these systems. The tracheal system is an invaginated extension of the general epidermis and will be considered first.

The Tracheal System

The tracheal system is composed of two cell types: the tracheal cell and the tracheolar cell. Essentially, the difference between a trachea and tracheole is that the former is formed as a layer of cuticle by the tracheal epithelium, while the tracheole is terminal and formed intracellularly by a single cell. At one time it was thought that tracheae possessed the spiral or annular thickenings, whereas the tracheoles did not. However, electron microscopy has shown that tracheoles do also.

A physiological distinction made between tracheae and tracheoles is that little gaseous exchange occurs through the tracheal wall, exchange being mainly by way of the tracheoles at the tissue level. It will be seen below that the metamorphosing pupa may be an exception to this generalization. The pupa has respiratory problems that are unique in that a single system laid down at the larval-pupal molt has to tracheate tissues that are destined to disappear during the pupal stage, tissues that persist through from larva to adult, and more significantly, yet other tissue that develop during the pupal stage and are not in existence at the time that the tracheal system of the pupa is laid down. These demands on the tracheal system will be taken up following a general survey of the development of the system from first instar to adult. Details can be obtained from Buck (1962) and Miller (1964) on the large amount of work that has been done on the tracheal system; however, problems of development are relatively unexplored.

In general, it has been thought that at each molt the tracheal lining is replaced but that the tracheolar lining is retained throughout the life of the insect. However, the situation is much more complex than this, particularly in the metamorphosing insect where drastic changes are occurring in other systems. A schematic drawing of the types of development that can occur in individual tracheae. In general, the insect tracheal system is divided into 10 pairs of tracheal metameres, occurring in segmentally repeating order and, in the pterygote insects, joined laterally and longitudinally at junction points of *tracheal nodes*. The original segmental pattern can be seen at each molt when the continuous tracheal tube linings "break" at the tracheal nodes, and each segment is removed through its respective spiracle. This occurs even in forms in which the spiracles are physiological closed an not functional as far as respiration is concerned. The degree of change or growth in the tracheal system from instar to instar is determined by the physiological requirements, so that as far as the metamorphosing pupa is concerned, there is a direct correlation between the degree of metamorphic change in, say, the muscular system and that in the tracheal system.

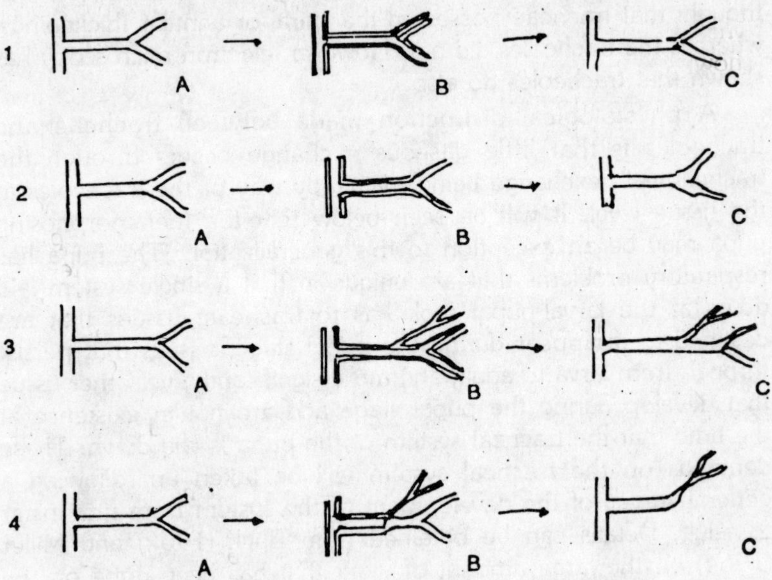

Fig. 3.6. Diagram showing the different types of fate that an individual trachea can undergo during development.

The pattern of development in a generalized dipteran, and members of the nematoceran Bibionidae would follow this closely. Here the larval basic pattern is taken right through from larva to adult, and the only drastic change is in the thorax where the tracheation to the new flight muscles, wings, halters, and legs is superimposed on the larval pattern by growth of new tracheal branches at the end of the last larval instar. In contrast, the sequence of events in *Drosophila melanogaster* where there is considerable change from larva to adult, correlated closely with the drastic changes in other systems and with the virtual reduction of the pupa to an embryonic system.

At the larval-pupal molt the posterior half of the larval system does not secrete pupal cuticle and is subsequently histolyzed as are the ganglionic trachea. Later in the pupal stage there is new growth of adult tracheal epithelium in this region, but the last metamere is nerve replaced. Between these two extremes there are all degrees of relative histolysis in the abdominal system, and these are highly consistent within dipteran groups of close affinity. This whole subject is as intriguing and probably as important in terms of the control of cell death as are problems of muscle cell death, particularly as tracheae are generally considered to be noninnervated. Without a nerve supply, why would an individual trachea or a whole tracheal segment cease to secrete cuticle and the cells subsequently die when its neighbour remains and functions normally? Perhaps the answer lies in physiological changes in the organ that is tracheated and destined to die. However, the converse cannot apply to the development of new tracheal branches to tissues that are not yet developed in the pupa.

It has been said that the elaborate tracheal development of the thoracic region of adult systems bears little relationship to that of the earlier larval system. This is not so. From a survey by the author of the development of the adult system within some 20 families of Diptera, it is evident that the very elaborate system of tracheae or air sacs represents branching of the tracheal epithelium of the earlier basic larval pattern. Superficially, they look very unlike each other and their corresponding larval systems. However, close inspection shows the same basic pattern of transverse and longitudinal elements, with very constant positions of origin of leg, wing, and halter tracheae. No comparable survey is apparently available for other endopterygote orders.

Some Unique Characteristic of Pupal Tracheal systems

The pupal system of the cyclorrhaphous Diptera, including forms like *Drosophila*, *Sarcophaga* and *Calliphora*, is characterized by grape like bunches of tracheoles suspended in the hemocoele. These arise by mitosis of cells of the tracheal system, and after the larval - pupal molt there is often extension of the tracheal cell cytoplasm toward various organs, including the developing epidermis. The pupal stage has been seen to be characterized by the absence of basement membranes : there is no penetration of individual tissues by the pupal tracheoles which lie essentially within the hemocoele. Some of these tracheoles. They are particularly interesting in that respiratory exchange is apparently occurring between the tracheole and the hemolymph.

The lining of cytoplasm, particularly on one side, is infinitesimally thin and so is any basement membrane. These pupal tracheoles are seen to move in later pupal stages to come in closer proximity to developing adult organs such as the reproductive system. For example, an early pupal testis or ovary has no tracheation, whereas later pupal gonads have functional pupal tracheoles associated with them, as well as the as yet nonfunctional adult tracheae and tracheoles. The condition in the Diptera Cyclorrhapha is extreme, but the same condition prevails in at least the thorax of more generalized forms where there is extensive development of muscles in the later pupal stages. The situation diagrammatically represents in an organism such as *Drosophila*.

In the present case it is the tracheal cell cytoplasm that extends towards the tissues, whereas in the bug *Rhodnius*, Wigglesworth (1959a) described the drawing of tracheoles, by epidermal cell extensions, to areas requiring oxygen. Epidermal cells at quite considerable distances away would send out processes towards tracheoles, "capture them", and draw them to the deprived area. By comparison with other epidermal cell types the tracheal cell shows remarkable uniformity, the only distinction being between tracheal and tracheolar. Whether the cells at the tracheal nodes are structurally different seems never to have been investigated. That they are important physiologically in connection with gradients has been shown by Locke (1964).

Cells of the spiracular region are not included in this tracheal system since they are most easily thought of as belonging to the general body surface. It would be of considerable interest to understanding of cell differentiation to know whether the same tracheal cell ever produces specific larval, as opposed to pupal and pupal as opposed to adult, tracheal cuticle. Whether the air sac cuticle is specifically different in an adult is not known.

Metamorphosis of the General Epidermis

The rest of the organ systems differ from the epidermis in proceeding more gradually from larva to adult except in cases where metamorphic changes are involved. The epidermis is, however, characterized by periodic molting and secretion of new cuticular layers, and often the cuticular layers of different instars are characteristically different from those of others instars. Thus, newly hatched *Thermobia* (firebrat: Thysanaura, Apterygota) that undergoes no metamorphosis during development, is scaleless whereas later preadult and adult instars have scales; a fly pupal cuticle is smooth, while that of the adult consists mainly of small and large bristles and their sockets. There has been a tendency in recent years to talk of the cuticles of metamorphosing endopterygote insects as "larval", "pupal", or "adult" and to relate the appearance of the specific type to the hormone balance between juvenile hormone and molting hormone at a particular period of development. Thus, the larval cuticle has been set to be produced in the presence of high levels of juvenile hormone and low ecdysone; the pupal cuticle is produced in the presence of high level of ecdysone and lower juvenile hormone; and finally the adult in high ecdysone and virtually no juvenile hormone.

In the case of the non-metamorphosing *Thermobia*, a piece of scaleless cuticle from a newly hatched larva, when implanted into an adult, is found to molt and prematurely produce scaled cuticle. The larva has a high and the adult a low concentration of juvenile hormone (Watson,, 1963). While there are in most Endopterygota considerable differences between larval, pupal, and adult cuticles, one cannot say, for instance, that bristles are confined to adult cuticles. A particular insect type will have characteristic cuticles at different stages, but whether a larval cuticle is ultrastructurally diagnostic has not been shown for any insect.

Certainly the tracheal cuticle appears to be essentially similar at all stages, and the epidermal cells secreting this are every bit as exposed to hormone fluctuations as are the cells of the general hypodermis. That changes are not simple responses to different hormone titers has recently been shown by Krishnakumaran and Schneiderman (1664). They questioned whether perhaps the cells of the lepidopteran adult differ from cells of the pupa in response to juvenile hormone, since it was seen in experiments that adult abdomens could be induced to molt again when joined to pupae, but that the second adult cuticle, although distinctiy adult, seldom produced scales. A much older adult did produce normal scaled adult cuticle. A high level of juvenile hormone did not seem to influence the development of the adult cuticle since experimental results were similar with abdomen containing high levels of juvenile hormone (e.g., *cecropia*, *cynthia*) and others containing low levels (*polyphemus*).

Krishnakumaran and Schneiderman quote a comparable phenomenon from the work of Piepho and his students, who showed that when fragments of adult cuticle were implanted into larvae they did not secrete larval cuticle immediately but required several molts before doing so. It may be of interest to observe in this context that the cells of the imaginal discs in the fly (which will be seen below to be determined but not differentiated), secrete first pupal cuticle and subsequently undergo extensive mitoses followed later by differentiation into adult cuticle secreting cells. These disc cells in an adult host continue mitoses indefinitely, but implanted into a metamorphosing larva they first secrete pupal cuticle, then further divide, and finally differentiate and produce adult cuticle.

Metamorphosis of the Epidermal Cells

The degree of change occurring in individual epidermal cells varies from insect to insect. In the more generalized endopterygotes or even the more generalized members of the specialized orders, e.g., the Nematocera of the Diptera, the epidermis passes more or less unchanged from larva to adult. On the other hand, in the higher flies, the Cyclorrhapha, there is said to be a complete replacement of larval epidermis by adult epidermal cells that originate from areas of undifferentiated imaginal cells—those of leg, wing, halter, genital, labial, antennal and eye discs, and paired groups of cells segmentally arranged on the abdominal segments.

The pupal cuticle is secreted by the larval epidermal cells of the abdomen and small areas of the thorax, and the remainder is secreted by the undifferentiated cells of the different imaginal discs. Only in the cyclorrhaphan tracheal system might it be possible to follow individual epidermal cells through from larva to adult, whereas in the Nematocera this should be possible for general epidermal cells also. The metamorphosis of the epidermis may best be looked at now from the viewpoint of "imaginal disc" development, since a tremendous amount of thought—provoking and fascinating work has been done on this subject. Following this, the fate of a few individual cells can be traced through from the time of the larval -pupal molt to the time of adult emergence. This will demonstrate the complexities on the one hand and the great potential on the other that the metamorphosing insect epidermal cells have in helping to solve some of the most puzzling of biology's and life's problems—those of nuclear-cytoplasmic relationships and of cell and tissue differentiation.

The insect epidermal cell has been the fascinating subject of several articles including that of Stern (1954) and of Wigglesworth (1959b). There have been new developments involving the epidermal cell, that include gradient studies (Locke 1964) and the discovery of giant polytene chromosomes in the dipteran epidermis: these, along with the fact that the epidermis is undoubtedly the main "target organ" of the molting hormone ecdysone, give the epidermal cell greater importance than ever.

Metamorphosis of the Imaginal Discs

In the more generalized Diptera the imaginal discs of wings and legs lie beneath the epidermis and are not withdrawn into the body as are those of the specialized Cyclorrhapha. Also, the cavities connecting with the general epidermis are never closed and the eyes do not develop from invaginated buds. Developing pigmented eyes can often be seen, as in the mosquito, through the semitransparent head capsule of a late larva. The imaginal discs of a cyclorrhaphan fly include genital and leg discs, genital and first and second leg discs, wing, halter, labial, eye and antennal. The groups of cells between duct and larval salivary gland, those at the junctions of foregut and midgut and hind gut, the mesodermal genital duct rudiment of *Sarcophaga*, and the hypodermal discs of the abdominal segments, are also imaginal rudiments.

It is difficult to draw a hard and fast line of distinction between some of these cell groups and those associated with individual systems, such as the tracheal cells that will suddenly at puparium formation undergo intense mitosis to produce the bunches of specialized pupal tracheoles, or the sheath cells of the peripheral nerves, which presumably functioned in the embryo to secrete the larval nerve sheath and now at puparium formation are stimulated into intense mitotic activity to give rise to the cells that form the skeleton of the adult peripheral nerves. Studies on the leg or wing or genital imaginal discs, are essentially studies on epidermal cell lines.

Embryos of endopterygotes that undergo complete metamorphosis, such as *Drosophila*, possess two cell types. Firstly, there are those cells that are determined to form the larval body and, in these, differentiation immediately sets in. Secondly, there are those cells that are determined but are set apart and remain in an embryonic state until the time of pupation (Hadorn, 1965).

Cells of the imagir al discs come into this category. They do divide and the discs grow considerably in size from their earliest appearance to the end of the larval stage, but they remain undifferentiated. Due largely to the classic earlier works such as those of Bodenstein using transplantation techniques and more recently the fascinating works of Hadorn and his collaborators, some amazing phenomena have come to light with respect to potentialities of imaginal disc cells in the metamorphosing insect.

It would seem that the cells of these discs are already determined for "leg" or "wing" or "eye" early in embryonic life, as early as the first to fifth hour, at a time when the imaginal discs are not yet visibly set apart within the still apparently homogeneous blastoderm. There is even determination for particular areas of these discs. There is *determination* at this time but no *differentiation*. This occurs only at metamorphosis. As stated by Hadorn, the change from undifferentiated to differentiated state in holometabolous insects is controlled by hormones. The juvenile hormone somehow in the larval instars blocks any differentiation; a drop in the level of juvenile hormone at the onset of metamorphosis results in a "deblocking" of readiness for differentiation present in the discs throughout larval life. *Perhaps the juvenile hormone acts as a repressor of the sets of genes needed in differentiation* but here Hadorn warns that much more information is needed before the happenings that clear the path for differentiation in a determined cell can be interpreted in molecular terms.

Until details are clearer of the responses by particular cells of known fate and history, it would be good to extend such reservations to work on hormone action, particularly with respect to the influence of the balance between juvenile hormone and ecdysone on the production of specific "larval", "pupal", or "adult" cuticle. The same reservations may be extended to the effects of hormones on chromosome puffing sequences. The results with larval salivary glands cannot necessarily explain hormone action on tissues that will undergo metamorphosis. The exciting results of Hadorn and co-workers involve the maintenance of imaginal discs within adult hosts, where they increase in size by extensive mitosis but do not differentiate, Removal from the adult and transplantation into a metamorphosing larva brings about cessation of cell division and differentiation into adult structures.

Of particular interest is the process of "transdetermination" whereby after some eight or nine transplants (from one adult host to another) the disc when allowed to determination" whereby after some eight or nine transplants (from one adult host to another), the disc when allowed to differentiate in a metamorphosing larval host, differentiates into an adult structure different from the original implant. Thus, a genital disc after many transplants may give rise to leg or wing or thorax. Transdeterminations occur in the probability order of genitalia, antenna, leg, palpus, wing, and thorax the latter occurring in the oldest transfer generations. Reversals were also found. No explanation can as yet be given for these amazing results, but it is suggested by Hadron that possibly a degree of dilution reached by cell cultures might have a feedback action on gene activation. Instead of the genes that formerly determined the original organ, other genes come into play, resulting in the transdetermination. Some of the transdetermination effects are reminiscent of some Drosophila mutants such as aristapedia were a tarsus is formed on the antenna. However, most of the transdetermination effects are not known as mutants.

Metamorphosis of an Imaginal Disc: The Leg Disc

The type of development undergone by individual elements of an imaginal disc can be seen in the case of the leg. Details have been followed for *Sarcophage bullata*, the flesh fly, but development is essentially similar in other Cyclorrhapha, including *Drosophila*. Event followed at light and electron microscope levels demonstrate the complex changes and co-ordinated development of the different components: various epidermal cells, tracheoles, and phagocytic blood cells. Some of these events have already been discussed in relation to hemocyte activity during metamorphosis. At this point the main concern will be with the development of individual epidermal cells. The mass of thousands of determined yet undifferentiated epidermal cells (stage 1) first secrete the pupal cuticle of the leg disc. Retraction of the epidermal cells from the pupal cuticle follows and cell division continues (stage 2). Five cells forming the future end of the claw and four future giant cells become recognizable at stage 3 while cell growth and continued mitoses occur among "mother" tenent hair cells whose descendants will occupy the future ventral surface of the foot. Growth and cell division thrust the five claw cells into the tip of forming claw, and the foot pads take shape by

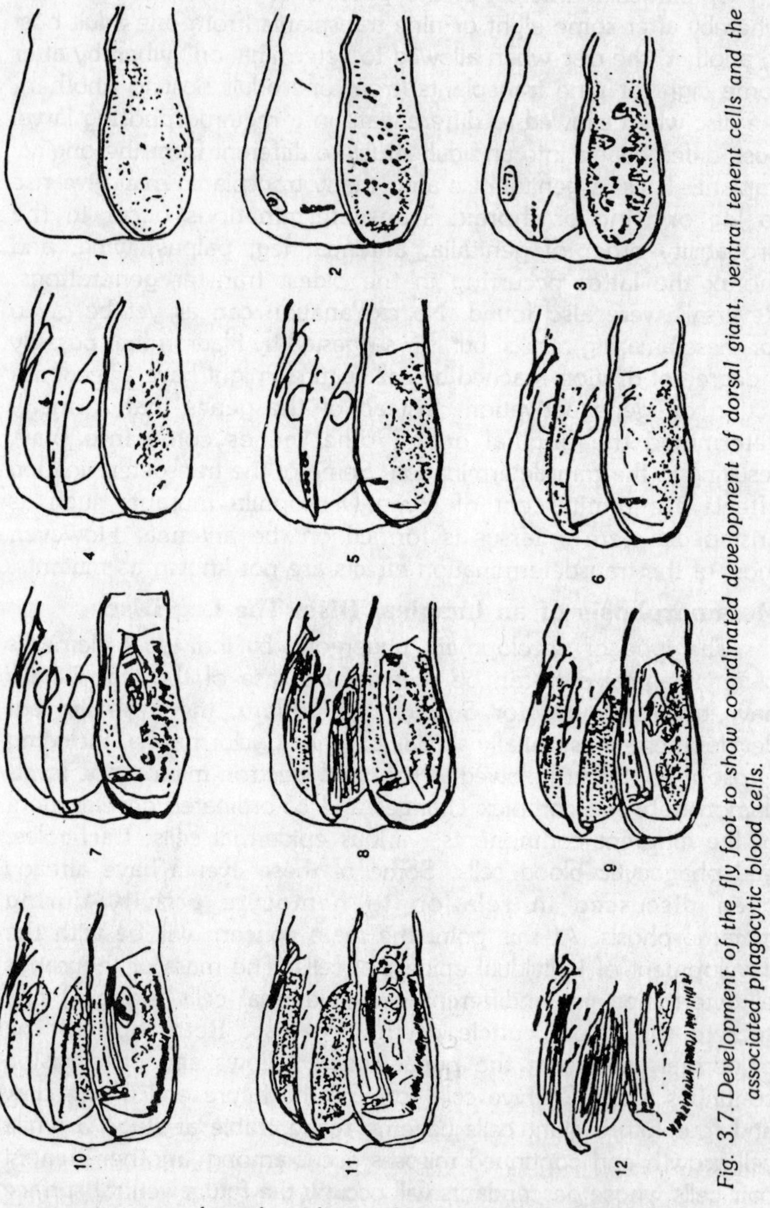

Fig. 3.7. Development of the fly foot to show co-ordinated development of dorsal giant, ventral tenent cells and the associated phagocytic blood cells.

progressive growth and median indentation. Dorsally, the two giant cells extend over each foot pad; their nuclei are characterized by giant polytene chromosomes.

The tenent cells give out very fine cytoplasmic processes, while the dorsal cells grow in width as broad, flat cells. The tarsal hair cells at this sametime from, like the tenent cells, cytoplasmic processes, varying in size according to whether a large or small hair is to be produced. Stage 4 marks the end of growth and differentiation, yet at this time no cuticle is as yet deposited. Thus, the future shape of a hair is laid down by cell growth (c.f. Lees and Picken, 1945) and is completed by the time cuticle secretion begins. This is of interest in questions of abnormal bristle development in mutant *Drosophila*. Cessation of growth and attainment -of- cell maturity is followed by a well-spaced sequence of cuticle deposition accompanied in the giant cell by constant and characteristic patterns of chromosome puffing activity. First is secreted an *ecdysial membrane* (1), followed by a cuticulin layer (2), a layer of dense exocuticle (3), which later becomes darkened and sclerotized, a homogeneous inner layer of exocuticle (4), and a narrow layer, of "mesocuticle" (5), each layer is indicated by an arrow. By stage 9 these sequences are completed in giant, tenent and trichogen cells. Cytolysis is now observable within the tenent cell nuclei and cytoplasm. This is seen at the electron microscope level, in which centers of focal degradation can be clearly seen.

The hemocytes, which at the time of first evagination of the legs were gorged with larval tissue fragments, have subsequently been digesting them and passing on the degradation products to the developing epidermal cells. These phagocytic blood cells now, at stage 9, embark on a second phase of phagocytosis. By stage 10 the phonetic tenent cell nuclei are nearly all engulfed. In fact, the exact age of the foot can be determined by three means: (a) by close inspection of the extent of the darkening process and actual shade of specific areas, (b) by close investigation of the relative degree of phagocytosis of tenent cell nuclei, and (c) by close inspection of the puffing pattern of the chromosomes. All are equally exact and reliable.

The tracheoles and giant cells remain unattacked at this time. Endocuticle secretion within the giant cell cytoplasm occurs after this at a time when darkening of the exocuticle is occurring over the ridges of the dorsal cuticle. Endocuticle (layer 6) secretion seems to occur by a combination of intracellular secretion and retraction of cytoplasm (and nuclei and blood cells) to the base of the foot, so that the space formerly occupied by giant and tenent cells is now occupied by endocuticle. The position of blood cells in relation to that or giant nuclei and the former position of tenent nuclei. Even at the light microscope level, association of blood cells with the giant cytoplasm can be seen. The association becomes more close, and at emergence the giant nuclei and cytoplasm also become ingested by the blood cells. The blood cells finally come to lie as a layer over the inner border of the foot pad cuticle, at the base of the pad. There is no innervation of individual cells in the foot, though possibly a basal stimulus is received that can be transmitted from cell to cell. What determines that cell division should cease in the giant cells and cells of the future claw at a time when mitoses are still occurring among the future tenent cells? What determines the regular pattern and uniform deposition of the different cuticular layers in the various epidermal cells? That the layers are similar and are deposited at comparable times in tenent, trichogen, and giant cells.

The biochemistry must be similar, yet the spatial distribution of the various layers is different, and this is essentially what gives each product its characteristic "differentiation". What initiates cell death among the tenent hair cells? Is it simultaneous or is there a focal point at which the process starts? Focal degradation begins before any phagocytosis on the part of the blood cells is to be observed. Does this prompt the blood cells to attack the tenent cells and not the giant or the tracheolar cells? Interestingly, the final deposition of endocuticle by the cytoplasm of the tenent cells is carried out in the absence of their nuclei. What of the chromosome activity while these sequences of growth, cuticle secretion, and finally death are occurring in the giant cells? For the first time, we seem to have a system where one should be able to correlate closely chromosome activity and puffing, sequences with the activity of a single cell through growth, various phases of clearly defined activity, and finally cell death. Detailed analysis will be made in the vast amount of information that has accumulated as a result of work carried out on dipteran salivary glands and to a lesser extent on the chromosomes of cells of the Malpighian tubules, gut, or seminal vesicles.

The Malpighian tubules, persist through pupation but are of small dimension while the salivary glands degenerate shortly after pupation. Since epidermal cells are specifically the "target organs" for the molting hormone ecdysone repetition of much of the work carried out on salivary glands should produce particularly interesting results with foot pad cells, particularly work such as that of Karlson and his colleagues (e.g., 1965). It would be exciting to see if and how hormone changes can affect puffing activity and in what way they may alter the biochemistry and ultrascopic structure of the cell. These discussions might seem to have taken us far from considerations of the whole metamorphosing insect, but in fact it is visible changes in the epidermal cells and their products that constitute external evidence of metamorphosis.

What determines the difference between the various cells of the epidermis of the developing insect? They are variants of the same cell type—the epidermal cell. When the layers of cuticle that are secreted are identical in two or more variants, what determines the spatial distribution of these cell products, and what determines the cell shape? If one could answer these questions one would be nearer to answering the problems posed by the phenomena of transdetermination of Hadorn and of the problems of patterns and prepatterns discussed at length by Sondhi (1963) and in a broader context by Waddington (1962). It will be seen from the foregoing necessarily superficial survey of insect metamorphosis that each system in turn has its own fascinating problems, which, to name but a few, include general considerations of body change, diurnal rhythms, problems of cell determination, cell differentiation, cell death, and cell interaction. Many of the problems are common to those found in other metamorphosing animals, as will be seen in succeeding chapters. Nevertheless, in the insects alone one cannot but be impressed by the infinite number of problems that the one phenomenon of metamorphosis poses to the inquiring developmental biologist.

4

Biochemical Changes during Insect Metamorphosis

The metamorphosis of the higher insects embodies the most profound reorganization of a grown animal that is known. It represents advanced polymorphism, for a single genome determines the development of a differentiated free-living organism which, when fully grown, is largely destroyed and reconstituted in a new form adapted for a totally different life. Such a phenomenon, under the control of hormones, should provide prime material for the study of the regulation of genetic expression and its role in morphogenesis. This challenge has inspired biochemical study of insect metamorphosis for many years. However, since all organ systems of the insect are to a greater or lesser extent modified in structure and function during metamorphosis, some change may be expected in almost any biochemical feature that is examined.

Accordingly, the literature of the subject is extensive and diffuse, and reviews of it have tended to give weight to each author's own predilections (Gilbert and Schneiderman, 1961; Karlson and Sekeris, 1964; Agrell, 1964). The contemporary emphasis on molecular genetics and nucleic acid and protein synthesis seems most likely to yield understanding of the fundamental mechanisms in metamorphosis, but investigation of these fields has not yet progressed far with insects. The earlier attention to nutrient reserves and pathways of energy conversion now seems less close to the central question. Nevertheless, these studies have contributed to knowledge of the milieu in which the genes are working and of the nature of the biochemical systems for which they are responsible, and we are not yet in a position to say confidently what is essential and what is not. Therefore, I shall attempt to summarize the present state of understanding in several biochemical areas where enough work has been done to permit some coherent discussion. Other subjects will be arbitrarily omitted; among the more important of these are the development and functions of the silk gland and the ovaries, although much work has been done on both. References to the literature will necessarily be selective, and most examples will be drawn from two insect groups which have received a great share of research attention.

The higher Diptera undergo the most complete metamorphosis among insects, replacing most of their organ systems. Some are easy to raise in the laboratory with a short generation time, but they have the disadvantage of small size which has discouraged biochemical work with discrete tissues (except to some extent flight muscle). In the Lepidoptera, there is far less abandonment of larval cellular structure, but this order includes many species with the experimental advantage of large size, and more work has been done with discrete tissues. Biochemically, there is probably unity in the underlying mechanisms of histolysis and of histogenesis in all insect groups. Processes which are not strictly part of metamorphosis may be relevant—e.g., the degeneration and reconstruction of muscles in the larval moulting cycle of the bug *Rhodnius* or in the adult diapause of the Colorado potato beetle. Metamorphosis is not wholly confined to the transformation from larva to adult, as there is often a small degree of change in form at each larval molt.

Use and Interconversion of Reserves

During their last larval stage, in preparation for metamorphosis, holometabolous insects accumulate substantial nutrient reserves. These are chiefly deposited as fat and glycogen in the fat body, a cellular tissue which is a center both for storage and for intermediary metabolism and which proliferates greatly during the last larval instar (Kilby, 1963). In the larva of the wax moth, *Galleria mellonella*, this preparation for metamorphosis can be detected early in the last instar, which differs metabolically from the preceding larval stages (Janda et al., 1966). Peak values for glycogen content are often in the neighbourhood of 10 percent (but sometimes as high as 30 percent) of total body dry weight, and the lipid reserves are generally greater, values of 20 to 50 percent of dry weight before metamorphosis being common. Analytically studies on changes and conversions in the insect pupa have been stimulated by the fact that it is a closed system which exchanges only gases with the environment.

A body of the early work is summarized and discussed in an admirable review by Needham (1929), but few conclusions can be drawn from it because of inadequacies in the analyses. In particular, trehalose was not recognized as an ubiquitous sugar in insects until 1957, and the extent to which the chitin and protein of the cuticle are resorbed in the molt and reutilized was generally underestimated or ignored. References to more recent studies can be found in several reviews (Wigglesworth, 1965; Gilbert, 1967; Wyatt, 1967a). Among the most comprehensive data on a single species are those recently obtained by Birt and colleagues on the sheep blowfly, *Lucilia cuprina*. Several conclusions may be drawn. The principal source of energy is fat, the content of which at adult emergence is about one third of that in the mature larva. The rate of fat oxidation is more rapid at the beginning and the end of metamorphosis than in the middle, in accord with the changing rate of respiration.

At the beginning of metamorphosis before the pupal molt, there is a fall in total protein and in free amino acids, which are partly oxidized and partly converted to carbohydrate and fat. The utilization of protein leads to accumulation of much uric acid in the early fly pupa (Russo-Caia, 1960). There is also a fall, before the pupal molt, in chitin as the inner layers of the larval cuticle are digested and resorbed; some of this is apparently deposited as glycogen, which is later reconverted to chitin in the growing adult cuticle. Thus, when chitin is included, the total carbohydrate content undergoes only a slight net fall throughout metamorphosis. Since files depend upon carbohydrate for flight energy, its conservation during their development is advantageous.

A slight temporary rise in total carbohydrate which is observed just after pupation is attributable to glyconeogenesis from amino acids. These conclusions differ in several respects from those from earlier, less complete analyses on related species, which had suggested synthesis of glycogen at the expense of fat, followed by substantial oxidation of carbohydrate during metamorphosis. The

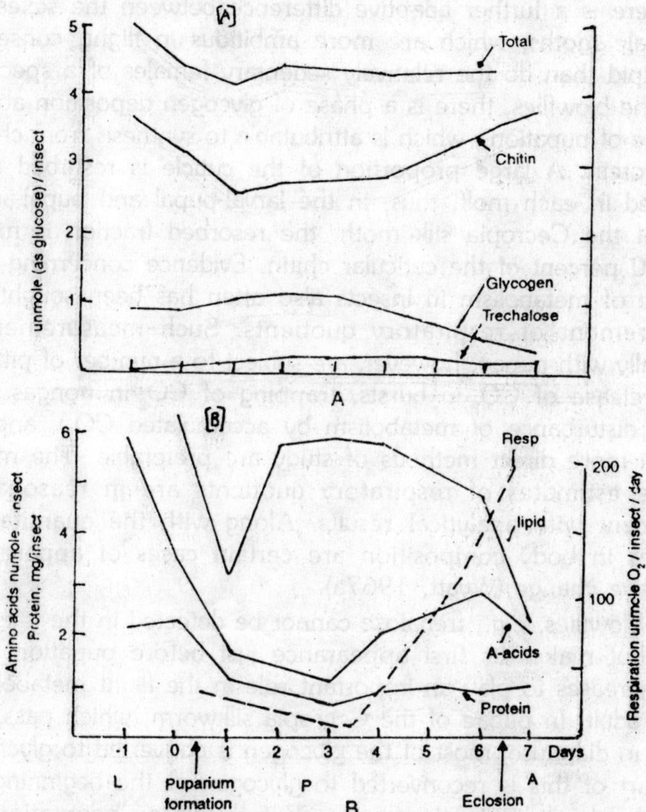

Fig. 4.1. Changes in composition and respiration of the blowfly, Lucilia cuprina, during metamorphosis. A—Carbohydrates. B—Respiration, neutral lipid, free amino acids, and soluble protein.

net conversion of fat to carbohydrate is not known to be possible in insects or vertebrate animals. Analyses on metamorphosing Lepidoptera, such as the oriental silkworm, *Bombyx mori* (Niemierko et al. 1956; Zaluska, 1959), and the American saturniid silk moth, *Hyalophora cecropia* (Gilbert and Schneiderman, 1961; Bade and Wyatt, 1962; Domroese and Gilbert 1964), present a generally similar picture. Again, fat is an important energy source. However, the extent of net oxidation of carbohydrate in passing from caterpillar to moth is greater than in the case of the fly, perhaps because moths use fat as their fuel for flight and so its conservation rather than that of glycogen, is advantageous.

There is a further adaptive difference between the sexes in that male moths, which are more ambitious in flight, conserve more lipid than do the relatively sedentary females of a species. As in the blowflies, there is a phase of glycogen deposition about the time of pupation,, which is attributable to synthesis from chitin and protein. A large proportion of the cuticle is resorbed and reutilized in each molt; thus, in the larval-pupal and pupal-adult milts of the Cecropia silk moth, the resorbed fraction is more than 80 percent of the cuticular chitin. Evidence concerning the balance of metabolism in insects also often has been sought by measurement of respiratory quotients. Such measurements, especially with pupae, however, are subject to a number of pitfalls (e.g., release of CO_2 in bursts, trapping of CO_2 in nongaseous forms, disturbance of metabolism by accumulated CO_2), and in general more direct methods of study are preferable. The more reliable estimates of respiratory quotients are in reasonable agreement with analytical results. Along with the quantitative changes in body composition are certain cases of apparently qualitative change (Wyatt, 1967a).

In blowflies, e.g., trehalose cannot be detected in the feeding larva, but makes its first appearance just before pupation and then increases to play an important role in the flight metabolism of the adult. In pupae of the Cecropia silkworm, which pass the winter in diapause, most of the glycogen is converted to glycerol, and part of this is reconverted to glycogen at the beginning of adult development in the spring. But this is an adaptation to survival in cold weather rather than a process characteristic of metamorphosis, for glycerol is accumulated by many insects which survive cold winters, and the life-cycle stage in which it appears is always that which passes the winter, be this the embryo, larva, pupa, or adult.

In short, the use and interconversions of reserves in the metamorphosis of different species is variously adapted to their respective needs. Little is known about the regulation of these conversions. A recent review on metabolic control mechanisms in insects cited much evidence for the existence of control but few instances where enough is known to justify designation as a mechanism (Harvey and Haskell, 1966). Fat and carbohydrate metabolism, like other aspects of insect development, must be more or less directly under control by hormones, and cases can be cited in which metabolic change follows change in hormonal state (Wyatt, 1967a). Certain rapid adaptive changes can result from hormonally released activation of preexisting enzymes, such as glycogen phosphorylase. The longer term changes in metamorphosis are presumably associated with changing enzyme levels, but there exists little satisfactory biochemical evidence on this point. One example is the elevation of chitinase, a participant in digestion of the cuticle, at each molt in the silkworm, *Bombyx mori* (Jeuniaux, 1961).

Respiratory Metabolism

Respiration and Energy Supply During Metamorphosis

One of the earliest and most frequently recorded metabolic observations concerning insect metamorphosis is that the rate of respiration changes in a manner that can be described as a U-shaped curve. In the earlier papers, the significance of this general course of respiration was much discussed. The first suggestion was that it reflected successive histolysis and histogenesis, the rate of respiration changing in accord with the amount of organized tissue present at any time, but it then became clear that the histological picture often did not support this simple view. When the activities of respiratory enzymes, such as cytochrome oxidase and dehydrogenases, were measured, a somewhat better correlation with respiratory rate was found (Agrell, 1949; Sacktor, 1951). No more than a rough agreement between respiratory rate and activities can be expected from such measurements, however, for the course of change during metamorphosis varies for different enzymes and different tissues. In any case, it is now clear that tissue respiration is generally not limited by the levels of respiratory enzymes.

The capacity for respiratory depends on the content of mitochondria and their electron transport system, which is relatively low in tissues such as fat body and high in others such as muscles. But the actual respiration is generally substantially less than this capacity, as a result of respiratory control by the coupled phosphorylating system, which, is subject to the levels of ADP, ATP, and inorganic phosphate in the cell (Harvey and Haskell, 1966). Thus, the rate of respiration adjusts itself to supply the demand, in the form of high-energy phosphate, of the work done by the tissue, which includes biosynthesis, active transport, muscular activity, and so on. In the pupa and the pharate adult insect, where there is little muscular activity, one may expect protein synthesis to be a major energy user, and indeed the rates of respiration often show some correlation with those of protein synthesis. Some evidence on the balance of energy metabolism in insect metamorphosis can be obtained by analysis of metabolite and coenzyme levels.

It was suggested as early as 1893 by Bataillon, on the basis of crude carbohydrate analyses, that the metabolism of the silkworm pupa might somehow be anaerobic, and similar ideas have recurred in various forms (Wyatt, 1963). Several compounds are now recognized as products of experimentally induced anaerobic glycolysis in insects: lactate, alanine, pyruvate, and glycerol-1-phosphate (Gilmour, 1965), but recent analyses show none of these accumulating during metamorphosis of the blowfly (Crompton and Birt, 1967). There is, in fact, much more lactate in the actively respiring larva than in the pupa. In addition, analyses of pyridine nucleotide coenzymes show that the ratio NADH/NAD is somewhat lower in midpupal life than in the active larval and adult stages, whereas anaerobic mechanisms would be expected to lead, on the contrary, to accumulation of the reduced form (Birt, 1966). Thus, there is no evidence for anaerobic processes in the blowfly pupa.

Indeed, it would be surprising to find them, for the supply of oxygen is ample to support the depressed rate of metabolism, and aerobic processes give more efficient use of reserves which the pupa needs to conserve for the construction and initial activity of the imago. The levels of "high-energy" and related phosphate compounds are also of interest. In the blowfly, ATP and the phosphagen, arginine phosphate, show a decrease in the midpupa, and inorganic phosphate shows an almost reciprocal increase.

The doubtless corresponds to changes in the body content of tissues which are characteristically rich in high-energy phosphate, notably muscle. The rise in AMP presumably represents degradation of ATP and ADP from such tissues. That the ATP/ADP ratio remains high through metamorphosis, and that phosphagen is never exhausted, show that the ATP-generating processes in the pupa are at all times fully able to keep up with the energy demand.

Respiratory Metabolism and Diapause

When the pupal stage enters the resting state of diapause, the depression of metabolism is extreme and prolonged. This is the case with *Hyalophora cecropia* and some related saturniid moths, which have received much attention directed toward testing the relationships between respiration, respiratory enzymes, tissues growth, and hormone action (reviewed by Shappirio, 1960; Harvey, 1962; Wyatt, 1963: Harvey and Haskell, 1966). Since diapause is terminated and adult development is initiated by the prothoracic gland hormone, ecdysone, this system has been used for analyzing its action. A provocative observation was that respiration of pupae in diapause is highly resistant to the cytochrome oxidase inhibitors carbon monoxide and cyanide, a fact previously noted for the embryonic diapause of the grasshopper, *Melanoplus differentialts*. In spectroscopic and enzymic assays, the cytochrome content of tissues in diapause was found to be much reduced—cytochrome oxidase is present at low levels, but cytochromes *b* and c could not be detected at all.

Accordingly, it appeared possible that terminal electron transport in diapause might proceed by some pathway other than the usual cytochrome system and that resumption of growth might depend upon the resynthesis of cytochromes which in turn might be closely linked to the action of the hormone. Further experiments showed, however, that when appropriately tested, diapause respiration is indeed sensitive to cyanide and carbon monoxide and since the latter inhibition can be reserved by light, participation of cytochrome oxidase is indicated. Apparently, the tissues contain substantially more cytochrome oxidase than is needed to support the normal rate of electron transport so that the oxidase can be largely inhibited with no effect on net pupal respiration. When the demand upon this enzyme is increased, however, either by lowering the oxygen tension or by enhancing the rate of respiration (e.g., with 2, 4-dinitrophenol), inhibition of the pathway becomes demonstrable. This conclusions raises the question: How is diapause respiration held to its low rate? Since cytochromes b and c were not detected spectroscopically, it seemed at first that a limiting step might exist here.

Oxygen uptake by the pupae, however, is strongly stimulated by dinitrophenol, which shows that the capacity of the electron transport enzymes exceeds their normal load and indicates respiratory control via the coupled phosphorylation system, as is usual in active tissues. In addition, analyses of adenine nucleotides in diapause and developing tissues show that the ATP/ADP ratio changes little, even though the turnover rate is much higher in development, so that a nice co-ordination between phosphorylation and dephosphorylation is indicated, just as in the nondiapausing blowfly pupa discussed above. The ATP level in diapause tissues is high enough so that deficiency of metabolic energy can scarcely be responsible for restraint of growth. The suggestion has also arisen that diapause pupae possess partially anaerobic metabolism as a consequence of restricted electron transport. This seemed to be supported by the gradual conversion of glycogen to glycerol in the Cecropia pupa.

It now appears, however, that the characteristic glycolytic products glycerol-1–phosphate, lactate, and pyruvate are not elevated in the normal diapause pupa, though they are produced in experimental anaerobiosis. In addition, the NADH/NAD ratio is not elevated in diapause. The production of glycerol, therefore, is probably subject to special regulatory mechanisms as an adaptation to cold, and the diapause silkmoth pupa is no more anaerobic than is the blowfly pupa. The present indication from these extended studies on the extremely suppressed metabolism of diapause is that the pathways do not differ qualitatively from those of other tissues, but there is a co-ordinated decrease in activities.

The breakdown of the cytochrome system is impressive, yet the actual rate of respiration appears to be limited by energy demand which, in the pupa, must largely represent biosynthesis of protein and other tissue constituents. Furthermore, resynthesis of cytochromes, though necessary, is not sufficient to release resumption of development, for injury to diapause pupae is followed by the former but not the latter. Certain problems remain: e.g., cytochromes b and c have not been detected in tissues other than muscle of the diapausing Cecropia pupa, and the evidence for their participation in electron transport remains indirect. The measurements of respiration all have been made with whole pupae, and it is not known how much is contributed by muscle, epidermis, fat body, and other tissues whose activities and enzyme complements are quite distinct.

Respiration and the Corpus Allatum Hormone

In addition to the effect of ecdysone in increasing respiration when diapause is broken, enhanced respiration is often attendant upon the action of the hormone of the corpora allata. Increased respiration has been noted both during the "juvenile" action of this hormone, when corpora allata are implanted into larval insects, and during its gonadotrophic action in the adult. When corpora allata were implanted into *Galleria* larvae of different ages, however, and development was modified to various extents, the effect on respiration was always correlated with the degree of morphogenetic change. It appears that this hormone, like ecdysone, may influence respiratory metabolism indirectly by inducing growth and structural change in the tissues (Sehnal and Slama, 1966).

Development of Flight Muscle

A picture of the co-ordinated synthesis of enzymes in the construction of a specialized tissue has been gained from studies on the biochemical development of flight muscle. The indirect flight muscles of insects are of the fibrillar type. Between the bundles of large fibrils are packed rows or giant mitochondria ("sarcosomes") which may be 1 to 3 μ in diameter and make up 30 to 40 percent of the weight of the tissue. These are directly supplied with oxygen by tracheoles. The flight muscle has almost no lactated dehydrogenase and in its normal activity is totally aerobic. The contents of soluble NAD-linked α-glycerophosphate dehydrogenase and of mitochondrial flavoprotein α-glycerophosphate oxidase are usually high and provide a shuttle for the oxidation of extramitochondrial NADH by passage of glycerol-1-phosphate and dihydroxyacetone phosphate in and out of the mitochondria (Sacktor, 1965). Flight muscle is thus constructed for sustained aerobic activity with exceptionally high energy output.

Insect leg muscle, such as the jumping muscles of the locust, on the other hand, has a smaller content of mitochondria and possesses lactate dehydrogenase in adaptation for sporadic violent activity followed by rest. The most comprehensive picture of the structural and biochemical development of flight muscle is that due to the work of Bucher and his colleagues on the locust, *Locusta migratoria* (Bucher, 1965). The muscle develops during a period of rather more than two weeks starting from a small anlage, or precursor muscle, that is present just after the fourth molt. Even this precursor muscle resembles flight muscle in having only a low content of lactate dehydrogenase. Four phases of growth are described: (1) larval growth, with increase in size (but not number of nuclei) with little differentiation; (2) a phase about the time of the final molt when the larval tracheal supply breaks down and new tracheoles penetrate into the muscle and a transient elevation of lactate dehydrogenase may indicate some temporary anaerobiosis; (3) a phase of differentiation, during the first days of adult life, in which the characteristic enzyme pattern of flight muscle is established by elevation of mitochondrial enzymes and degeneration of lactate dehydrogenase and glucose-6-phosphate dehydrogenase; (4) finally, further growth, with approximate doubling of all components. Thus, most of the growth of the mitochondria and synthesis of the eventual enzymes takes place after emergence of the adult insect, and the locust cannot fly until this maturation is completed.

There are "constant-proportion groups" of enzymes, so that the mitochondrial oxidative enzymes (represented by a-glycerophosphate dehydrogenase) increase co-ordinately with one another and, similarly, a group of soluble glycolytic enzymes represented by glyceraldehyde phosphate dehydrogenase) retain an approximately constant mutual relationship. The existence of such constant proportionality within groups of enzymes, despite wide variations in absolute levels, extends to other tissues than flight muscle and presumably indicates some common regulation of their synthesis. Although the flight muscle of the locust is somewhat less specialized than that of holometabolous insects such as flies and bees, the biochemical development of the latter is probably essentially similar.

A greater proportion of the enzyme synthesis may occur before eclosion than in the locust, but there is always period of maturation, involving further mitochondrial growth and enzyme synthesis, in the emerged adult. In saturniid silk moths, cytochrome *c* first appear in measurable amounts in the flight muscle on about the sixteenth day of the 21-day period of adult development, and its synthesis thereafter is very rapid (Chan and Margoliash, 1966). In the blowfly, *Phormia*, the cytochrome content rises about fourfold during the first week of adult life, which reflects an increase in the size of the mitochondria while their number remains constant (Levenbook and Williams, 1956). Growth of the flight muscle mitochondria is also reflected in a 50 percent increase in phospholipid content of the thorax of *Lucilia*, during the weak after emergence (D' Costa and Birt, 1966).

In *Lucilia*, the distribution of enzymes and protein in particles of various sizes has been determined, and in as much as the specific activities of several respiratory enzymes relative to mitochondrial protein change greatly during development of the fly, it is proposed that the respiratory enzymes and the nonrespiratory, structural protein follow different courses of synthesis (Lennie and Birt, 1967). This is quite likely, since it appears that the nonextractable, structural protein of mitochondria may be synthesized within them, whereas synthesis of the extractable enzymes is a function of extramitochondrial ribosomes.

In the honeybee, the mean volume of a flight-muscle sarcosome increases from 0.015 μ^3 in the 10-day pupa (pharate adult) to 0.041 μ^3 on the day of emergence and to 0.52 μ^3 in the 20-day-old adult. The increase in content of several cytochromes, determined by the sensitive technique of spectroscopy at the temperature of liquid nitrogen. It is evident that their synthesis is by no means synchronous. Cytochrome b_5, which occurs in the microsomal fraction and is believed to participate in biosyntheses, is found in the pupa and declines sharply in adult development. Among the mitochondrial cytochromes, although the general pattern is similar, there are distinct differences in time of first detectability and course of synthesis, so it is clear that constant mutual proportionality is not a universal rule among these electron carriers.

Sclerotization of the Cuticle

A characteristic feature of insects is their tough external cuticle. This is secreted at each molt by the epidermis and consists chiefly of protein and chitin, plus smaller amounts of specialized lipoidal substances. Unlike the cuticle of Crustacea, which contains inorganic deposits, it is hardened by organic tanning or cross-linking of the protein in a process known as sclerotization. It has been accepted for many years that sclerotization involves quinones derived from tyrosine, but the chemical nature of the tanning agent was established only recently, and the details of its reactions with the protein are still obscure. The biochemistry of sclerotization of the blowfly puparium has recently been intensively studied by Karlson, Sekeris, and coworkers because of its connection with the action of the hormone, ecdysone. This subject is reviewed by Cottrell (1964), Brunet (1965), and Sekeris and Karlson (1966). When the blowfly maggot is full grown and ready to pupate, its soft, white cuticle is inflated into a barrel shape and then becomes hard and dark brown, as the puparium, before the insect molts to a pupa within it. The principal tanning agent has been identified as N-acetyldopamine its acetylation has the chemical role of preventing cyclization into an indole, and when the phenolic ring is further, oxidized to an o-quinone, it is reactive with protein amino groups. In the early third-instar larva, metabolism of tyrosine is largely by a degradative pathway that starts with transamination.

During the instar, the level of tyrosine mounts until it is one of the most abundant free amino acids in the hemolymph. Then, at the time of puparium formation, it is largely used in the production of N-acetyldopamine by the enzyme-catalyzed steps shown. The switch in metabolic fate of tyrosine is the result of

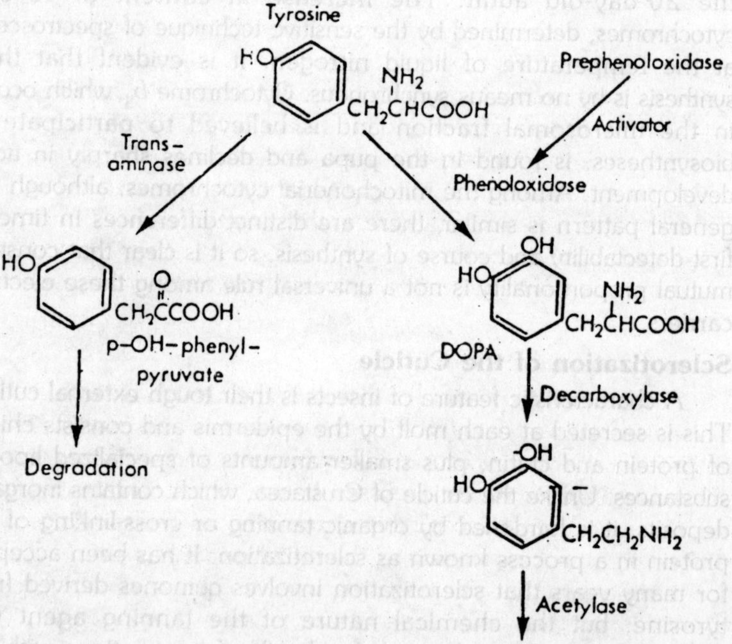

Fig. 4.2. The metabolism of tyrosine in the tanning of the blowfly puparium.

changed enzyme activities: tyrosine transaminase apparently decrease, prephenol oxidase accumulates in the hemolymph and its activator protein builds up in the integument, and a specific dopa decarboxylase is synthesized. The acetylase, on the other hand, is present throughout the instar. The decarboxylase, whose activity shows the greatest changes, is believed to be the controlling enzyme.

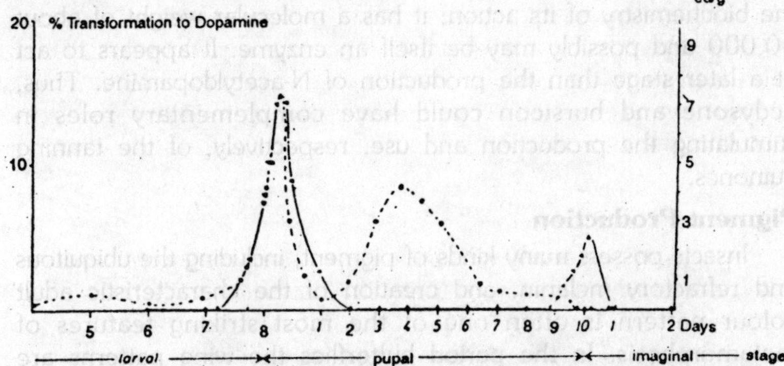

Fig. 4.3. Ecdysone titer and dapa decarboxylase activity during development of the blowfly Calliphora erythrocephala. Ecydosone titer (–) is expressed Calliphora-units per gram, and enzyme activity (—) as percent transformation of dopa to dopamine.

The whole process is set off by the increasing level of ecdysone at this stage of the life cycle; it can be prevented by ligation and induced by injection of the hormone, and this forms the basis of the usual assay for ecdysone. Because of the close correlation between production of dopa decarboxylase and ecdysone titer, and other evidence to be discussed below, it has been proposed that the hormone "induces" this enzyme by an effect on nucleic acid metabolism. The tanning of the dipteran puparium just before the pupal molt is a special case in insect ontogeny, as sclerotization commonly affects newly deposited cuticle just after ecdysis. In several such cases recently examined in Karlson's laboratory (the, blowfly adult, the locust, and the mealworm, *Tenebrio molitor*) the mechanism appeared to be similar, for dopa decarboxylase was active, and tyrosine was converted to N-acetyldopamine, close to times of sclerotization.

In the blowfly adult, a further source of tanning agent is apparently the hydrolysis of N-acetyldopamine -4-0-β-glucoside, in which form excess N-acetyldopamine is conserved after puparium formation. After the adult molt, ecdysone titer is low, but there is now recognized another hormone specifically associated with cuticular hardening and designated bursicon (Cottrell, 1964; Fraenkel, Hsiao and Seligman, 1966). It is required for sclerotization of the cuticles of newly emerged flies and newly molted cockroaches and will effect the darkening of discs of colourless integument from the latter in vitro. Little is known about the biochemistry of its action; it has a molecular weight of about 40,000 and possibly may be itself an enzyme. It appears to act at a later stage than the production of N-acetyldopamine. Thus, ecdysone and bursicon could have complementary roles in stimulating the production and use, respectively, of the tanning quinones.

Pigment Production

Insects possess many kinds of pigment, including the ubiquitous and refractory melanin, and creation of the characteristic adult colour pattern is often one of the most striking features of metamorphosis. In the period butterflies the wing patterns are due chiefly to pteridines, and the several pteridines of *Colias eurytheme* wings follows characteristic courses of synthesis, beginning on the third day of pupal life (Watt, 1967). But little is yet known about the regulation of such processes. One instance in which there is a relation to hormone action involves certain ommochromes. These are phenoxazone compounds derived from tryptophan via kynurenine, which occur widely in the eyes, integument, and internal organs of insects. They are of historic interest because it was experiments with certain eye-colour mutants of *Drosophila* and *Ephestia* blocked in tryptophan metabolism that some 30 years ago first demonstrated the association of specific genes with successive steps in a metabolic sequence.

The structure and distribution of ommochromes was worked out in a monumental series of investigations in Butenandt's laboratory. In the hawk moth, *Dicranura vinula*, the green caterpillar develops a characteristic red-brown pattern six days before pupation as a result of the deposition of xanthommatin and rhodommatin. Ligation prevents this change in the abdomen, and it can then be induced by injection of small doses of ecdysone. The enzymic basis of the change has not been established, but the hemolymph levels of tryptophan and 3-hydroxykynurenine rise sharply at the time of pigment synthesis, and it is proposed that ommochrome formation may serve in the removal of tryptophan liberated by proteolysis (Buckmann, Willig and Linzen, 1966).

Biosynthesis and Degradation of Macromolecules

The Proteins of Larval and Adult Insects

A question of fundamental importance is to what extent to protein constitution of an adult insect differs from that of a larva of the same species. No all-out attack on this problem has been described, but there has recently been a good deal of skirmishing with it. Most attention has been given to insect blood (hemolymph), because this protein solution is easy to obtain and it undergoes large quantitative changes during ontogeny (Wyatt, 1961). The usual pattern with respect to protein is a relatively low level often 1 to 2 percent), during the earlier larval stages, then a rapid rise just before pupation (often 6 to 8 percent, but 18 and 20 percent have been recorded), then there is some decline in the pupa and a further fall during the later part of adult development. Qualitatively, a number of analyses by electrophoresis on filter paper (summarized by Wyatt, 1961) gave too little resolution to be of much significance.

More recently, starch gel and acrylamide gel electrophoresis have been applied to the hemolymph of a number of species, and between 10 and 20 bands can generally be recognized. These have no correspondence with the components of mammalian serum and differ characteristically with the species. There is often a yellow or green chromoprotein band and a dense band of low mobility which may represent the major hemolymph component of very high molecular weight (more than 500,000) which has been detected in several insects by other methods.

In the transformation from larva to adult, most of the electrophoretic components generally persist, but there are certain qualitative differences. Thus, in the tent caterpillar *Malacosoma americanum*, out of 17 bands detected in the full grown larva and early pupa, two to five were apparently missing in young larvae and in adult moths, and the adult did not a possess any obvious new components (Loughton and West, 1965). When blood from fourth-instar tomato hornworms was concentrated fourfold before electrophoresis, several otherwise undetected components were revealed, and in this insect the adults apparently lack some components characteristic of larvae and contain several fresh ones (Hudson, 1966).

A study on the blowfly, *Phormia regina*, in which polyacrylamide gel was used with equal amounts of protein from each developmental stage, led to the conclusion that "the newly emerged adult fly has an entirely different picture" from the larva and pupa (Chen and Levenbook, 1966a), though inspection of the published diagrams suggests that about the half of the bands in the adult pattern may correspond to larval components. Bands observed in zone electrophoresis are, of course, rarely pure proteins, nor is their mobility an adequate criterion of identity.

The technique of immunodiffusion, in which antigen and antibody diffuse toward one another in an agar gel, can assist with establishing identity, although when applied to whole hemolymph, it has generally revealed fewer components than has electrophoresis. In an immunological study on Cecropia silkworm hemolymph, Telfer and Williams (1953) observed nine distinct antigens and followed six of them quantitatively from the fifth-instar larva to the adult. Each followed its own course of variation, which shows that their levels are individually regulated. One protein found in female pharate adults was passed into the ovaries and their contained eggs, and was present only as traces in the male (Telfer, 1954). But the appearance of uniquely adult antigens, distinct from those of the larva, is not reported.

By a technique of immunoelectrophoresis, which should combine high resolution with definitive identification, Terando and Feir (1967) detected 11 antigens in hemolymph of the milkweed bug, *Oncopeltus fasciatus*, and finding no evidence for stage-specific antigens, arrived at the conclusion that nymphs and adults of both sexes showed only quantitative differences; but it should be noted that this concerns a hemimetabolous insect. Laufer (1963) has applied both electrophoretic and immunological methods to the examination of hemolymph and tissue protein of the Cynthia silk moth and several other species and concludes that there exist both stage-specific and tissue-specific antigens.

Enzyme activity (esterases and dehydrogenases) was found associated with a number of the protein zones, but as the enzyme assay is very sensitive it may reveal proteins other than those made visible by staining. Extracts of the cellular tissues would be expected to be more representative than hemolymph of the overall protein makeup of an insect. Butler and Leone (1966) have prepared whole-insect extracts of several developmental stages of the beetle mealworm *Tenebrio molitor* and examined these by several types of electrophoretic and serological test.

By immunoelectrophoresis, 10 precipitin zones were obtained; in several zones, larval and adult protein were distinguished serologically despite similar electrophoretic mobility. These authors conclude that during ontogeny there are gradual and sequential qualitative and quantitative changes in protein complement. The electrophoretic pattern of basic proteins extracted from an acetone powder with acid also changes during blowfly metamorphosis. From these various studies, it is very difficult to feel confident in any conclusions as to the extent of change over from a larval set of proteins to an adult set in metamorphosis. For the enzymic requirements for their metabolism, insect cells, like other cells, must contain at least several hundreds of distinct proteins. Yet the methods that have been used have detected at most 20.

These are no doubt the most abundant or the most antigenic components but are not necessarily the most significant in ontogeny. In only a few studies has there been any attempt at quantitation, and in view of the limited sensitivity of the detection methods, it is very difficult to distinguish a qualitative from a quantitative change. There is suggestive evidence (with respect to hemolymph proteins) that, as might be expected, some correlation exists between the completeness of metamorphosis in an insect group and the extent of change in its proteins. Thus, a bug shows no qualitative change, a moth changes a small proportion of its complement, and flies exhibit extensive change.

In all cases, however, a proportion of the demonstrable protein components are found at all stages of development. A quite different, and more demanding, approach to the question of identity of larval and adult proteins is to isolate homologous proteins from each stage and compare them chemically. This has been done with tropomyosin, which was obtained crystalline and apparently pure from both larvae and adults of the blowfly, *Phormia regina* (Kominz et al., 1962). The two preparations were very similar but showed slight differences in amino acid and peptide composition and physical properties. These small differences could conceivably result from impurity or secondary modifications of the protein rather than genetic difference, and thus the question of genetic identity is unfortunately unresolved. Another indication of similarity of larval and adult proteins, though not conclusive evidence of identity, comes from low-temperature spectra of mitochondria from larvae and adult thoraces of *Drosophila*; the two show no significant differences, while both differ characteristically from mammalian mitochondria.

Free Amino Acids and Peptides

A peculiar feature of insects that always has to be taken into account in a discussion of their protein metabolism is their high level of free amino acids. This is highest in the blood, and the tissue pool is generally somewhat less concentrated and differently constituted with respect to individual amino acids. In the more advanced insects, the hemolymph amino acid level is commonly in the range 5 to 20 mg/ml-many times that mammalian plasma. In addition to the amino acids of proteins, some less usual ones have been identified from certain insects, including β-alanine, tyrosine-0-phosphate, methionine sulfoxide, D-alanine, D-serine and others. The literature, of this subject, as well as other aspects of amino acid and protein metabolism in insect development, is reviewed by Chen (1966).

The significance of the high amino acid level is not well understood, although it provides a pool for protein synthesis and appears to have an osmotic role. During metamorphosis, while individual amino acids change in concentration, the variation in total free amino acid pool is remarkably small when one considers the extent of tissue reconstruction that is taking place. Along with free amino acids, insects also exhibit a variable content of peptides. In *Drosophila* larvae, intensive chromatographic analysis led to the conclusion that more than 600 peptides were present, and these incorporated injected C^{14} glutamate much more rapidly than did the total body protein (Mitchell and Simmons, 1962). In *Phormia*, a group of peptides is transiently abundant during the early third instar; these are mostly composed of only two amino acids one of which, in each that was analyzed, was either lysine or histidine. Their metabolism and significance is unknown.

The Biochemistry of Histolysis

It repeatedly has been observed since the early histological studies on insect metamorphosis that at certain developmental stages the fat body, and to a lesser extent other tissues, become loaded win protein granules or "albuminoid spheres". These have been regarded as a form of protein reserve and appeared to constitute an exception to the generalization from mammalian biochemistry that animal tissues have no specifically storage proteins. Their probable nature is now becoming apparent with the use of the electron microscope and with growing understanding of the nature and functions of lysosomes.

In Lepidoptera, such as the skipper *Calpodes ethlius*, protein granules appear in the fat body about one day before pupation, at the time when the larval cuticle is being resorbed. Since injected radioactive amino acid showed little incorporation into these granules, whereas injected plant peroxidase (detectable histochemically) appeared specifically in them, it was inferred that they contain protein sequestered from the hemolymph and not newly synthesized. Similar uptake occurs in other tissues, particularly the epidermis and especially at times in the molting cycle when there is active protein synthesis. The granules are surrounded by membranes and are believed to transform into multivesicular bodies in which digestion takes place, presumably by addition of lysosomal enzymes, as a source of amino acids for new protein synthesis.

Cell organelles such as mitochondria due for destruction, are believed to be isolated in a similar way. The uptake of hemolymph proteins into fat body, gut , and to some extent other tissues of Lepidoptera is independently demonstrated by immunodiffusion tests. Which show several blood antigens to be present in them during the prepupal and early pupal stages. As an example of programmed histolysis, have studies the breakdown of the abdominal intersegmental muscles of saturniid silk moths, whose function is to pump blood into the wings of the newly emerged moth and which lyse when this role is completed. The trigger for lysis is apparently the cessation of nervous motor stimuli to the muscle. Lysosomes like bodies are observed to undergo change at this time.

Proteolytic activity of a catheptic type (pH optimum 4.0) builds up is these muscles at first gradually, then rapidly, during the 3 weeks of pharate adult development, and is associated with lysosome like particles. It can be activated and partially released from the particles by detergent treatment, but in the emerged moth the activity and the proportion of soluble enzyme are naturally enhanced. Similar catheptic activity is found also in the far body, but not in thoracic muscle (which, of course, is not destined for dissolution). The intersegmental muscles and fat body also contain an acid phosphatase which shows greatest activity in advanced adult development. These enzymes are considered to be directly responsible for the destruction of the intersegmental muscles. It is remarkable that, despite the dominant place of histolysis in insect metamorphosis, there is almost no other information on the enzymic basis of the process. Not much can be concluded from several reports of hydrolytic activities without evidence as to their source in tissues or cytological fractions.

Rate of Protein Synthesis and Turnover

Linked to the question of changing protein composition of an insect during metamorphosis is that of the rates and extent of protein degradation and synthesis. A number of measurements have been reported of the incorporation of injected amino acid into the proteins of hemolymph or tissues of insects at selected stages. It is found that incorporation rates rise sharply when adult development is initiated in diapausing silk moth pupae, and that fat body qualifies kinetically as a source of hemolymph proteins. Similar results have been obtained with isolated insect tissues incubated in vitro, a technique which avoids complications due to the blood amino acid pool. But the systematic quantitative study of protein metabolism during metamorphosis seems to be confined to the recent work of two laboratories. Using *Drosophila melanogaster*, Boyd and Mitchell (1966) fed C^{14} amino acids to larvae, then after appropriate times for incorporation, extracted body fluid (chiefly hemolymph), and separated radioactive protein fractions from it by acrylamide gel electrophoresis. Then, their turnover was measured by reinjecting into larvae or pupae the isolated radioactive protein fractions and, after incubation, subjecting hemolymph samples to the same kind of analysis.

The redistribution of activity was extensive and varied in different fractions: One was stable in the larva, but after pupation had a half-life of 13 hours; another was too stable for the half-life to be measured. Specificity is also indicated by the fact that injected mammalian hemoglobin did not turn over. When enough nonradioactive valine was injected to increase the blood valine pool about thirty-fivefold, the redistribution of activity from valine-labeled protein during 42 hours was cut from 56 to 39 percent, which suggests that free valine, but not the valine of the hemolymph, may be an intermediate in the interconversion of hemolymph proteins. The authors conclude that hemolymph proteins "may be an important channel through which the massive protein turnover of the metamorphosing insect passes". Levenbook and colleagues have examined the question of protein synthesis and turnover in *Phormia regina*.

The hemolymph protein is shown to be synthesized chiefly in the larva; at pupation there is a rapid fall in blood protein content, and there is little synthesis from injected amino acids or change in level thereafter. Radioactive hemolymph protein was prepared from C^{14}-labeled larvae, and after injection into larvae, pupae, and pharate, adults, its fate was followed. Very little was converted to C^{14} O_2, although injected free amino acids were rapidly oxidized, and conversion of hemolymph protein to free amino acids was also slow. It is inferred, in contrast to the conclusions of Boyd and Mitchell, that the hemolymph proteins are relatively stable. When mature larvae were injected with labeled hemolymph protein and kept until they emerged as flies, however, about there quarters of the recovered radioactivity was found in protein in the tissues.

Since the blood free amino acids clearly cannot be intermediates, some form of direct transformation of blood proteins into tissue proteins is suggested. The data seem only to rule out *hemolymph* free amino acids as intermediates and do not oppose degradation to free amino acids which remain within the tissues and are rapidly used in resynthesis. In a related study, Dinamarca and Levenbook (1966) examined the dynamic state of the free amino acid pool and the rate of protein synthesis from it in *Phormia regina* at different developmental stages. C^{14}-alanine and C^{14}-lysine were injected, and their conversion to C^{14}-O_2, total disappearance from the pool, and incorporation into protein were measured, as well as the pool size for each amino acid in each insect used. Corrected values for the rate of incorporation into protein could then be calculated and were found to follow a U-shaped curve, closely similar for the two amino acids, except in the adult. The importance of the procedure became clear, for lysine always yielded more of the injected dose into protein than did alanine, but this was a consequence of the more rapid turn-over of alanine. With the assumption of a single homogeneous pool of each amino acid in the insect, the further calculation could be made of the total amount of new protein synthesis during metamorphosis, and this came out at the surprisingly small value of 0.16 mg, or about 2 percent of the total protein of the insect.

As Dinamarca and Levenbook recognize, the assumption is a great oversimplification; nevertheless, they suggest that the total amount of new protein synthesis during metamorphosis is less than has been generally assumed. The chief conclusions from these studies are, perhaps, to confirm that the hemolymph amino acids (which make up more than half of the total pool) are not in free exchange with those in the tissues and to indicate that the greater part of the protein turnover involves only local intracellular pools. This would be in accord with the cytological evidence that proteins are engulfed from the blood and digested within various tissues that are also synthesizing protein. Calculations based on the total pool of the animal might then fail to give any indication of the true amount of protein synthesis. It seems that still more refined techniques, perhaps involving isolation of specific proteins from individual tissues, are required.

Enzymes Associated with Protein Synthesis

The amino acid activating enzymes, responsible for the first step in protein synthesis, have been assayed and to some extent characterized in different developmental stages of the blowfly, *Lucilia cuprina*, and the fat body of *Sarcophaga bullata*. Activity was found for almost all of the protein amino acids, and where it was not found, this may have been due to inactivation. The total soluble enzyme activity in both series of experiments was minimal in midpupal life and rose to a maximum just before emergence of the adult fly, thus paralleling the changes in rate of protein synthesis in a blowfly pupa.

In *Lucilia*, during the early pupal stage, some amino acid activating enzymes, particularly for tyrosine, are bound to a class of cell particle, and it has been suggested that these may have some role in storage of tyrosine for use in cutincular synthesis and tanning. The activity of glutamate-aspartate transaminase in the housefly declines from a high value at pupation to a minimum just before adult emergence and then rises steeply. A relation to protein synthesis is suggested, but the curve appears more closely correlated with that to be expected of protein degradation. In the Cecropia silk moth, however, this enzyme rises strikingly in thoracic muscle during late adult development, which is certainly an active site of protein synthesis.

Ribonucleic Acid and the Machinery of Protein Synthesis

Increasing attention has been directed recently to the nucleic acids as determinants in the specificity and possible regulations in the rate of protein synthesis, but there is not yet a great deal of information relevant to insect metamorphosis. Measurements of total RNA during metamorphosis of several insect species reveal little change, which is consistent with the histological observation that histolysis and histogenesis are concurrent rather than consecutive, so that the total amount of cellular tissue does not change much. Some changes reported in RNA fractions differing in extractability with phenol from *Calliphora erythrocephala* are very difficult to interpret. When a single tissue is examined, more significant changes may be seen. Thus, in the fat body of the saturniid silkworm, *Samia Cynthia ricini*, during pupation there is a great decrease in cytoplasmic RNA, which is shown by P^{32} incorporation to have been synthesized early in the fifth larval instar.

In the Cecropia silk moth, when adult development begins after the pupal diapause there is no discernible net increase in fat body RNA, but there is an increase in turnover, associated with renewed activity in protein synthesis. Other recent studies on RNA in insect metamorphosis have concentrated on particular developmental processes related to hormone action, Karlson, Sekeris and colleagues have examined changes in fully grown *Calliphora* larvae, at the stage when formation and tanning of the puparium follow upon the increasing titer of ecdysone. After injection of ecdysone into larvae taken about two days before pupation, the incorporation of P^{32} into RNA is elevated. With 1.5-hour pulse corporation, the maximal stimulation of 80 percent over controls was observed at 4 to 7 hours after giving the hormone. A greater effect was observed in RNA from the epidermis, a target tissue for ecdysone, and in RNA prepared with hot phenol from epidermal nuclei there was some stimulation within 1 hour of giving the hormone. The nuclear RNA stimulated amino acid incorporation in microsomal preparations from rat liver or *Calliphora* epidermis.

It has been reported from the same laboratory that ecdysone added to isolated epidermal nuclei stimulates their incorporation of C^{14}-uracil into RNA and that RNA prepared from pupating larvae when added to rat liver microsomal preparations acts as a template in the synthesis of dopa decarboxylase, but fuller data are required before one can interpret these observations with confidence. The results obtained with this system have been presented in support of the view that ecdysone and other hormones act by directly causing derepression of DNA in the cell nucleus and release of specific messengers. The evidence, however, does not eliminate the presence of further links in the chain of hormone action.

The synthesis of RNA under the stimulus of ecdysone has also been investigated in the pupal wing epidermis of *Hyalophora cecropia*, during the initiation of adult development after diapause. During natural development, in pupae incubated after a period of chilling, the synthesis of RNA begins well before morphological change, and by the second day of visible adult development the content of RNA in the wing tissue is about fourfold that in diapause. After initiation of development by injection of ecdysone into pupae whose endogenous endocrine sequence is blocked by removal of the brain, stimulation of RNA synthesis is detectable at 8 to 10 hours, intense (about sixfold) at 24 hours, and falls off thereafter. Some effect on protein synthesis begins to appear at about the same time, but the rise is more gradual and more prolonged.

The RNA produced under the influence of ecdysone has been characterized in several ways. Its base composition (by pulse labeling with P^{32}) is intermediate between the bulk RNA and the DNA of the species and could be accounted for by synthesis of a mixture of ribosomal RNA and DNA-like RNA. Fractionation of the RNA, after pulse labeling with uridine, showed incorporation in the 4s, 17s, and approximately 30s regions (corresponding to transfer, ribosomal, and ribosomal precursor RNA, respectively) and also in the residue from cold phenol extraction, which would contain nuclear RNA. The extent of stimulation by ecdysone was rather similar in each of these fractions. When assayed for stimulation of amino acid incorporation in an E. coli ribosomal system, the RNA produced early in Cecropia wing development was found to be highly active, but there are reasons for believing that such stimulation in a heterologous system may not detectable in the epidermis in diapause, become abundant early in development.

It is concluded from this work that new RNA synthesis forms an important early part of the action of ecdysone and contributes to building the stage for increased protein synthesis. But the connection between hormone action and RNA synthesis, and the question of specificity in template RNA production, require further study. The picture in a tissue being awakened from diapause may, of course, differ from that in a tissue which, when already metabolically active, is influenced by ecdysone. But the observations on the Cecropia wing epidermis are consistent with those on the action of various developmental hormones in vertebrate material.

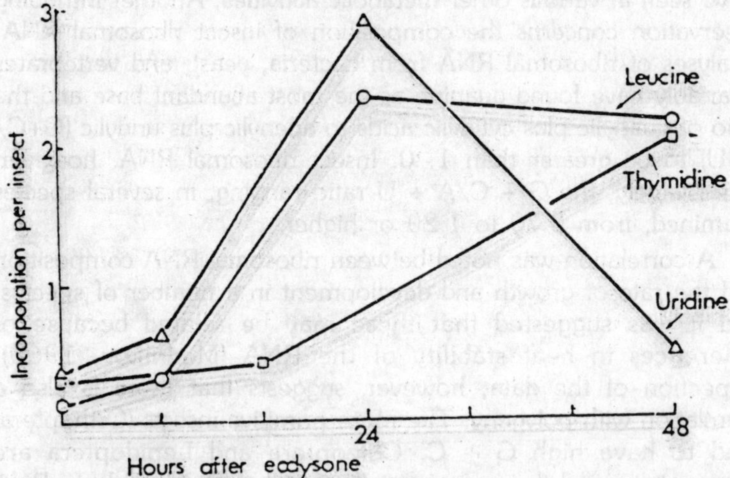

Fig. 4.4. The effects of ecdysone on macromolecular synthesis in wing tissue of Cecropic silk moth pupae. Into pupae from which the brains had been removed, sufficient ecdysone was injected to cause visible initiation of development. Then, of the intervals shown, radioactive leucine was injected as a precursor of protein uridine as a precursor of RNA, and thymidine as a precursor of DNA. After 4 hours for incorporation, the pupal wings were removed and prepared for counting.

The juvenile hormone of the corpora allata, acting in the presence of ecdysone, permits growth and molting in insects but modifies their course, so that the characteristics of immature stages are retained. A plausible basis for this would be a regulatory effect upon RNA synthesis so that larval-specific templates are produced and adult templates repressed, but there is as yet no biochemical evidence for this. When juvenile hormone is injected into *Antheraea polyphemus* pupae at the beginning of adult development, the rate of RNA synthesis is increased, but this may be a result of stimulated output of ecdysone from the prothoracic gland rather than a direct effect on other tissues. A provocative observation for which no interpretation can be offered concerns RNA-methylating enzymes.

In *Tenebrio molitor*, the enzymes active in adding methyl groups to *E. coli* tRNA are more active (with a characteristic day-to day pattern) in the early pupa than in the larva or pharate adult, thus portraying the approximate reverse of the U-shaped curve seen in various other metabolic activities. Another intriguing observation concerns the composition of insect ribosomal RNA. Analyses of ribosomal RNA from bacteria, yeast, and vertebrates invariably have found guanine as the most abundant base and the ratio of guanylic plus cytidylic acids to adenylic plus uridylic (G+C/A+U) to be greater than 1, 0. Insect ribosomal RNA, however, varies widely, the G + C/A + U ratio ranging, in several species examined, from 0.70 to 1.20 or higher.

A correlation was noted between ribosomal RNA composition and the rate of growth and development in a number of species, and it was suggested that these may be related because of differences in heat stability of the RNA (Mednikov, 1965). Inspection of the data, however, suggests that there is also a correlation with polygeny. The more primitive insects (Orthoptera) tend to have high G + C, Coleoptera and Lepidoptera are intermediate, and the most specialized group (Diptera) have RNA distinguished by low G + C content. Studies on the properties of ribosomes with such differently constituted RNA would be of interest. Understanding of the control mechanisms of rate and specificity of protein synthesis in insect metamorphosis will undoubtedly depend on the development and study of cell-free amino acid incorporating systems of insect origin. In such a system for *Tenebrio molitor*, the ratio of incorporation of tyrosine to that of leucine rises greatly during the last two days of adult development, which is believed to reflect the high tyrosine content of the cuticular proteins then being synthesized.

Deoxyribonucleic Acid

When one considers the pivotal position of DNA in determining the characters of organisms, and the wealth of recent critical and ingenious research on this nucleic acid in various living systems, the paucity of information on DNA in insect metamorphosis is remarkable. For very few insect species do we even have reliable estimates of the changing quantity of DNA during development. Nucleic acid determinations with insect material require special attention to technique because of various complications that may arise, e.g., difficulty in securing quantitative extraction, the presence of compounds which interfere in colorimetric reactions, and uric acid which interferes in the ultraviolet range. In *Lucilia cuprina*, the total body DNA falls to about one half just before pupation and then rises rather steadily during the pupal stage and adult development, presumably reflecting breakdown of larval cells, followed by cell multiplication in the imaginal *anlagen*.

In the mosquito, on the other hand, there is apparently little change in DNA during metamorphosis, although in larvae there is the unusual situation that up to 30 percent of the total DNA is found in the soluble supernatant fraction of homogenates. Patterns of DNA synthesis in different tissues are readily surveyed by autoradiography after incorporation of H^3-thymidine, and this technique has been applied to the development of saturiniid silk moths. In the larval molting cycle there is a burst of synthesis in most tissues centered about 3 days before the molt, but in the hemocytes and midgut, synthesis continues at a reduced rate at other times, and in the imaginal wing buds synthesis is apparently unrelated to the molt cycle.

Shortly after pupation, in diapausing species such as *Hyalophora cecropia*, DNA synthesis stops in all tissues except the hemocytes and spermatogonia, in which a significant rate of incorporation continues. In a species without diapause (*Samia Cynthia ricini*), incorporation in most tissues comes almost to a stop at the time of pupation, but then resumes within a day with the onset of adult development. At the conclusion of a pupal diapause, DNA synthesis appears to be correlated with the stimulation of growth by ecdysone. Adult development of moths is blocked by mitomycin C, an inhibitor of DNA synthesis, and it is clear that DNA replication is an essential part of ecdysone-stimulated development. The cecropia wing epidermis developing after the pupal diapause has also been examined by biochemical methods (Wyatt, 1967b).

When incorporation of thymidine is measured at intervals after injection of ecdysone, clear stimulation of DNA synthesis is not detected until about 48 hours—much later than the effect upon RNA. Thus, the stimulation of DNA synthesis seems to be separated from the earliest effects of the hormone. During the 21-day period of natural development to the moth, two cycles of DNA replication centered at about the second and sixth days can be detected in the wing tissue. In a quest for the enzymic basis of the cessation of DNA synthesis in diapause, the crucial enzymes thymidine kinase and thymidylate kinase were assayed in silk moth pupal wing epidermis. Both increased manyfold at the initiation of adult development, yet diapause tissue did contain appreciable activity, so that the lack of these enzymes could not be responsible for the total cessation of synthesis.

Inorganic Ions

One further subject which must be mentioned is the remarkable inorganic ion content of many insects. In contrast to the usual composition of body fluids of vertebrates and most other animals, the higher insects, particularly phytophagous species, contain much potassium and little sodium, so that the Na^+; K^+ ratio is far below unity. Further, the level of magnesium is often extraordinarily high (often up to 30mM and occasionally to 100 mM, though part of this is chemically complexed). This situation is regarded as an adaptation to a vegetarian diet high in potassium and magnesium. Normal ion levels are maintained with the aid of active transport mechanisms, including the transport of K^+ through the gut wall. In metamorphosis, the inorganic levels may be changed Thus, in the silkworm, *Bombyx mori*, falls in sodium and magnesium at pupation have been reported, though the data of different authors are not consistent. Since the ionic proportions found in different insect species vary greatly, their having an important role in metabolic regulation is perhaps unlikely. Little is known, however, about the distribution of ions between hemolymph and intracellular fluids in insects, and it is interesting that an effect of ion levels on protein production by blowfly larval fat body in vitro recently has been reported. The matter deserves further attention, in view of the indications that potassium can have roles in the regulation of gene activity in insects and in protein synthesis in various living systems.

5

NUCLEUS IN DEVELOPMENT

A living cell is made of a nucleus and cytoplasm. There is a partnership between the nucleus and the cytoplasm. The nucleus, with its store of hereditary factors, the *genes* is the conservative member of the partnership. It controls the syntheses which go on in the cell. Hence, it directs that most amazing of all syntheses. The course of development. Hence also it determines the species and individual characteristics which are the outcome of development. On the other hand it is primarily the cytoplasm which develops and is the progressive member of the partnership. It begins in relative simplicity and then, by differentiation and molding becomes the mature organism in all of its structural and functional complexity. For the smooth functioning of the cell and also for the continuation of life both the nucleus and the cytoplasm are needed. Many experiments have been performed in which the nucleus of a cell has been removed. For example, when an amoeba is cut into two parts, the part containing the nucleus heels and continues to live. The part which lacks a nucleus may carry on for a time, but ultimately it wears itself out and perishes, for it cannot continue to rebuild its substance.

It is reported that a sea urchin egg from which the nucleus has been removed may cleave times, but it will not continue to develop. It is even more obvious that a nucleus cannot live and carry on its functions without the cytoplasm. In particular, the nucleus depends upon the cytoplasm for its raw material and energy. It is probable that only in the cytoplasm do the oxidations take place by which energy-rich molecules (ATP) are synthesized.

The Life Cycle of the Nucleus

The story of the nucleus is a repetitive one and begins at fertilization when the chromosomes of the two germ cells gametes), an egg and a sperm, come together in the nucleus of the fertilized egg or zygote. The story ends and is ready to start again, at least as far as the contribution of the individual to the race is concerned, when the new individual has matured and has produced gametes of its own. Between fertilization and the maturing of the germ cells, the story of the nucleus is one of alternating growth and division by mitosis. It is responsible for growth repair and inheritance.

Fertilization

It is the union of the male and female gametes. From the standpoint of the nucleus, fertilization is the coming together of

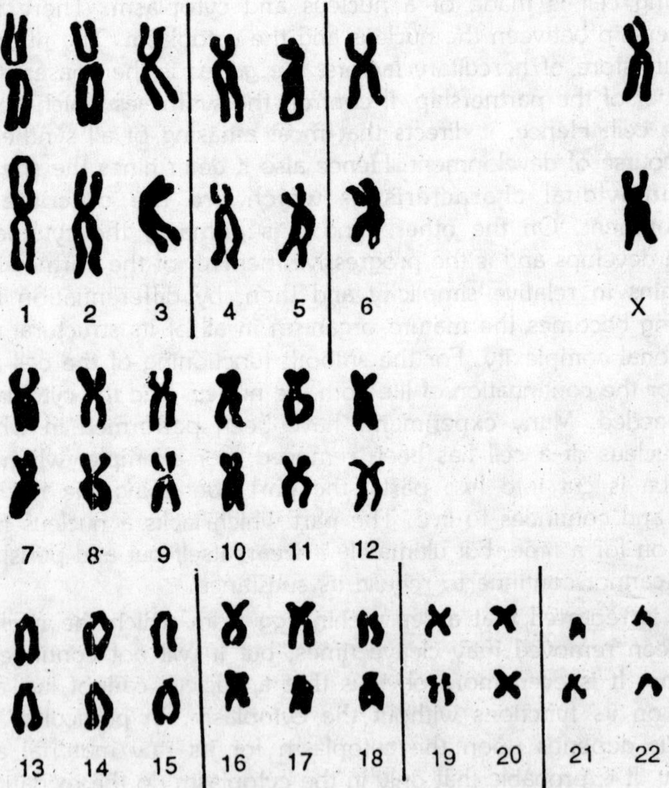

Fig. 5.1. The 46 human female chromosomes.

the chromosomes of the egg and the chromosomes of the sperm in the same nucleus, the zygotic nucleus. One set of chromosomes, the haploid number, is present in the egg. Another set is present in the sperm. This makes two sets the diploid number present in the zygote. In man the haploid number is 23. Hence the human diploid number is 46. Each chromosome of a haploid set has its own individuality that is, it is of its own kind. It becomes visible at the beginning of cell division and disappears at the close of cell division in its own characteristic manner. Very significantly, its genes are arranged within it in a definite sequence. Moreover, it affects development in its own specific way. Each chromosomes of a diploid set usually has a mate. Yet although the two mates

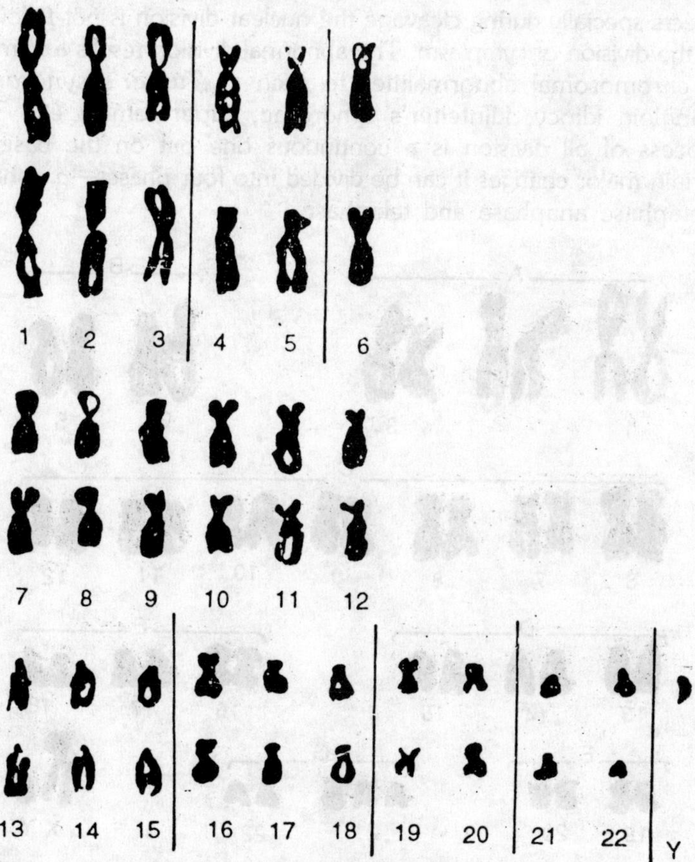

Fig. 5.2. The 46 human male chromosomes.

are within the same nucleus, with rare exceptions each remains independent of the other throughout the cell divisions (mitoses) of development.

Mitosis

The coming together of the male and female nuclei in fertilization is followed by *mitosis*, the process by which one cell becomes two cells in so complicated but precise a manner that each daughter cell possesses just the same two sets of chromosomes that the mother cell possessed. It is the mitosis only that convents a single called zygote into a multicellular organism with billion cells. In most of cases the *Karyokinesis* which the division of nucleus is followed by *cytokinesis*—the division of cytoplasm. But in group insects specially during cleavage the nuclear division is not followed by the division of cytoplasm. The abnormal division results a number of chromosomal abnormalities to such as *Turner's syndrome*, Mangloid idiocy, klinfelter's syndrome, super female etc. The process of all division is a continuous one but on the basis of certain major changes it can be divided into four phases—prophase, metaphase anaphase and telophase.

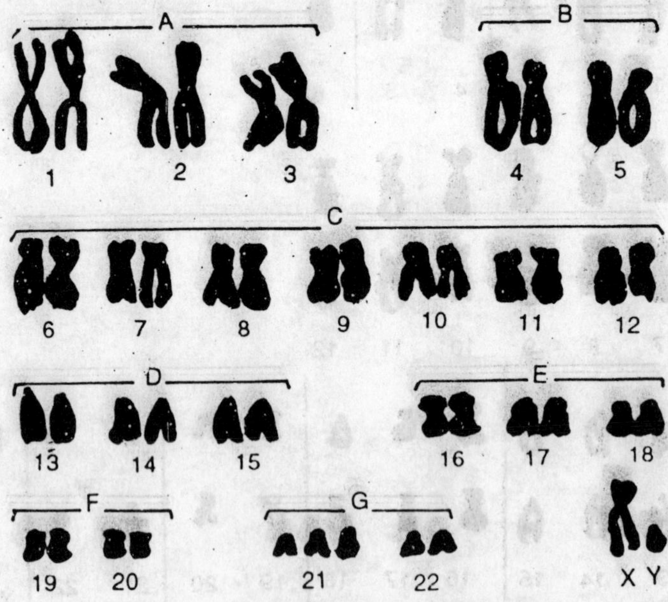

Fig. 5.3. Chromosomes of a mongoloid idiot.

The interphase is the stage between telophase of one division and the prophase of the next division. All necessary biochemical changes occur during interphase at which the cell is metabolically most active. It is during the interphase that the genes self-duplicate and carry on their function of supervising syntheses. Following self-duplication the chromosomes continue to be stretched out. Usually they are so attenuated that their boundaries cannot be seen with the light microscope. The extended chromosomes are able to carry on their chemical functions in an efficient manner. "Nuclear sap," or karyoplasm fills the interstices between the chromosomes. One or more rounded bodies, the *nucleoli*, are usually present in the interphase nucleus. The visible boundary between the nucleus and the surrounding cytoplasm in the *nuclear membrane*. In the cytoplasm adjacent to the nucleus there is a body, the central body, which consists of two granules, or after each granule has replicated, of two pairs of granules the *centrioles*.

Prophase

The events in the nucleus and in the cytoplasm which lead to the division of the cell constitute the prophase. The following is a generalized account.

1. According to some accounts chromosomal fibers grow out from particular locations in each chromosome until they reach the poles of the spindle. The locations from which they grow out are known as *centromeres* or kinetochore. The chromosomal fibers seem to consist of a contractile fibrous gel.
2. A fusion body, the *mitotic spindle*, now forms between the pairs of centrioles, sometimes in the cytoplasm sometimes in the substance of the nucleus. The spindle's poles are anchored in the region of the centrioles of the cytoplasm. The electron microscope shows that it consists of hollow fibrils (microtubules). It elongates as the centrioles move apart. Possibly this is a result of inhibition of water or incorporation of other molecules, or changes in the shape of its constituents proteins. The amphiaster and the spindle together form the "achromatic figure," so called because it does not stain deeply with basic dyes.
3. In the cytoplasm the pairs of centrioles move apart, and a sphere of gel rays the *aster*, grows around each pair. Together the two asters constitute the *amphiaster*. It is of interest that

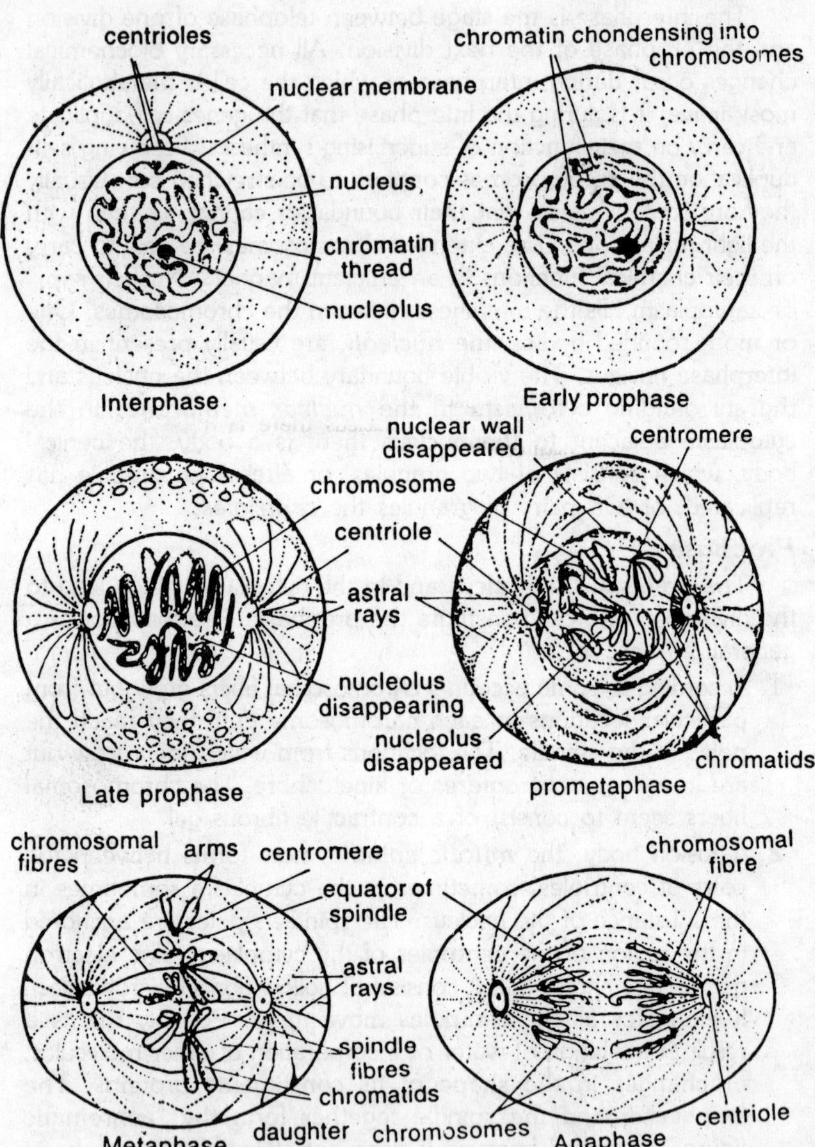

Fig. 5.4. Mitotic cell division in animal cells.

the centrioles are short cylindrical bodies each consisting of nine rods (or double or triple rods). In this they are structurally comparable to a cilium or flagellum. (They lack the two central tubules of a cilium) In a typical case, the centrioles of a pair lie at right angles to each other and to the axis through the pairs.

4. The nucleolus also disintegrates, and its substance along with the nuclear sap, passes into the cytoplasm.
5. Toward the end of the prophase, the nuclear membrane disappears.
6. The chromosomes condense into tight spiral coils. Under the right microscope these appear to be rods. But when the material is suitably fixed and strained, each chromosome can be seen to be a double structure each half of which is a chromatid.

Metaphase

The chromosomal fibres appear to contract and pull the chromosomes to the equator of the spindle (prometaphase). When thus spread out and seen in polar view, their number can be counted.

Anaphase

Continuing or resuming, what appears to be contraction the chromosomal fibres seem to drag each chromatid toward one pole of the spindle. Actually two processes may be at work a continuing enlargement and elongation of the spindle and a shortening of

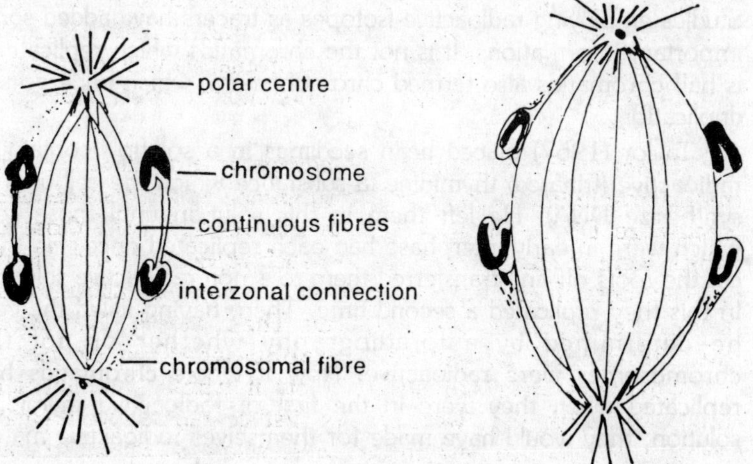

Fig. 5.5. Two types of Anaphasic spindles. A—Direct types, B—Indirect type.

the chromosomal fibres. Since the fibres shorten without becoming thicker the process probably involves the removal of water or other molecules from the fibres. As the chromatids move apart, they are often V or J-shaped depending on the location of the centromeres to which the chromosomal fibres are attached. A band of "interzonal fibers" is often seen for a time after the separation has been accomplished, connecting the chromosomes which have pulled apart, and often including a remnant of the spindle.

Telophase

Having reached the poles of the spindle the chromatids swell and disappear from view as seen with the light microscope. Although usually called "daughter chromosomes," they are single chromatids. The nuclear membrane reappears, and shortly one or more nucleoli reform. Of the mitotic apparatus little more than the centrioles of the cytoplasm remain.

Interphase

The daughter cells have now returned to the interphase or "resting stage" expect for one feature: each chromosome still consists of only one chromatid. Investigations by a staining technique known as the Feulgen reaction (specific for DNA) have made it clear that the restoration of chromosomal material takes place during the interphase. At that time each chromatid becomes two chromatids and the double nature of each chromosome is restored. Studies employing radioactive isotopes as tracers have added some important informations. It is not the chromatids which replicate. It is half chromatids also termed chromonemata, which undergo self duplication.

Taylor (1957) placed bean seedlings in a solution containing radioactive (tritiated) thymidine (a substance which the cell uses to synthesize DNA). He left them in this solution until those cells which were in early interphase had each replicated once. He then cut the roots off and transferred them to a non-radioactive solution. In this they replicated a second time. Then, having killed the cells he determined by autoradiography whether or not the chromosomes were radioactive. Now if whole chromatids had replicated when they were in the first or radioactive thymidine solution, they would have made for themselves radioactive mates.

After cell division half of the daughter chromosomes would have been radioactive and half namely, the original chromatids would have been non-radioactive. But Taylor found that *all* the daughter chromosomes were radioactive! Why? The answer is that half-chromatids replicated, so that each chromatid now consisted of a new radioactive half-chromatid and an original non-radioactive half-chromatid. At the second replication this time in non-radioactive solution each half chromatid whether radioactive or non-radioactive mate. Therefore following the next cell division, the daughter chromosomes (chromatids) were half radioactive and half non-radioactive.

Similar results were obtained by Prescott and Bender using cells of hamster issues raised in tissue culture and also human leucocytes. Experiments of this sort emphasize the remarkable stability of the half-chromatids. Although most of macromolecules within the cell are in a state of flux, continually breaking down and being replaced the half-chromatids are passed on intact from one cell generation to the next. And this has been going on for countless generations! There is evidence that the centrioles may self-duplicate even earlier than the chromosomes in some cases as early as the preceding anaphase. At this time each centriole (there are two at each pole) makes for itself a partner, not usually by longitudinal splitting, however, as might be expected, but apparently by a process of lateral budding. The introduction of mitosis in the history of life on the earth was one of the grand innovations of all time. Possibly it took place only once, for which variations, it occurs in all plants and animals except bacteria and blue-green algae. Even in these there must be a process akin to mitosis, for without some such process the transmission of hereditary characters to each new generations in a balanced fashion would not be possible.

Meiosis

In 1884 Van Beneden described the processes of maturation, fertilization and cleavage as they take place in the parasitic roundworm of the horse. *Ascaris megalocephala*. He demonstrated that the egg and sperm contribute an equal number of chromosomes to the offspring. This led, a few years later, to the discovery of *meiosis*, the process by which the number of chromosomes if reduced to one half when the germ cells ripen. Each body cell and each unripe germ cell or *gonium* has two sets of chromosomes, the *diploid number*. Each of the chromosomes has a mate (except in the case of the unpaired sex chromosomes).

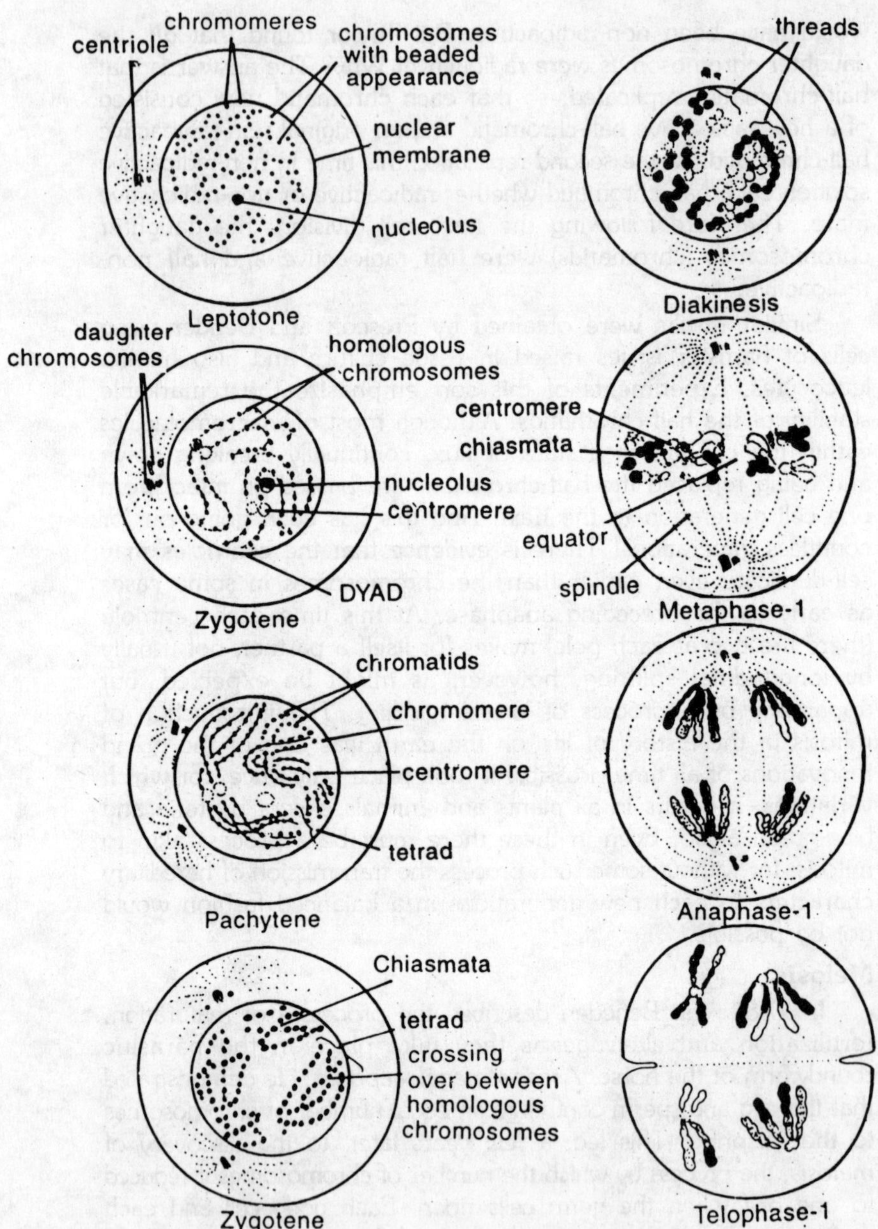

Fig. 5.6. Diagrammatic representation of different stages in the first meiotic division.

When meiosis takes place, an oogonium or spermatogonium, as the case may be gives away one set of its two sets of chromosomes; that is, it gives away one chromosome of each pair of chromosomes. The result is that each ripe egg or sperm has left only one set of chromosomes, the *haploid number*. Now, meiosis unlike mitosis (1) takes place only in gonads, (2) occurs only in those unripe germ cells which at the time are in the process of ripening and (3) although it begins earlier, reaches completion only during the period of sexual maturity of the plant or animal. In most cases some unripe germ cells (residual gonia) remain in the gonad where they multiply by mitosis and so produce more gonia. In female birds and mammals however, the period of multiplication of oogonia comes to an end about the time of hatching or birth. At this time several million gonia may be present; yet only a very few ever ripen into ova-about 400, in the case of women. The gonia are thus lineal descendants by mitosis of the original fertilized egg or zygote. Hence, they are cousins of the body cells, and generally speaking they possess the same two sets of chromosomes which the body cells possess. Meiosis has been compared to two mitotic divisions, for two successive spindles are formed and two separations of chromosomes take place.

Yet there is only one complete prophase, and this differs from a typical prophase in that it involves a pairing of the chromosomes side by side each with its mate. The process of pairing is known as *synapsis*. In a sense this pairing of homologous chromosomes completes the coming-together of the germ cells at fertilization; for all through the mitotic cell divisions of development the chromosomes of the egg and those which came from the sperm have remained separate. Now each chromosome joins its mate. The stages of the prophase of meiosis have been given names; leptotene, when the chromosomes become threadlike; synaptene when they come together in pairs; pachytene, when they contract and become tightly coiled; diplotene, when they begin to separate except for locations when "crossing over" between synapting chromatids has taken place; and diakinesis, when again they become compacted prior to separating. Note that at the beginning of meiosis each chromosome consists of two chromatids as in ordinary mitosis, so that, when, pairing takes place, the result is bundles of four chromatids. These are known *bivalents* because each consists of two chromosomes. They are also known as tetrads because each is four chromatids.

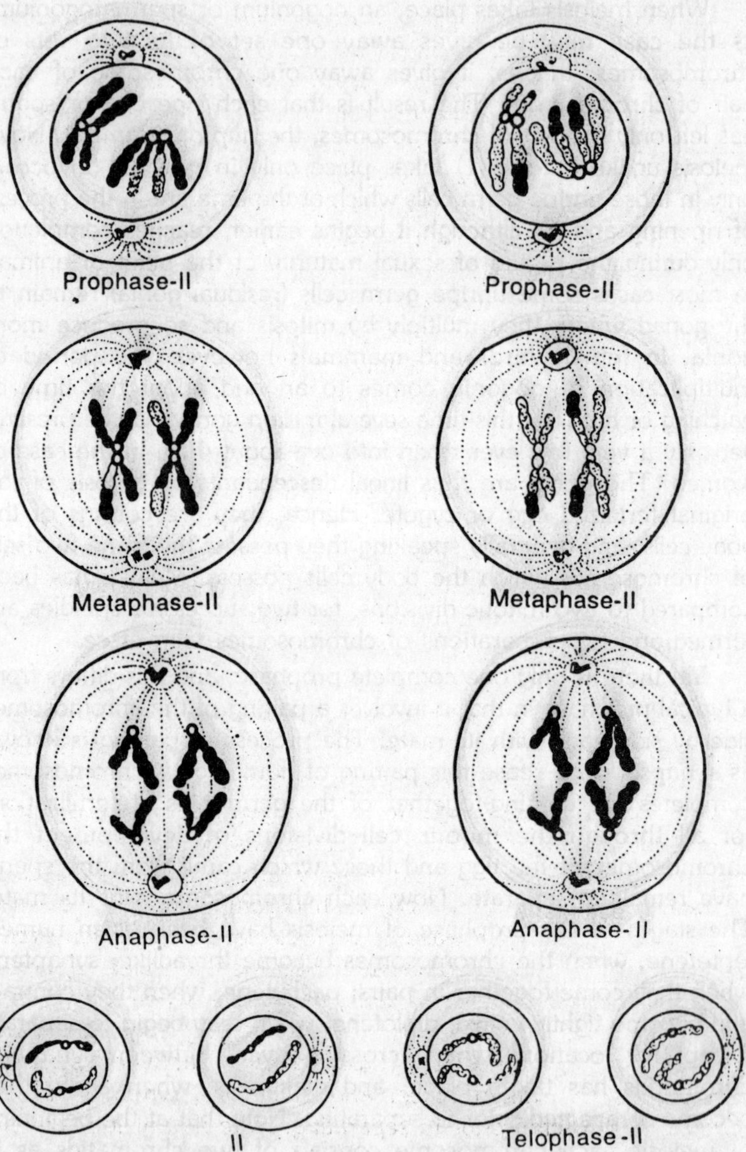

Fig. 5.7. Diagrammatic representation of homeotypic division of meiosis.

During this period preparatory to the first meiotic division, the ripening germ cell is called a *primary gametocyte*, primary spermatocyte or oocyte, as the case may be. The diplotene stage of oocytes is of particular interest, not only because it is long drawn out (it may be as long as 40 years in the human female), but because it is a period of great growth. The nucleus enlarges greatly due to the accumulation of nuclear sap and is often referred to as the *germinal vesicle*. The chromosome become fuzzy objects which have been compared to lampbrushes.

First meiotic division

The stages of the prophase which have just been described end as the metaphase approaches. The bivalents move to the equator of the first meiotic spindle. At the metaphase they split; and at the anaphase each bivalent divides and *monovalent chromosome* or dyad (pair of chromatids) goes to each pole of the spindle. The monovalents are actually the original chromosomes which united in synapsis expect that some crossing over has usually taken place. The products of the first meiotic division are known as *secondary gametocytes*.

Second meiotic division

Usually the telophase of the first meiotic division is brief and is quickly followed by the much reduced prophase of the second meiotic division. In some instances, the anaphase of the first meiosis leads directly into the metaphase of the second meiosis. In any case, the monovalent chromosome (dyads, two chromatids each) take a position at the equator of the second metaphase spindle. At the anaphase the chromatids or *daughter chromosomes* (*monads*) move apart to the poles of the spindle. The resulting cells are known as *tids*, ootids or spermatids, as the case may be. Note that a reduction in the number of chromosomes has taken place. The gonia had two sets (diploid number) of chromosomes.

The primary gametocyte had one set (haploid number) of bivalent chromosomes (tetrads). The secondary gametocytes had one set of monovalent chromosomes (dyads). Now the tids have one set of daughter chromosomes, that is, one set of chromatids (monads). But this is important the set which each tid had is a complete set with one chromosome present to represent each pair of chromosomes of the original diploid set. In the case of spermatogenesis, each primary spermatocyte divides twice equally and produces four equivalent spermatids. In oogenesis, however, although the meiotic divisions of the nucleus are equal those of the cytoplasm are grossly unequal.

At each division most all the cytoplasm goes to only one of the daughter cells and thus is conserved to supply the substance of the embryo. The result is that the first meiotic division gives rise to one secondary oocyte and one small *first polar body*. The second meiotic division similarly produces one ootid and a small *second polar body*. The spermatids undergo metamorphosis and become sperm cells or *spermatozoa*. The ootids, on the other hand, as a rule need only to burst from the ovary to cause them to ripen and become ready for fertilization. It is a strange fact that the stage in meiosis which is attained before the fertilizing sperm enters the egg is not the same for all species. In sea urchin eggs meiosis is complete and both polar bodies have been formed before the egg is receptive to the sperm. This is rather rare.

In vertebrates the first polar body has been given off and the second meiotic division has progressed to the metaphase before

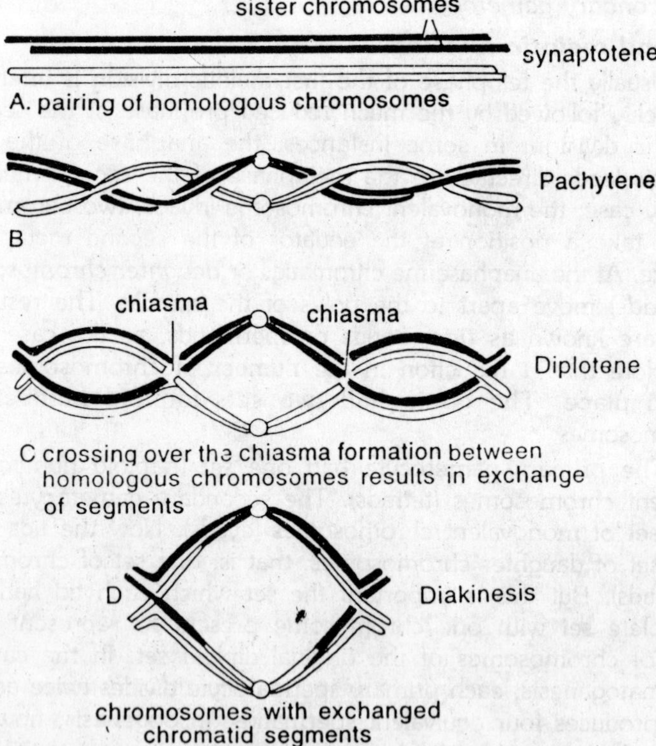

Fig. 5.8. Mechanism of chiasmata formation and crossing over.

fertilization takes place. The extreme case is the parasitic roundworm, *Ascaris*, in which the sperm enters the cytoplasm of the egg before even the first meiotic spindle has formed. It remains inactive in the centre of the cytoplasm while the egg completes meiosis. In most animal species therefore the meiosis which close the nuclear cycle of the egg overlaps to a greater or lesser extent the entrance of the sperm which begins the nuclear cycle of the new individual. Why does this complicated process of meiosis exist? The answer was pointed out in 1903 by Walter Sutton, who showed the parallelism between the story of hereditary factors (later termed genes) as worked out by Mendel and other students of plant and animal breeding and the behaviour of chromosomes in fertilization, mitosis and meiosis. Meiosis, Sutton showed is nature's way of reducing the number of chromosomes in each

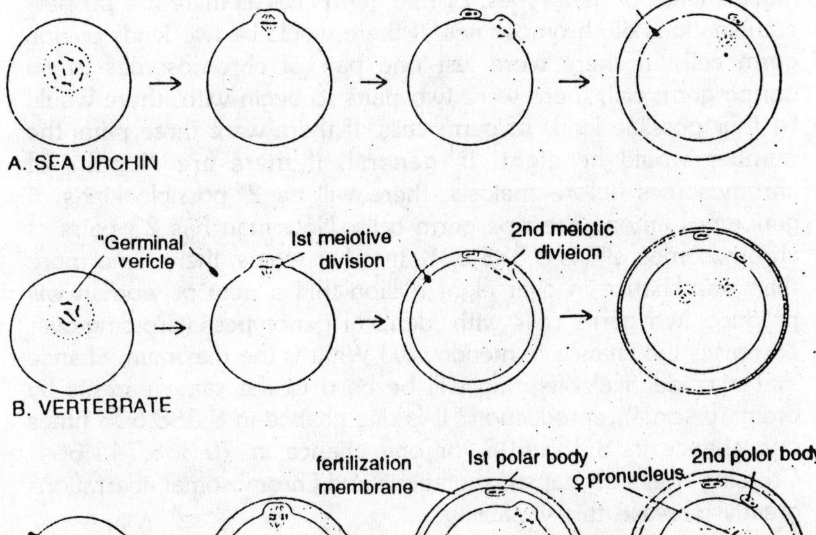

Fig. 5.9. Fertilization (entrance of sperm) takes place at different stages of meiosis in different animals: A—after meiosis is complete in the sea urchin; B—midway in the second meiotic division in most vertebrates; and C—before meiosis in the parasitic roundworm Ascaris.

germ cell so that at fertilization, the normal diploid number of chromosomes will be restored. But it is more than this.

It is nature's way of seeing to it that the germ cells are all different. Let us assume that the chromosomes of different pairs are different and also that the chromosomes has some mutant genes which make it unlike its mate. In nature this is probably always true. Since all the cell divisions during development are made by mitosis, every gonium will have the same two sets, or diploid number, of chromosomes and genes. Then, when meiosis takes place, each ripe germ cell retains only one chromosome of each pair of chromosomes, one gene of each pair of genes.

Moreover it is a matter of chance which chromosome of a given pair of chromosomes the cell retains. Thus, from the standpoint of heredity, there can always be as many different genetic kinds or *genotypes*, of ripe germ cells as there are possible combinations of chromosomes. If there would be two kinds of ripe germ cells. If there were just one pair of chromosomes in an unripe germ cell, there were two pairs to begin with, there would be four possible kinds of germ cells. If there were three pairs the number would be eight. In general, if there are N pairs of chromosomes before meiosis, there will be 2^N possible kinds of genotypes among the ripe germ cells. Now man has 23 pairs of chromosomes. 2^{23} is 8,388,608. In other words, there is no more than one chance in over eight million that a man or woman will produce two germ cells with identical genotypes. (Crossing-over decreases the chance tremendously.) What is the maximum chance that two identical offspring will be born to the same parents by ordinary sexual reproduction? It is one chance in 8,388,608 times one chance in 8,388,608, or one chance in 70,368,744,664. Crossing over at synapsis, mutations and chromosomal aberrations greatly increase this variability.

The Role of the Nucleus

The account of fertilization, mitosis and meiosis which has just been given is similar to that usually found in textbooks of biology. It stops short however, of explaining how the nucleus controls development. Indeed, it avoids the problem, for the compacted chromosomes of fertilization mitosis, and meiosis are not active in development. It is only when they are expanded as at the interphase, that they carry on their functions in growth and development. (The chromosomes of oocytes are expanded also during the long prophase.) Furthermore, the chromosomes presumably are alike in every cell of the body of an embryo. Now, how can chromosomes which are everywhere the same account for the origin of differences within the embryo? In order to approach this problem we must consider the chemistry of the nucleus.

The Chemistry of the Nucleus

The distinctive chemical substances of the nucleus are long chain molecules known as *deoxyribonucleic acid*, or DNA for short. Each such molecule is a gene; more likely it is a string of gene, the material units of heredity. A closely related type of nucleic acid is *ribonucleic acid*, or RNA. Both DNA is in the chromosomes. RNA is the chromosomes, but especially in the nucleoli, and it is also present in the cytoplasm. Now, nucleic acids have three unique and remarkable properties:

1. Nucleic acids (genes) occasionally mutate; that is, a short segment of a gene may change in composition. When the occurs the gene replicates according to its new or mutant form. If this were not so, life on earth would never have evolved and the ever increasing adaptation of living things to their environments would not have been possible. The shortest segment of a gene which may mutate is probably a single nucleotide unit. It is known as a *muton*.
2. They are self-duplicating molecules. Possibly they are the only truly self-duplicating molecules. Centrioles and plastids are morphologically self-duplicating, but it is likely that, in this case, nucleic acids are involved.
3. Nucleic acids supervise the synthesis of proteins within the cell. The proteins are the principal structural compounds of protoplasm and are the chemical basis of its functioning. They include the enzymes which control the innumerable chemical reactions of the cell and also structural proteins. In fact, a gene (cistron) has been defined as the template for the production of one polypeptide chain. It is probably true that in the absence of nucleic acids, no proteins are ever produced. These statements are oversimplified. The self-replication of a nucleic acid requires the presence of specific enzymes (polymerase) and numerous other molecules in the protoplasm. The process is therefore a circular one: nucleic acids bring

about the synthesis of proteins, and proteins are needed to make possible the replication of the nucleic acids. It is the entire protoplasmic system which endures.

Genes are nor invisible. Crossing-over of parts of genes may take place between synaptic mates during meiosis. The smallest part of a gene which may interchange thus with a mate has been termed a *recon*.

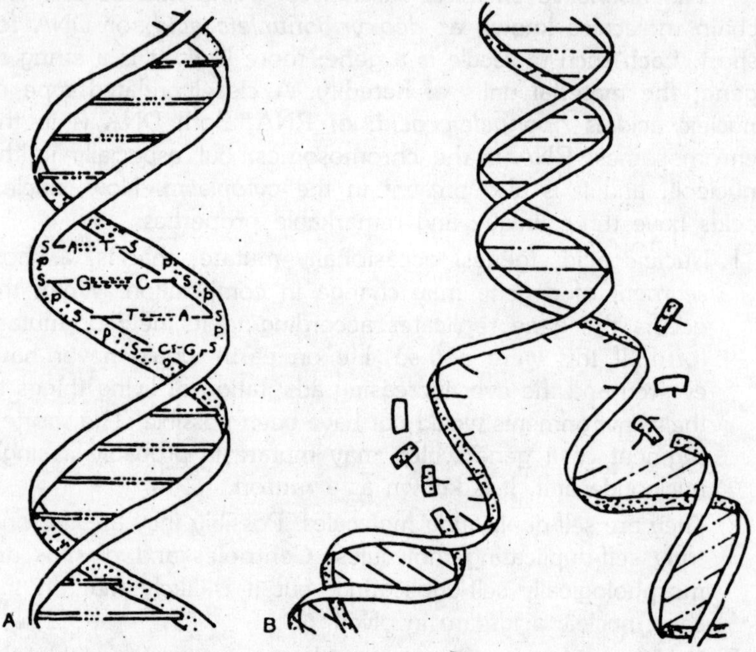

Fig. 5.10. The structure of DNA and the manner of its replication.

Great progress has been made in the last few years in the knowledge of the structure of nucleic acids and of the manner in which they duplicate. According to the Watson-Crick hypothesis, each DNA molecule is a long double chain of units known as nucleotides. It may be compared to a rope ladder twisted into a helix. Each single nucleotide consists of a sugar, deoxyribose (S), a phosphate group (P) and one of four different "bases." The bases are adenine (A, a purine), guanine (G, also a purine), thymine (T, a pyrimidine), and cytosine (C, a pyrimidine). The single chain consists of alternating sugars and phosphates, with a base attached as a side group to each sugar.

The precise sequence in which, the nucleotides are arranged in the long nucleic acid chain is the basis of the coded information which is passed on by heredity and which guides the course of development. In the double chain of the DNA molecule, each base of one chain is linked crosswise, by weak hydrogen bonds, with it mate in the other chain. The relations are precise.

Adenine is always bounded with thymine, and guanine with cytosine. There are no other combinations. Hence, the two chains fit each other exactly like a mold fits a model. When a nucleic acid (DNA) molecule self-duplicates, the chains supposedly separate from each other. Each single chain then makes for itself a replica of its former partner by attracting and assembling to itself appropriate nucleotides from the surrounding protoplasm. The result is that there are now two identical DNA molecules (double chains of nucleotides) where previously there was one.

One notes at once the comparison between the replication of the half-chromatids (chromonemata), which takes place during the interphase of mitosis and the replication of the single chains of DNA molecules. The nucleus also contains proteins of two sorts: histones (proteins rich in basic amino acids), which are associated with the genes; and nonhistones, which compose the larger amount of protein and may be a part of the metabolic and mitotic machinery of the cell. The relation between histones and DNA is loose, yet it is so consistent that the combination is usually referred to as nucleoprotein. A small amount of nonhistone protein, termed residual protein, presumably holds the genes in order and organizes them into the definite bodies, the chromosomes. Lipids are also present in the nucleus.

The Influence of the Nucleus on Development

The genes (DNA) do not directly influence development. Instead they "transcribe" their coded information to RNA, which then collects around the chromosomes and in the nucleoli. Apparently one of the two strands of a DNA molecule assembles ribonucleotides and makes for itself a complementary strand of RNA. (The other strand possibly self-duplicates.) The nucleotides (ribonucleotides) which compose the RNA, however, differ form those which compose DNA. The sugars are ribose instead of deoxyribose and one of the four nucleotides is uracil instead of thymine (it lacks the methyl group of thymine). Although these differences from a chemical standpoint appear to be slight, yet they account for important differences in function.

The RNA molecules are single chains and as such are not self-replicating. (RNA is self-replicating in some viruses.) They are dependent on DNA for their production. Moreover, the RNA molecules do not remain in the nucleus. From time to time they migrate in to cytoplasm where they function in the manufacture of proteins. Some of them serve as templates for the production of proteins, notably for the production of the enzymes which catalyze the chemical processes of the cell. Possibly this release of RNA is not a continuous process. An impressive release takes place at the prophase of the first meiotic division of oogenesis when the swollen oocyte nucleus (the so-called germinal vesicle) breaks down and its nuclear sap and the material of the large nucleolus mingle with the cytoplasm. A similar release occurs at the prophase each time a cell divides. In very active cells, such as growing oocytes and certain cells of larval insects, the nuclear membrane has been shown by the electron microscope to be perforated.

Nuclear material has been found to "bleb" through pores into the cytoplasm where it supervises the syntheses of proteins. There may be a delay between the time of the formation of the RNA in the nucleus and the time when its influence becomes apparent in cytoplasmic processes. The clearest evidence that this is so comes from experiments on species hybrids, that is, on eggs which have been fertilized by sperm of another species. In this case, the early events of development, for example, the pattern and rate of cleavage are determined by the egg cytoplasm alone, presumably by the RNA that was synthesized while the oocyte grew in the mother's ovary and under the influence of the mother's genes.

Generally speaking, the effect of the genes (DNA) brought in by the sperm is not apparent until about the time of gastrulation. But from then on, the genes of both the eggs and the sperm control the course of development. Even more crucial are certain experiments in which eggs were enucleated before they were fertilized by the sperm of another species, or in which the egg nuclei were destroyed immediately after the foreign sperm had entered. In such haploid eggs (termed androgenetic hybrids), the cytoplasm belongs to one species and the nucleus to another. Such combinations have not lived beyond the early embryonic stages, but the evidence indicates that, from the time of gastrulation onward, the influence of the chromosomes of the stranger nucleus is pronounced.

Do Nuclei Undergo Differentiation?

In the early days, Weismann and Roux believed that the nuclei of body cells become different during cleavage as a result of what were thought to be unequal divisions of the nuclei. The organization of the embryo, so they theorized, is the expression of the inherited organization within the chromosomes of the zygote nucleus. This view has long since been given up. There is organization within the cytoplasm of the uncleaved egg. There is strong evidence that the nuclei of every cell of the body contain initially the same set of chromosomes and genes, and that, during cleavage at least, mitosis is equal cell division in so far as the chromosomes are concerned. This fact was originally demonstrated in an experiment by Spemann in which he constricted an uncleaved amphibian egg by tightening a hair noose around it. The egg was pinched into two halves, with only a narrow neck of protoplasm between them. One half contained the nucleus; the other half lacked a nucleus.

After several cleavages had taken place in the nucleated half, one of the descendant nuclei, migrated across the isthmus of cytoplasm into the non-nucleated half (delayed nucleation). Cleavage followed and both halves (provided the needed cytoplasm was present) became whole embryos of half size. This proves that a cleavage nucleus can do everything that the original zygotic nucleus can do. It is an exact copy of its original. The fact that the daughter nuclei are equal during early development has been clinched by some remarkable experiments by Briggs and king. They removed the nucleus of an uncleaved frog's egg, and substituted a nucleus from a cell of a blastula or an early gastrula. With refined techniques, many such operations were performed and were successful. Normal embryos developed.

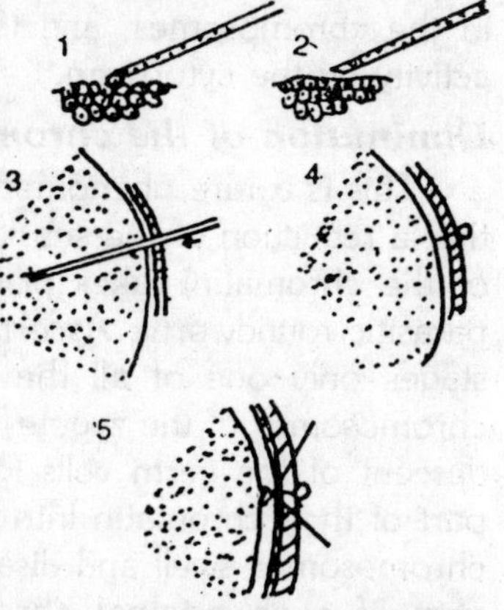

Fig. 5.11. Method used by Briggs and King when they substituted nuclei from older embryos for the nuclei of uncleaved eggs of the frog.

Here again, it is demonstrated beyond question that at this early stage of development when the future fates of the regions of the embryo are being determined. The nuclei, i.e., the genes of the cells of the different tissues, are equivalent. Since the nuclei are alike, it therefore must be the cytoplasms which are different. Gurdon and others have extended this work and have found that some nuclei of stages as fate as swimming larvae, when transplanted to enucleated eggs, are capable of supporting a complete development. Some of the tadpoles even live to metamorphose into frogs and become sexually mature. This, however is not the whole story.

It is a common observation that the nuclei of cells of different tissues do differ in size and shape. Somehow during development nuclei do change. The cytoplasm, which supplies them with substance, energy, and necessary distinguish them. It does more than this it affects their activity; and, at least in some cases, it affects their composition. The nuclear changes which arise during development are of three sorts, and they arise in three distinct ways: (1) by the diminution of the chromatin, (2) by stable changes in the chromosomes, and (3) by the epigenetic control of gene activity by the cytoplasm.

Diminution of the chromatin

This is a rare phenomenon. Boveri observed many years ago that a reduction of the substance of the chromosomes (diminution of the chromatin) takes place during the early cleavages of the parasitic roundworm, *Ascaris*. At the two, four, eight, and 16-cell stages only one of all the blastomeres retains the typical four chromosomes of the zygote. In each case, this cell is in the line of descent of the germ cells (germ line). The other cells cast off a part of their chromatin into the cytoplasm; that is, the tips of the chromosomes swell and disappear. At the same time, the central part of each original chromosome breaks into several small segments which remain as separate chromosomes. Other examples of the diminution of the chromatin in the development of body cells have been described, especially among the insects. Sometimes this process takes place early in development, sometimes late. But it never occurs in the germ line.

We may conclude that, in the body cells of these embryos, the chromatin (DNA) has already performed its function by producing RNA. The RNA is stable and functions during development. The DNA (chromatin), no longer needed in the development of the body cells, is discarded. Experiments have shown in *Ascaris* that a certain substance localized in the cytoplasm prevents the diminution of the chromatin. Normally, this material is passed on to only one blastomere during each of the first four cleavage divisions. But if, as a result of manipulation (centrifuging), this material becomes divided between two daughter cells, then diminution of chromatin does not take place in either of them. This is a clear case in which substances in the cytoplasm control the nucleus.

Changes in the chromosomes

Diminution of the chromatin is not known to occur in vertebrate development; yet there is evidence that changes in the genome or sum total of genes may take place. In their studies of transplanted nuclei, Briggs and King found that when a nucleus from a late gastrula or neurula of an amphibian is substituted for the nucleus of an uncleaved egg, normal development does not usually take place, although it may take place. When examined under the microscope the cells often do not have the normal chromosomal equipment. It is not clear whether changes of this sort take place in normal development. Possibly the nuclei of differentiating cells become increasingly sensitive to experimental manipulation.

Possibly the transplanted nuclei are forced by the mitotic apparatus to divide before they are fully ready. Possibly, also when cells differentiate the mitotic cell divisions becomes slovenly. In any case, in these experiments the daughter cells did not always receive the normal complement of chromosomes. The nuclear changes which take place in this manner are stable. Briggs and King have demonstrated that when the nucleus of a defective embryo is transplanted to an uncleave egg, the result is an embryo with a defect similar to the defect of it nuclear parent. This transfer of nuclei may repeated over and over with the same result (nuclear cloning).

For a time it was suggested that the changes in the genome which Briggs and King had discovered might be related to the normal differentiation of cells. For example, the nuclei derived from the endoderm seemed to favour the development of endodermal organs. But this interpretation has not been borne out. Defects of a similar sort have also been obtained when the nuclei are taken from the neural tube. On the other hand, some nuclei are capable of supporting a normal development even when taken from the gut of a swimming tadpole. However, one thing is certain. Any changes which may take place in the genome of differ entiating cells always occur *after* the cytoplasm has already become specialized and committed to its fate.

The epigenetic control of the genes by the cytoplasm

Whether or not nuclear substance differentiates during development, it is certain that the same genes are not equally active in every cell and at every place, nor are they equally active at all times. Genes represent potentialities are realized depends upon the conditions within the cytoplasm which surrounds a given nucleus the genes off and on like *faucets*. To change the metaphor, a nucleus is like a stockroom of a foundry full of patterns. From its vast store, patterns are requisitioned when where, and as they are needed. No casting can be poured at the foundry for which a pattern is not provided and of course the form of the casting depends upon the shape of the pattern. In the same way, no synthetic chemical process can take place in a cell unless a gene is present to supply the pattern (i.e., template) for the enzyme which mediates the process.

Genes are templates in stock. But the decision as to which genes are active at a given time and place-this depends upon conditions in the cytoplasm. And these conditions change epigenetically as development goes on. It is a mistake to pass over lightly the epigenetic factors which turn genes on and off, for they, and not the genes, ultimately control the time and place of developmental processes. Direct evidence that this is true is found in the behaviour of the giant chromosomes of certain insect larvae as they approach metamorphosis.

Such chromosomes consist to multiple strands of chromatids (so-called polytene structure). They are the result of self-duplication of the chromatids without accompanying separations. For example, the chromatids of salivary glands of fruit fly larvae multiply nine or ten times without separating, with the result that each giant chromosomes consists of 512 to 1,024 chromatids. These compound chromosomes show bands (possibly representing genes) of DNA which stain darkly and are large enough to be seen under the light microscope. They also show swellings, or "puffs," at definite loci. These seem to be the result of RNA and protein accumulating and forcing the strands apart. The puffs may expand until they become what are known as Balbiani rings. Beermann has studied the puffs and rings in the larvae of the midge, *Chironomus*, and has shown clearly that they occur at different locations in the chromosomes of different tissues and at different times. The explanation seems to be that the puffs and rings are centers of intense synthetic activity in particular tissues, in preparation for the role which that tissue will play in the metamorphosis.

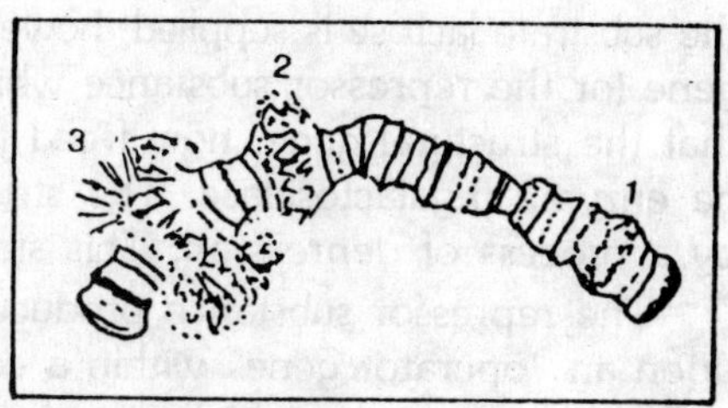

Fig. 5.12. Polytene chromosome of the salivary glands of the midge, Chironomis, showing stages of "puffing."

How are the Genes Controlled

The analogy of a nucleus to a stockroom of pattern, at a foundry has one obvious flaw. A stockroom has a stockkeeper who delivers the patterns as they are called for. The nucleus of a cell, however is automated so that, cued by conditions in the surrounding cytoplasm, it delivers the templates when they are needed. Lately developmental geneticists have been much interested in the problem of how the genes are activated. We shall make reference to the closely reasoned concepts of Jacob and Monod of the Pasteur institute who studied enzyme induction in the bacterium, *Escherichia coli*. Normally this microbe produces no enzyme to metabolize the nutrient sugar lactose.

Yet, when lactose is supplied in the nutrient medium, the enzyme β-galactosidase, which is able to metabolize it, appears in abundance. The presence of the substrate (substance acted upon) serves as an "*inducer*" and leads to the production of the enzyme. The authors reasoned that no enzyme could have been produced if a "structural gene" for its production had not already been present in the germinal material. (Genes which determine the structure of a protein are called structural genes.). If this is so, then why was the enzyme β-galactosidase not produced until lactose was added?

Genetic studies by Jacob and Monod on mutant strains of *Escherichia coli* supplied them with the answer. The gene for the production of β-galactosidase is inactive because another gene, so-called "regulator gene," is present which produces a *repressor substance* that inhibits the activity of the structural gene. When the substrate lactose is supplied, however, it inactivates the regulator gene (or the repressor substance which it produces). The result is that the structural gene, now freed from repression, gives rise to be enzyme β-galactosidase. The structural gene is thus activated by a process of *depression*. This statement is oversimplified.

The repressor substance produced by the regulator gene acts upon an "operator gene" within a complex of structural genes. It is the operator gene which directly controls the structural genes, turning them on and off. An operational complex of genes of this sort was called an "operon" by Jacob and Monod. Such a complicated mechanism would be unbelievable if its several steps had not been experimentally verified.

Generally speaking, a substrate acts as an inducer. Its presence leads to the production of the enzyme which metabolizes it. In the language of modern communication theory, this is a case of "positive feedback". In a similar manner, the products of a reaction are inhibitors which repress the production of the enzymes which produce them. This is a "negative feedback." Embryologists are much interested in concepts such as this one of Jacob and Monod because they suggest how epigenetic factors in the cytoplasm (represented in the above case by the substrate lactose) can control the activity of the genes. There is, however an important difference between the actions of genes in bacteria and in the development of higher organism. The genes of bacteria are immediately responsive to their environment, and the mRNA which they produce by transcription is immediately active in protein synthesis. It is also short-lived. The genes which control the development of higher organisms on the other hand, commonly produce "masked" mRNA, which may be stored for a time until the time and place for its activity have arisen.

6

Cytoplasm in Development

Cytoplasm is the part of protoplasm present between nuclear membrane and the plasma membrane. Nucleus alone fails to regulate the activities. It the mutual affords of nucleus and cytoplasm that regulate all the developmental processes. The genes of the nucleus control the course of development, but it is mainly the cytoplasm which develops. The genes are the "physical basis of heredity," but principally it is the cytoplasm which expresses this heredity in so far as it is expressed. There is interaction between the genes and the cytoplasm, but their roles are different. The genes largely control the species characteristics of the new organism and to a great extent its individual characteristics as well; but the characteristics which distinguish one tissue from another, or one organ from another *in the same individual*, are decided initially by processes which take place in the cytoplasm.

The history of the cytoplasm is indeed a true eoigenesis; that is, it is a straightforward coming-into-being, of molecules, structures, and functions which did not at first exist. It starts in the relative simplicity, or at least plasticity, of a single cell, the unripe egg. It moves forward irreversibly in ever-increasing complexity through the stages of the life cycle; embryo; larva (if any), juvenile, adult, and finally the senescent individual. The ultimate end of development is death unless, as in some of the lower organisms which are capable of regeneration, the story is able to begin again. Embryology is therefore, for the most part, a recital of processes which take place within the cytoplasm. The cytoplasm is a colloidal aqueous system of proteins, lipids, sugars, and numerous other organic molecules plus various mineral salts. It has a structure, an intimate, precise anatomy such as no system of a merely chemical nature ever had. Its proteins and lipids are organized into membranes filaments, and granules, at the surface of which chemical reactions take place. These are the organelles which carry on the many processes of living.

The precise patterns of the protein surfaces are the principle basis of the specificity of the cell's chemical reactions for the proteins include the enzymes which catalyze and control the rates of chemical processes. They also include non-enzyme proteins. A very few years ago cytoplasm was described as a homogeneous, transparent gel or sol which was optically "empty." The various formed bodies which were seem under the microscope were spoken of as being within the "ground cytoplasm"; or they were interpreted as artifacts produced by fixation and staining. An intricate morphologic structure of cytoplasm has been studied with the help of phase-contrast microscope, the electron microscope, and the ultracentrifuge. There is a variety of cells and the cytoplasms of different kinds of cells have specialized functions (no single type of cell can be said to be "typical"), yet the

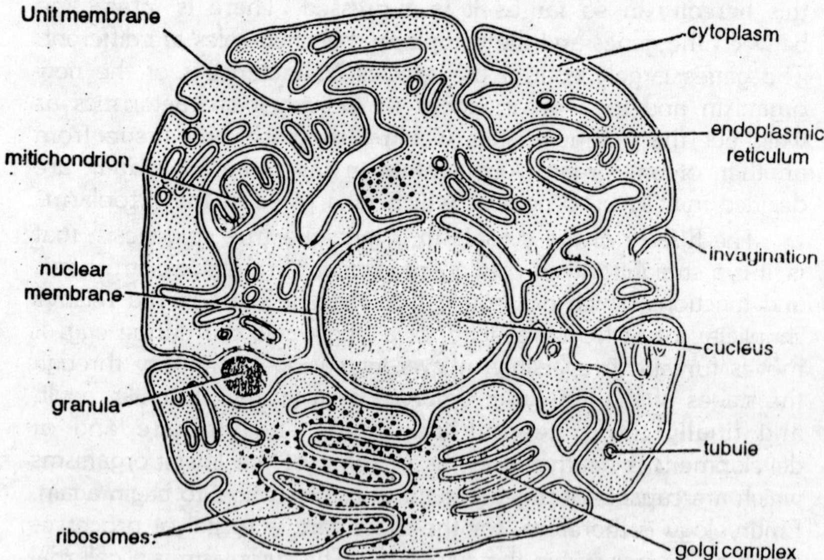

Fig. 6.1. Diagram showing the origin and development of membranous structures of the cell from the unit membrane.

cytoplasms of most cells posses certain organelles in common. These include: a plasma membrane (cell membrane of plasmalemma, the outer layer of the living cytoplasm); inner membranes, which form vacuoles, tubules, and lamellae; a nuclear membrane (actually a part of the cytoplasm); and formed bodies such as ribosomes and mitochondria.

Each of these organelles plays of role in the physiological life processes of the cell. Specialized cells have filaments and granules of various kinds which perform distinctive functions. Muscle cells have contractile fibrils; nerve cells have conductile fibrils; secretory cells have secretory granules; certain other cells produce pigment granules; while still others store reserve food materials such as yolk platelets, glycogen granules, and fat globules. The embryologist sees these various structures appearing in the cytoplasm as development proceeds, each at the proper time and in the proper place. It is believed that they result from the synthesizing activities of distinctive enzymes which are released in the cytoplasm, and that in turn each enzyme obtains the "information" (specific pattern) for its production from a particular gene or genes in the nucleus.

Ribosomes and the Synthesis of Proteins

Protein is the building stone and it is to be synthesized to make the developmental growth possible. Beginning in 1941 and 1950 with the demonstration by Caspersson and Brachet that ribonucleic acid (RNA) is involved in the synthesis of protein, a great deal of information has been accumulated. As noted in the last chapter, the genes of the nucleus (DNA) "transcribe" their specific genetic "information," coded in the sequence of their nucleotides, to RNA molecules. These accumulate in and around the chromosomes and ultimately enter the cytoplasm. Apparently the DNA produces RNA of three sorts; *ribosomal RNA* (rRNA), *messenger RNA* (mRNA), and *transfer* or *soluble RNA* (tRNA). The ribosomal RNA associates basic proteins to itself and accumulates in nucleoli within the cell nuclei. Later this material (rRNA plus protein) enters the cytoplasm where it appears as ultramicroscopic bodies, the *ribosomes*. As much as 30 per cent of the dry weight of cytoplasm may be ribosomes, and about one-half of each ribosome is rRNA. From time to time messenger RNA (mRNA) also leaves the nucleus and enters the cytoplasm.

There it becomes closely associated with the ribosomes and serves as templates, at the surfaces of which the polypeptide chains

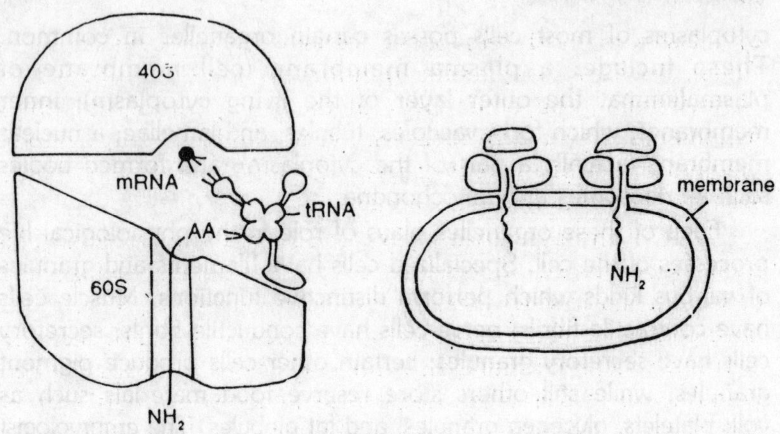

Fig. 6.2. Diagram of ribosome showing the two sub-units and the probable position of mRNA and tRNAs.

of amino acids are build up. These chains become the proteins. How is this "information" which is brought from the nuclei by the mRNA transferred ("translated") to protein molecules? It is carried in the form of an alphabetic code of four "letters," the ribonucleotides. These differ from the nucleotides of DNA in that the sugar is ribose, instead of deoxyribose, and the bases are adenine, guanine, cytosine, and uracil (which substitutes for thymine).

The precise sequence of the nucleotides is the most important feature, for it determines the sequence of the amino acids in the polypeptide chain. But since there are only four kinds of nucleotides and there are twenty or more kinds of amino acids in a polypeptide chain, it must take a "*word*" or *codon* of at least three nucleotides to pick or choose one amino acid. The third sort of RNA is transfer or soluble RNA (tRNA). It consists of smaller molecules, each of which bears, possibly at one end, a triplet of nucleotides which corresponds specifically to one amino acid. Since there are twenty kinds of amino acids, there must be at least twenty specific kinds of tRNA. According to present evidence, the "translation" of information from mRNA to the polypeptide chain takes place mostly in the cytoplasm in the following manner: (1) First the amino acids in solution in the cytoplasm become activated with the aid of energy-rich molecules, such as ATP, and activation enzymes. (2) Then the activated amino acids, still attached to the enzymes, are "recognized" by specific tRNA molecules. By recognition we mean that each amino acid becomes attached to its proper tRNA molecule. (3) Having accomplished this, the tRNA-amino-acid combinations separate from their enzymes, and each adheres one by one in sequence to a specific locus on a messenger RNA molecule as it moves across and past a ribosome.

Thus, in turn, the amino acids, having been brought close together, link up, and a growing polypeptide chain is produced. A polymerizing enzyme is required for the linking-up process. What is the role of the ribosomes in the synthesis of proteins? It seems not to be specific. Cell-free solutions in which mammalian mRNA have been combined with bacterial (!) ribosomes (plus a "pool" of amino acids, transfer RNA, and ATP) are capable of synthesizing mammalian proteins. One writer has referred to ribosomes as "workbenches" on which the polypeptide chains are forged. Ribosomes are said to "read" the message brought in by mRNA as it moves past. It may help us to visualize the process if we think of a ribosome as a sort of automatic machine which is controlled by a punched tape, the mRNA. The machine (ribosome) receives, as raw material a mixture of activated and recognized amino acids (amino acids bound to tRNA molecules) and attaches particular combinations as the mRNA tape dictates.

The amino acids link up and an ever-lengthening polypeptide chain is produced is released to be used again). When the tape runs out, the polypeptide chain is complete and moves away. The entire process takes but little over one minute. The machine is also free to start over again with the same or another tape. Since the tape is long, it may be moving through or past several ribosomes at the same time, each of them in the process of producing a polypeptide chain. Such a series of ribosomes, connected by a thin thread of mRNA molecules, can be demonstrated with the electron microscope. They are known as *polyribosomes* (polysomes). There is some evidence that ribosomes are not all the same in their nucleotide composition, even that they differ in the cells of different tissues. Still, they may all be just work-benches. The "*code*" by which mRNA transfers its information to polypeptide chains has been broken. This was accomplished first by students of microorganisms. Bacteria were homogenized without destroying the ribosomes. The resulting cell-free system was supplied with amino acids, a source of energy

(ATP), and artificially synthesized RNA chains of known composition. The polypeptide chains which were produced were then analyzed.

The breakthrough came in 1961 when Nirenberg "fed" his system an artificial RNA consisting of the nucleotide urocil only. He obtained polypeptide chains composed of the amino acid phenylalanine. From that point on it was only a matter of time until the entire code was broken. At first the ribosomes in the cytoplasm are free and scattered. Then, about the time that newly synthesized proteins begin to be secreted away from the cell, the ribosomes become attached to the endoplasmic reticulum within the cytoplasm. This work on RNA and the ribosomes is of great interest to embryologists because it is possible that differentiation manifests itself first quantitatively in the distribution of the ribosomes.

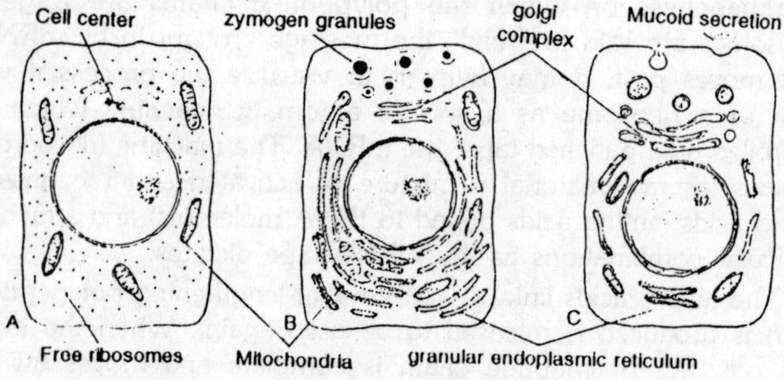

Fig. 6.3. Three types of cells engaged in synthesis (idealized). A—A protein-retaining cell with free ribosomes. B—A protein-secreting cell with an extensive indoplasmic reticulum coated with ribosomes. C—A mucus-secreting cell with a well-developed Golgi complex.

At any rate, gradients in the arrangement of ribosomes exist early in development, and any manipulation which disturbs. In cytoplasm there are several cell organelles like mitochondrias, Golgi body, Endoplasmic reticulum, ribosomes lysosomes etc. These organelles play their allotted roles there by helping the cytoplasm during development. As usual mitochondria provides energy to perform various metabolic processes possible. Golgi body secretes all necessary enzymes and endoplasmic reticulum transports the various products from one place of all to another wherever they are required. Ribosomes play is the synthesis of protein as explained above.

The Cell Cortex

The from and activity of a cell are intimately dependent on the properties of the outer zone (or zones) of the cytoplasm, namely, the *cell cortex*. The outer layer of the cortex is the *plasma membrane*. It directly relates and cell to its environment. Nutrients must enter through it, and unwanted substances must be kept out. Wastes are eliminated by it. Stimuli, except perhaps those of light, affect it first. The plasma membrane is of the order of 100 Å thick (=100 × 10^{-7} mm). It has properties of semi-permeability, but is must not be thought of as a passive sieve. On the contrary, it is an extremely active membrane which produces and retracts minute processes (microvilli) from its external surface, and forms microtubules and vesicles that push inwardly into the interior of the cytoplasm. By this means the cell feeds (phagocytosis) and drinks (pinocytosis). The particles and droplets which thus enter the cell do not directly penetrate the plasma membrane. Rather, the membrane moves before them and surrounds them as they advance into the cell. Later the membrane may dissolve, or the contents of the vacuole may be digested and absorbed into the cytoplasm. In addition to the plasma membrane, the cell cortex commonly includes a somewhat thicker layer of cytoplasm which is from 1 to 5 μ thick (=1 to 5 × 10^{-3} mm). This layer, sometimes referred to as ectoplasm or plasmagel, is usually gel-like in its physical structure, although it may change state from gel to sol and back again.

When photographed with the electron microscope, the plasmagel often differs very little, if at all, from the more liquid internal cytoplasm (endoplasm, plasmasol). Indeed, it may not be present at all. But in other cases, it is specialized by the presence of filaments or granules which are not found elsewhere. The various responses by which a cell changes its shape are often cortical reactions. Development, also, is to a remarkable degree a matter of cortical response. This is obvious in the case of the extrusion of the polar bodies, the rising-up of the fertilization membrane, and the division of the cytoplasm during cleavage. It is not so obvious that the folding of the cellular layers by which the gut and later the nervous system are formed are also to a large extent cortical reactions.

Protozoologists long ago became aware that the form and much of the activity of one-celled animals are properties of the cell cortex. Consider, for example, the ciliate, *Stentor*. The body of this animal is conical in shape. The sides are marked by a hundred or so longitudinal rows of cilia separated by intervening pigmented stripes. The base of the cone (anterior surface or "head") is a disc with spiral rows of cilia. The head is surrounded by an almost complete circle of membranelles. At the clockwise end of the arc of membranelles are the structures used in feeding, namely. The oral pouch, cytostome, and gullet. Nearby is the anal pore. The apex of the cone (posterior end) is specialized as an organ of temporary attachment, the holdfast. All these structures are specializations of the cell cortex. Tartar has performed numerous surgical operations on *Stentor* and has repeatedly observed its extensive capacity for regeneration and reorganization. The controlling feature, he finds, is the presence and orientation of the cortex. A tiny bit of *Stentor*, only 1/123 the size of a normal animal, is capable of regenerating a whole *Stentor* provided that a little of the cortex and some nuclear material are present. But a far larger piece of *Stentor*, lacking a bit of cortex, does not regenerate even though adequate nuclear material is included.

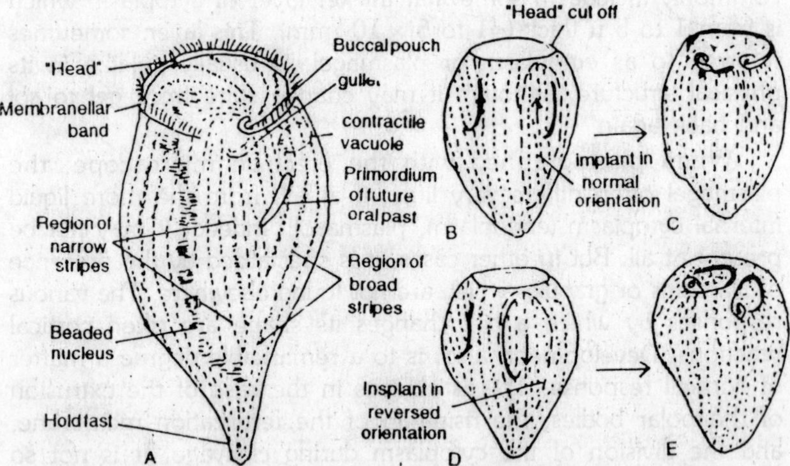

Fig. 6.4. A—Schematic diagram of the ciliate protozoan Stentor. B and C—Tartar's experiment in which a region of the cortex with narrow strips was implanted in a region of broad stripes in normal orientation. Result—a "doublet" Stentor. D and E—A similar experiment but with the implant revolved 180°. The implant developed a set of oral parts with reversed asymmetry.

Undoubtedly the nucleus is necessary to control the processes by which substances of the cell are synthesized; but Tartar concludes, the polarity and asymmetry of the body are transmitted from one generation to the next solely by the cortex. In a somewhat similar fashion, the cortex of a metazoan egg transmits the pattern which largely controls the polarity, symmetry, and asymmetry of the future embryo. For a time these features are labile to varying degrees and subject to modification, by the environment, Raven has studied the role of the cortex in the eggs of molluscs. When such eggs are centrifuged, the visible materials of the cytoplasm are displaced either toward or away from the axis of rotation. They are displaced toward the axis if they are less dense than the fluid cytoplasm which surrounds them, and away from it if they are more dense.

In some cases, development proceeds normally except that the granules and droplets remain unnaturally distributed. In other cases, the displaced visible stuffs move back to their original locations. Now, since the internal cytoplasm is quite fluid and subject to flow (witness its movements during meiosis, fertilization and cleavage). It must be the cortex which remains fixed and so transmits the pattern of the developing organism. Extreme centrifugation liquefies the cortex and results in abnormalities.

The Adhesion of Cells

The cells of an embryo adhere to one another in varying degrees and in manners which are characteristic of the different tissues. At first the adhesions are weak. Cell easily dissociate when early embryos (blastulas and gastrulas) are placed in calcium-free solutions. As development proceeds, however, the adhesions between most cells become stronger. This is especially true in the case of epithelial cells, that is, in the case of cells which form layers and either bound the outside of the embryo or line cavities within it. Adhesion is especially close and strong immediately adjacent to where the contacts between cells border a cavity. In effect, an epithelium presents a continuous outer surface—Holtfreter thought of it as a "coat"—which is semipermeable and which serves to separate the external medium which surrounds an embryo (or fills its cavities) from the internal medium which bathes its inner cells.

Actually, it is not a continuous coat, but a network of close contacts between cells (often referred to as terminal bars). Except at these outer contact pockets of liquid may separate the plasma membranes. As development proceeds, adherent patches known as *desmosomes* develop between adjacent cells and strengthen the bonds between them. The electron microscope sometimes shows filaments extending from each desmosome into the cytoplasms on both sides. Other cells of the embryo, notably those known as mesenchyme, are less adherent to each other.

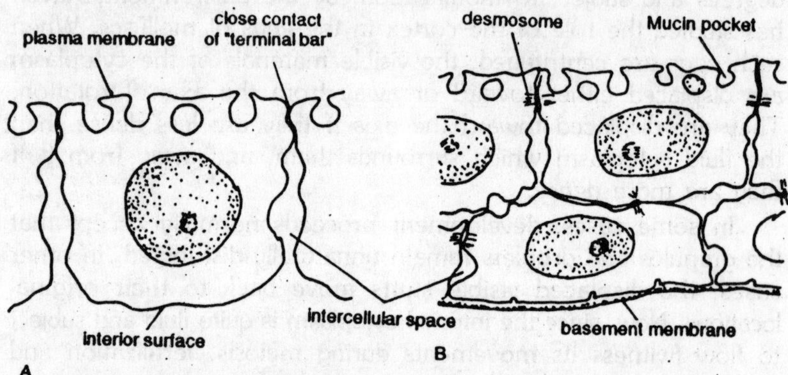

Fig. 6.5. A—Adhesion of epithelial cells. A early one-layered epithelium in which there is close contact next to the external surface. B—Later two-layered epithelium with desmosomes and a basement membrane.

Apparently they are free to wander independently; but they do not wander haphazardly. On the contrary, they move to appropriate positions in the embryo, and there they remain as if trapped by their surroundings. What guides mesenchyme cells? What finally brings their wanderings to an end? In general, what holds embryonic cells to their places in the embryo? Holtfreter has attempted to account for the direction of cell migration in terms of "contact guidance". Movement on a surface, he supposes, is controlled by the oriented macromolecular structure of the surface. He explains the final location of cells by the principle of "selective adhesion." That is, cells (or rather, cell surfaces) adhere most strongly to other cells for which they have an "affinity" possibly a comparable surface pattern). When a cell makes contact with a congenial cell, its wanderings cease, Abercrombie terms the behaviour "contact inhibition." These responses of the cell surfaces are markedly specific.

In some cases, they appear to be species-specific, as when an egg cell reacts to sperm cells of its own species, or when lymphocytes react to foreign proteins (antigens) by producing antibodies. In other cases—and this is especially characteristic of embryo—the specificity of the cell surface seems to be tissue-specific. When an embryo is experimentally dissociated into, its constituent cells and then the cells are permitted to reaggregate, they first adhere to each other and then tend to sort out and reform the tissues from which they were taken. Superficially, it seems as though each cell, when it contacts another cell, recognizes whether the other cell is of its own or another kind. Steinberg offers a less mysterious explanation of contact guidance and selective adhesion. He explains segregation in terms of quantitative differences in the strength by which cell surfaces adhere.

The principle is much the same as that which causes oil and water to separate; namely, the mean of the cohesive forces between like molecules (water to water and oil to oil) is greater than the adhesive forces between unlike molecules (water to oil). Similarly, intermingled cells sort out because like cell surfaces adhere to each other with more average strength than unlike surfaces adhere. Sorting out, therefore, is not so much a property of cells as individuals, as it is a property of cells in populations. A corollary of this is that those cells of a mixed population which have the strongest mutually adhesive properties tend to pull together and become encapsuled within an outer mass of cells with less adhesive tendency. The foldings and moldings of tissues during development (about which we shall have much to say in the following pages) have been interpreted in terms of programmed changes in the adhesive characteristics of cell surfaces. For example, a thickening of an epithelial layer of cells involves an increase in the areas of contact between cells. Presumably this results from an increase in the strength of their cohesiveness. A thinning of an epithelium is associated with a decrease in the areas of contact.

The Role of the Cytoplasm in Development

Many embryologists today are engaged in describing in details the chemical and cellular changes which take place in an embryo as development proceeds. This is the problem of *cyto-differentiation*, Notable progress is being made with the aid of sophisticated physical and chemical techniques such as electron microscopy, radioactive tracers, and immunological methods. As a result, today we know something about the sequence of chemical events in the embryo and when and where new cellular structures appear and disappear. But it is one thing to *describe* the physical and chemical changes which take place during development and quit another thing to give a *causal account* as to how these changes are brought about. The key question is: What controls differentiation? There is no doubt today that the cytoplasm controls the course of cell differentiation.

But there are two "levels" at which this control may be exerted: (1) it has been generally assumed that control takes place at the level of "transcription": that is, takes place in the nucleus when DNA becomes activated and transfers its information to messenger RNA. Some genes are "turned on" at any given time and place, while other genes (probably most of them) remain repressed. (2) There is, however, increasing evidence that some control by the cytoplasm takes place at the level of "translation," namely, when mRNA supervises the synthesis of protein. In this case the mRNA produced in the nucleus remains "masked" until it reaches the proper time and place in the living system. The classic example of control of differentiation at the level of translation is seen in the marine alga, *Acetabularia*. This is a giant cell, some two inches long, and shaped like a mushroom or parasol. Its apical end spreads out radially like a hat or cap. Its basal end forms rhizoids (holdfasts), one of which contains the large nucleus. The entire life cycle of *Acetabularia* does not concern us now.

Our present interest is in its capacity to regenerate a new cap when the old cap is cut off. Surprisingly, the regeneration of a cap in acetabularia. Hammerling, who studied these matters intensively, interpreted his findings, in terms of gradients in the distribution of morphogenetic substances, i.e., a cap-producing substance with its highest concentration at the apical pole, and a holdfast producing substances centered at the basal pole. His concept is that these gradients determine the location of the new parts. The place even when the nucleus has been cut away two or three weeks previously. Clearly the presence of nuclear DNA is not directly responsible for the restoration of lost parts. The role which the nucleus plays in regeneration must be an indirect one.

Presumably the genes of the nucleus produce the morphogenetic substances which then migrate within the cytoplasm and concentrate at the apex or base, as the case may be. There they stimulate regeneration and at the same time control the

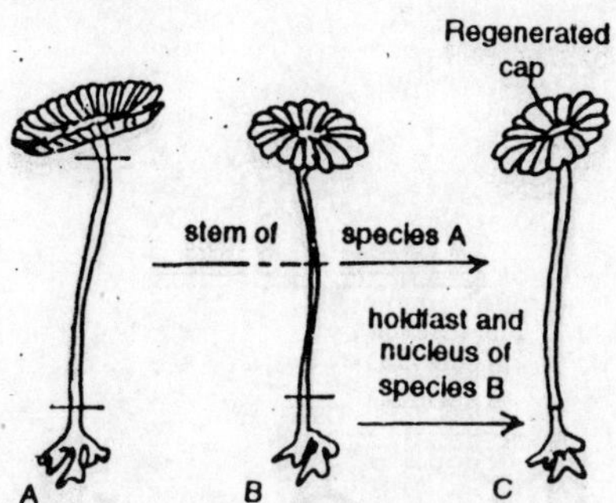

Fig. 6.6. Regeneration in the unicellular alga; Acetabularia. A—Acetabularia mediterranea ("species A"). B—Acetabularia crenulata ("species B"). C—Result of grafting a decapitated stem (no nucleus) of species A to a hold fast (with nucleus) of species B. The first cap which regenerated was cut off. The second cap to regenerate had the characteristics of species B.

character of that which regenerates. This interpretation is borne out by Hammerling's observation that, when a piece of the stem of *Acetabularia mediterranea* (we shall cell this "species A") is grafted by its basal end to a holdfast of *Acetabylaria crenulata* ("species B"), the cap which regenerates resembles the caps of species A. He interprets this to signify that, at the start, regeneration is under the control of the morphogenetic stuffs which were emitted by nucleus A before it was removed. But if this first cap is cut away, the second cap which regenerates has the character of species B. This confirms the view that the nucleus, which is located at a distance from the apex both in time and space, is the source of the morphogenetic stuffs.

In terms of present theory we may identify the morphogenetic stuffs (as Brachet does) with messenger RNA. They are produced by transcription in the nucleus. Under the influence of the cytoplasm. The cap-forming stuff migrates and concentrates at the apical end of the cell. Here translation takes place; that is the information brought by the mRNA is transferred in *situ* to newly synthesized enzymes and structural proteins. No theory of turning genes on and off is sufficient to account for this localized reaction. What brings about the concentration of morphogenetic stuff at the apex? Since the inner cytoplasm of *Acetabularia* is engaged in active streaming, some factor in the cortex must be responsible. But it must be admitted that, as yet, we know very little of how this influence of the cortex originates or is exerted.

7

CELL LINEAGES

The term-localization is used here to show the specification of cell fate according to the sector of egg cytoplasm inherited by an embryonic cell lineage. As the early blastomeres divide up the egg cytoplasm, they appear, in some organisms, to inherit "instructions" for various kinds of cell differentiation, including specific patterns of macromolecular synthesis. Among the main problems posed by the localization phenomenon are the molecular nature of the morphogenetic agents stored in the cytoplasm; the level(s) of control at which these agents might operate; their subcellular location, and the cytoskeletal structures to which they are anchored; and the means by which they are distributed to given regions of the egg or early embryo. The description by C.O. Whitman (1878) of cleavage and cell fate in the leech *Clepsine* was the first complete study of this kind, and greatly influenced the field. In this organism, as in many others shortly thereafter the subject of cell lineage analysis, the number of individual blastomeres at the cleavage and blastula stages is relatively small, and they can be distinguished by size, shape, and position, so that the developmental destiny of each cell and its lineal descendents can be followed.

Whitman showed that the mesodermal and neural "germ bands" of the leech embryo are budded forth from bilateral stem cells, the teloblasts (annelid teloblasts had first been noticed by Kowalevsky, (1871). Whitman's most important finding was that the ventral nerve cords of *Clepsine* develop from bilateral sets of four individual neuroblast stem cells, themselves deriving from early bilateral blastomeres. The mesoderm also forms from individual bilateral stem cells, the mesoblasts. Though the discussion in Whitman's 1878 publication is largely focused on the evolutionary affinities implied by his observations, their significance for the cell biology of development was widely recognized. This was the demonstration that highly differentiated structures, such as nerve cords, can be derived lineally from *specific individual blastomeres that arise only in certain regions of the egg, by a precise pattern of cleavage, and that normally inherit certain segments of the egg cytoplasm*.

In the following decades elegant and in some cases extremely detailed cell lineage investigations were carried out on animals belonging to diverse phylogenetic groups, including numerous molluscs and annelids, the nematode *ascaris*, and various ascidians, flatworms and ctenophores. In many cases the fates of *every* cell of the embryo were determined. Sufficient data were obtained for the annelid-mollusc group to support comparative analyses of the developmental values of homologous blastomeres by Wilson (1898); by Child (1900) in his study of cell lineage in the polychaete *Arenicola*; and by Conklin (1897), in his classic description of the cell lineage of the mollusc *Crepidula*. Many of the investigators who carried out cell lineage studies were deeply interested in the evolutionary implications of cellular homologies in the development process. Thus a by-product of these studies was the construction of an imposing morphological argument for the phylogenetic relationships of various protostome phyla, particularly polyclad worms, annelids and molluscs. For us, as for contemporaries focused primarily on the problems of developmental mechanism, the main significance of the cell lineage determinations lies in a different direction.

Conklin's description of the lineage of *Crepidula* in many ways illustrates the major theoretical impact of these studies. Conklin followed the fate of individual blastomere lineages beyond the 100-cell stage, from their initial appearance to their participation in differentiated structures, and thus defined exactly the embryonic cellular origins of all the organ systems and structures of the advanced veliger larva. Contemporary writers therefore stressed that the determinants in mosaic developmental processes are the maternal cytoplasmic substances, rather than being a particular property of particular embryonic blastomeres.

(i) Localization of in the Egg that Display Determinate

The cases of localization that most impressed classical experimentalists were those in which areas of future cell fate could be visibly mapped out on the uncleaved egg cytoplasm. A spectacular example is Conklin's (1905) study of development in the ascidian *Cynthia* (*Styela*), which greatly expanded and confirmed earlier experiments on ascidian embryos by Chabry (1887). In the egg of *Styela* pigmented areas of cytoplasm corresponding to a morphogenetic fate map for the cells of the early embryo can be distinguished. Some of Conklin's elegant hand-drawn figures are reproduced. These display the relationship between the various pigmented regions of the egg cytoplasm and the tissues ultimately formed from these regions. Five kinds of cytoplasm were noted by Conklin: a dark yellow granular cytoplasm eventually included in the tail muscles of the larva; a light yellow material later segmented into the coelomic mesoderm of the larva; a light gray substance inherited by notochord and neural plate cells; an opaque gray material segregated into the endoderm cell lineage; and a transparent cytoplasm later present only in ectodermal cells. The dark yellow granular cytoplasm referred to by Conklin as the "yellow crescent" is shown in recent photomicrographs of eggs that had been extracted with detergent.

As described in the legend, the persistence of the yellow crescent in these eggs indicates that the pigment granules that Conklin saw are embedded in the cytoskeletal matrix of the cytoplasm. Conklin focused on the striking correspondence between the sets of blastomeres that give rise to specific groups of differentiated larval cells, and the distribution of the five types of eggs cytoplasm that he could visually distinguish throughout cleavage. The definitive distribution of the "organ forming substance" only appears after fertilization. It is not a primordial property of the egg. The process by which this distribution develops was observed carefully by Conklin and some of his illustrations are reproduced. One axis of polarity is already present in the unfertilized egg.

The establishment of the anterior-posterior axis depends on the acentric movement of the male pronucleus. The fusion nucleus comes to lie near the future posterior pole, and there the yellow crescent cytoplasm later incorporated in the embryonic muscle cells is localized. Conklin regarded the ascidian embryo as a "mosaic-work," and as had the other observers cited above, he also argued that "the mosaic is one of organ forming substances rather than of cleavage cells." He concept of ascidian development was supported by an extensive series of experiments in which he determined the developmental potentialities of individual blastomeres and sets of blastomeres in embryos in which the other blastomeres had been killed. In general the surviving blastomeres were found to establish cell lineages the differentiate in their respective normal directions, though the overall organization of the embryo is of course affected. The drawings reproduced. That each embryo fraction contains the potentiality of forming certain presumptive tissue types, such as notochord, mesoderm, neural plate, gut, or ectoderm, even though the isolates were often not cultured sufficiently long to reveal completely their developmental capabilities.

The ascidian cell lineage includes some corrections of Conklin's original assignments, mainly obtained by the method of horseradish peroxidase injection into individual blastomeres of 8-cell embryos of *Ciona intestinais*, *Ascidia abodori*, and *Halocynthia roretzi* and into blastomeres of 16-and 32-cell embryos of *Halocynthia* for similar assignments obtained for some lineages of *Phallusia mammaillata* embryos by a different method. This lineage describes diagrammatically the ultimate capacities of the various early blastomeres *in situ*, and in general in partial embryos as well, and thus provides a developmental linkage between Conklin's "organ forming substances" and embryonic cell differentiation. However, the relation between embryonic cell fate and cell lineages is in some areas not as simple as implied by Conklin's (1905a) proposal that segregation of morphogenetic agents has been completed by the 64-cell stage, and thus that the progeny of each 64-cell blastomere contribute to only one type of tissue.

The lineage chart indicates that some functionally asymmetric divisions must occur later than the 64-cell stage, since certain 32-cell blastomeres contribute to at least three different types of tissue. Examples of clonal determinations by the 64-cell stage, and also of later determination, are to be found in the muscle cell lineages. The concept that early embryos are a "mosaic-work" of determined blastomeres was widely challenged. Beginning in the late 1880s, blastomeres were isolated and cultured from many different embryos, in order to determine whether their division products indeed develop *in vitro* as do their Normal lineages in the while embryo. Diverse results were obtained with different organisms, and a ferocious controversy over the basic nature of embryological development soon arose. The conflicting theories that followed are reviewed briefly in the next section, while in this we consider further examples of the evidence supporting one side of this controversy, viz., those isolated blastomere experiments that provided powerful evidence for early fixation of morphogenetic potential in localized regions of the embryo.

Consistent results were obtained with other isolated cell types and with partial *Patella* embryos. Among these were isolated one-sixteenth embryo macromeres, which produced endodermal gut rudiments, and isolated apical progenitors, which differentiated *in vitro* into apical sensory and ectodermal cells. The morphogenetic fates of the cell lineages descendant from early embryonic blastomeres were studied in marine annelids, such as the polychaetes *Sabellaria* and *Nereis* and an illustrative series of experiments was carried out by Penners (1926) on the oligochaete annelid *Tubifex rivulorum*. This animal develops essentially as described for *Clepsine* by Whiteman. Thus a neuroblast germ band and mesodermal germ band both derive from the D quadrant of the embryo (blastomere nomenclature here. At the 4-cell stage of the Tubifex embryo the D macromere is the largest, and with respect to both size and rate of division, its products remain distinct from those of the A, B and C quadrants.

Penners found that each of these blastomeres would continue its normal course of development even if all the others were killed *in situ* by UV microbeam irradiation. Thus if A, B and C are killed, the D macromere nonetheless adheres to its unique cleavage pattern. It gives rise to an ectodermal neuroblast stem cell, 2d and to a primary mesodermal stem cell, 4d. After this columnar ectodermal and mesodermal germ bands are produced by anterior budding. Penners showed that if the 4d cell is individually killed, the neuroblast germ bands still form, but the embryo lacks coelomic mesoderm. Recent investigation has demonstrated that the individual neuroblast stem cells of annelids each give rise to detain specific neurons. In the leech, for instance, ablation of a given germ band neuroblast results ultimately in the absence of one member of each of the three bilateral segmental pairs of larval neurons that contain serotonin, and ablation of a different neuroblast causes the absence of one member of each pair of segmental body wall neurons that contain dopamine.

(ii) The Early History

The initial discovery of determinate, asymmetric cell lineages was among the antecedents of the proposals that morphogenetic substances are localized in the cytoplasm of the very early embryo. There was also a parallel theoretical development, the concept of "precocious segregation" of maternal molecular determinants. The origins of this idea lie in the nineteenth century controversy between preformationist and epigenetic theories of development. The participants in the final stages of this controversy included many eminent biologists engaged personally in research on localization. Huxley (1878), Hertwig (1894), Bourne (1894), and Whitman (1895a) all published discussion of the implications of localization in which the main issue considered was whether the embryo actually increases in biological completely or, on the other hand, merely reveals progressively to the observer an organizational complexity that is already resident in the unfertilized egg. A major accomplishment of mid-19th century biology had been to establish the apparently epigenetic nature of early development.

The first investigations directly supporting epigenetic interpretations were those of Wolff (1759, 1768), which showed that embryonic chick blood vessels and gut develop from undifferentiated tissues; Pander's (1817) description of epigenetic development in the chick from primitive germ layers; and the studies of Von Baer (1828, 1837), as a result of which the germ layer theory was generalized to other animals. Von Baer showed that skin develops epigenetically from ectoderm, muscular and skeletal systems from mesoderm, etc. Kowalevsky (1867) suggested that the germ layers of many animal phyla are also formed epigenetically. These observations were regarded as crucial refutations of the traditional preformationist theories previously current, such as those of Bonnet (1762 see translation and discussion in Whiteman 1895b).

Bonnet had argued that little change incomplexity actually occurs in embryological development, and that "organized bodies pre-exist from the beginning," According to Bonnet's rather crude ovist theory of preformation (1746) a complete embryo is patterned in every egg, and each such embryo contains an ovary, with eggs that bear embryos within, etc. The matter seemingly settled by scientific investigation, it was thus a striking event when in the 1870's a novel and sophisticated new form of preformationism was proposed, by Wilhelm His and others. His was the teacher Johann F. Miescher, the discoverer of DNA, and was a proponent of the view that satisfactory explanations of biological phenomena can only be obtained at the molecular level. In 1874, several years before the publication of Whitman's *Clepsine* study. His suggested that the epigenetic character of early chick development is only apparent, the underlying phenomenon being the "coalescence of performed germ." The close relationship between these ideas and the subsequent concept of cytoplasmic localization is obvious. However, these passages were written a decade before it was first realized that the hereditary determinants are confined to the chromosomes of the cell nucleus.

To obtain from them the idea of cytoplasmic localization as we now think of it requires the conceptual separation of localized *cytoplasmic* determinants from the *genomic* determinants shared equally among all the blastomere nuclei. Roux proposed that two kinds of processes leading to embryonic cell differentiation occur in development, *self-differentiation* and *correlative differentiation*. The first of these describes the determinate behaviour of blastomeres and their cell lineages in mosaic forms of embryogenesis, and predicts that when isolated, embryonic blastomeres will continue to display their normal potentialities. Correlative differentiation was invoked by Roux to explain the regulative behaviour of partial embryos. Through their powers of correlative differentiation embryonic cells could secondarily modify their normally determined fate. Roux suggested that correlative differentiation is mediated by means of interactions between cells.

Unfortunately, Roux (1883) and Weismann (1885, 1892) initially attempted to explain blastomere self-determination as the consequence of the qualitative partitioning during cleavage of *nuclear* genetic determinants, a proposal that was almost immediately discredited. Experiments of Driesch on sea urchin eggs, Wilson on *Nereis* eggs, and Yatsu on nemertine eggs, among others, showed that treatments which *redistribute* nuclei destined for given blastomeres to other blastomeres nonetheless often result in normal development. The most convincing experiments were carried out by the transient application of pressure to cleaving eggs, e.g., by exposing them to the weight of a cover slip. The result is to inhibit normally occurring transverse cleavages, and to induce extra rounds of vertical cleavage. Three-dimensional development resumes on release of the pressure, and nuclei normally assigned to given blastomeres are not found to be located in different blastomeres. For example, in Wilson's *Nereis* experiments (1896a) the nuclei that would in normal embryos populate the entoblasts were shown to be able to direct the development of first quartet micromeres, and vice versa.

In other words, these nuclei must contain genes for endodermal as well as ectodermal differentiation. Generalizing, it follows that any cleavage stage nucleus contain all the zygote genes, and therefore that the differences in cell fate displayed by given lineages must be due to segregation of *cytoplasmic* factors. The promulgation of Roux's mosaic theory of development—ignoring the nuclear segregation portion-initiated a most interesting intellectual divergence which to some extent is still with us. While tests of the developmental capacities of isolated blastomeres in many cases directly supported the mosaic view of embryogenesis, some regulative capacity was often noted even in blastomeres from eggs that were described overall as "mosaic-works."

Most importantly, it was discovered that in some organisms isolated blastomeres could give rise to complete rather than partial larvae. The latter result was obtained with at least some of the early cleavage blastomeres of species of sea urchins, hydroid coelenterates, cephalochordates, nemertean worms, amphibians, and fish (for a complete summary of the classical partial embryo experiments). Proponents of various forms of correlative differentiation regarded the amazing regulatory behaviour revealed in these partial embryo experiments as a decisive blow to the mosaic theory that the embryo is the sum of its self-differentiating blastomeres.

The basic difference in attitude this related to the significance of specific substances in specific cells. Wilson and Conklin, for example, considered there to be convincing evidence that the fate of a given blastomere or set of blastomeres may be determined by its specific cytoplasmic contents. The focus on the overall organizational properties of the embryo, and on the role of *position* can be regarded as ancestral schools of embryology oriented toward analysis of organizing centres, morphogenetic fields, and gradients, and that more recently has given rise to the concepts of positional

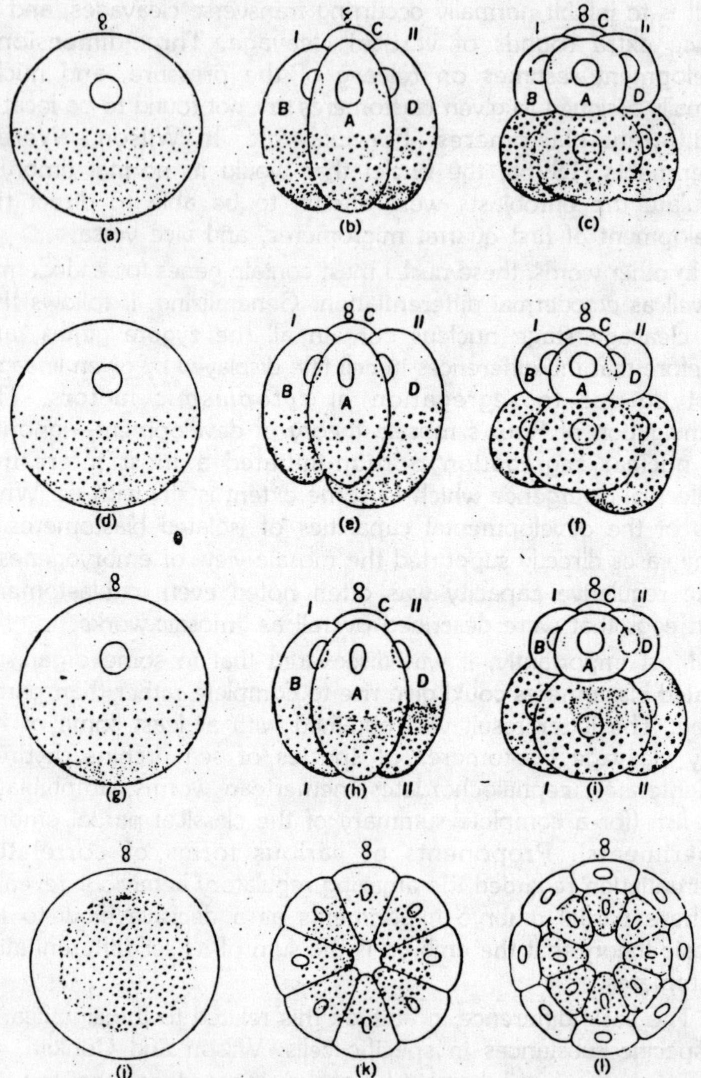

Fig. 7.1. Localization of morphogenetic substances in the early blastomeres of four types of egg.

information. The opposing *cellular* focus was soon bolstered by the rediscovery of Mendelism and the general acceptance of the presence in blastomere nuclei of genomic determinants for all the specific traits manifested by differentiated cells.

An event of considerable significance at the time was the demonstration of Boveri (1902) that blastomere gene action is specifically required for early development (reviewed in Davidson, 1985). This study concerned the causes of developmental arrest in dispermic sea urchin eggs. Boveri showed that the tetrapolar and tripolar mitoses that occur in these eggs usually result in aneuploidy, so that particular chromosomes are missing or present in greater than diploid number in some blastomeres. He demonstrated that the developmental potential of isolated blastomeres containing partial genomes is invariably incomplete, while in contrast, as Driesch had shown, complete plutei could be obtained from isolated blastomeres of normal embryos. The classical *cellular* theory of localization, and Boveri's essentially modern assertion of the role of the genome in differentiation, are the traditions that led to the persistent effort to understand embryo cell function, which today is largely molecular in nature. It is an interesting historical note that in the modern pursuit of the mechanism by which initial specification of embryo cell function occurs we once again find ourselves face to face with "the localization problem."

Determination of Specific Cell Lineages

Three examples of embryonic cell lineages that are apparently determined by processes of cytoplasmic localization are considered in this section. One difference between the classical studies just reviewed and the modern investigations we now review is the use of specific ultrastructural and biochemical markers as indices of differentiated cell lineage function. The first of these examples is the determination of the ascidian muscle cell lineage, for which current evidence on the role of localization is probably the most complete; the second is the determination of the gut cell lineage in a nematode; and the third concerns determination of germ cell lineages in certain nematodes, insects, and anuran amphibians.

The Gut Cell Lineage of *Caenorhabiditis Elegans*

It is a well-known fact that the development in the nematode *Caenorhabditis elegans* is a highly determinate process. The timing and orientation of each embryonic cell division is programmed genetically, and under normal circumstances the ultimate fate of each cell is rigidly fixed. The simplified diagram in Fig. illustrates the origin in the initial cleavages of the six embryonic founder cells. The first cleavage produces an anterior cell, AB, and a posterior cell, P1, and the latter then undergoes several unequal divisions, each resulting in a germ line precursor cell plus a primordial somatic cell. The endoderm derives from the Somatic founder cell EMS, the descendants of which give rise to endoderm, various mesoderm structures, and some neurons.

The endoderm lineage is definitively segregated out at the first division of EMS, with the formation of E (endoderm) and MS (mesoderm-stomadaeum) founder cells. Since these cells all arise from topologically specific regions of the egg, and since the destiny of the lineage descendant from them is fixed, a possible interpretation of the initial lineage assignments is that each is determined by the localized cytoplasmic elements that it inherits. In this section we consider evidence related to this proposition in respect to the gut cell lineage descendant from the E blastomere, and in the following, the homologous arguments that can be applied to the germ line lineage deriving from the P4 stem cell. Two markers that specifically indicate gut cell differentiation are the appearance of fluorescent particles that contain tryptophan catabolities, called *rhabditin granules*, and a specific esterase that can be identified by histochemical staining. These appear in the gut cells a few cycles prior to the terminal divisions of their lineage. Laufer *et al*. (1980) and Cowan and McIntosh (1985) used the appearance of rhabditin granules to determine the cellular location of gut cell determinants in embryos whose cleavage had been blocked with cytochalasin and colchicine, and similar experiments monitoring the appearance of esterase activity in cytochalasin treated eggs were reported.

The drugs were introduced after bursting the impermeable chorion by application of pressure, or shear forces. The results of these cleavage arrest experiments are essentially similar to those of the ascidian cytochalasin studies just reviewed. When cytokinesis is arrested at the 2-cell stage, rhabditin granules and esterase appear after a number of hours in the PI cell but not in AB; when it is arrested at the 4-cell stage the fluorescent bodies and esterase are observed in EMS, but in no other blastomere: and in arrested 8-cell embryos only the E cell develops these granules or gut esterase. By the time the differentiation reflected by the presence of the cytotypic rhabditin granules has occurred, the arrested blastomeres have accumulated an amount of DNA equivalent to many genomes. It is evident that gut differentiation does not require the appearance of the individual gut cell precursors normally produced during subsequent cleavage of the E lineage, and probably does not depend on intercellular interactions among them, or with other cells. Whether the genomic multiplication is necessary. However, unlike the to those 20th century case in ascidian embryos.

Cytochalasin-blocked zygotes are unable to produce rhabditin granules, though they do produce hypodermal marker. Nor do blocked one-cell embryos produce the gut esterase is shown to be an exclusive and intrinsic property of the E blastomere, and the EMS and PI blastomeres from which it derives. This is confirmed in studies on the development of burst partial embryos in which given blastomeres have lysed. Rhabditin granules are observed only in the E cell derivatives, and not, for example, in isolated AB, PI or P2 cell partial embryos. These results are consistent with a determinant role for cytoplasmic factors that are segregated into the E cell during cleavage. However, demonstration of such a role would require that gut differentiation occur in cells normally destined to develop into other structures as a consequence of the experimental introduction of E or EMS cytoplasm. One test of this possibility was carried out by fusing AB a blastomeres (i.e., the anterior daughter of the AB cell) to the contiguous EMS cell by puncture of the membranes separating them with a laser microbeam. Following the next cleavage, either the MS cell or the E cell was destroyed by UV irradiation.

In 90% of case when the E cell was intradiated no rhabditin granules formed, though they appeared in the remaining 10% of the experimental embryos, while most control embryos in which the E cell was preserved developed these granules. However, when cleavage was arrested with cytochalasin after the AB a cell. An alternative experimental design has yielded more positive results. At the 2-cell stage the P1 blastomere was punctured by laser microbeam and the nucleus extruded. In such embryos only the nucleated AB cell divides further and no rhabditin granules are ever formed. However, if nuclei of the progeny of the AB cell are exposed to the enucleate P1 cytoplasm, by laser puncture of the intervening membranes, rhabditin granules form amongst the AB cell progeny in 50% of cases. These blastomere fusion experiments thus provide preliminary support for the causal specification of the gut cell lineage by maternal cytoplasmic factors segregated into the P1, EMS, and then the E blastomeres.

The perfect precision, and the predictability of the developmental fate of each blastomere indeed suggest the solution, just as did the same features of many other mosaic embryos to classical writers, However, there is no evidence in this or the foregoing ascidian examples, as to the mode of action, or the directness with which such maternal factors may function. At one extreme. They could immediately bind to and activate gut-specific genes, and at the other, they could serve merely as the initial triggers for a cascade of subsequent regulatory events, an ultimate consequence of which is the activation of these genes.

(iii) Localization of Germ Cell Determinants

In most species that have been examined the definitive germ cell lineage can be observed after midembryogenesis, and thereafter be traced directly to the gametogenic cells of the larval or juvenile gonads. Primordial stem cells which serve as the exclusive progenitors of the germ line are known to arise in specific embryonic locations at the outset of development in some organisms, but in others, where germ line specification occurs later, the origins of the germ cell lineage remain controversial or obsecure. A unique, localized region of the egg from which the germ line arises can probably be excluded in the case of mammals, for example, since any of the first four or even eight blastomeres can give rise to a complete and fertile embryo.

Grafting experiments suggest that in the mouse determination of germ line stem cells occurs in the primitive ectoderm, but not until about 6-7 days of development, as the cells of this region remain totipotent until this time. It is interesting that very early germ cell determination from localized regions of the egg occurs in phylogenetically distant groups, e.g., nematodes, insects, and anuran amphibians. Thus specification of the germ line by means of a cytoplasmic localization process may be an extremely ancient mechanism, since it is found at least occasionally in both deuterostomes and protostomes.

Determination of the Germ Cell Lineage in Nematodes

Boveri and his associates described the Cytoplasmic localization of determinants for the primordial germ cells of the parasitic nematode *Ascaris*. In *Ascaris*, as in *Caenorhabditis*, the definitive germ line stem cell is the product of an early series of unequal cleavages. A special feature of *Ascaris* development not observed in *Caenorhabitis* is the phenomenon of *chromosome diminution*.

At the first cleavage one daughter cell retains the complete chromosome complement while the other eliminates a large fraction of the genome in deeply staining pycnotic granules. The stem cell not undergoing chromosome diminution retains the capacity to give rise to the germ cell lineage, while at each of the next several cleavages it produces one daughter which becomes a somatic cell. After fifth cleavage there are 31 somatic cells and one definitive germ line stem cell, the only blastomere that retains all of the zygotic DNA.

Boveri and Hogue (1909) showed that retention of the complete genome and hence the initial step in determination of the germ line, depends on a special sector of polar egg cytoplasm. In eggs undergoing abnormal cleavage induced by polyspermy or centrifugation, nuclei other than the normal germ line stem cell nucleus that are surrounded by the polar cytoplasm now behave like normal germ line nuclei, in that they are spared chromatin loss at mitosis, while all nuclei distributed into the remainder of the cytoplasm undergo diminution. As many as 3 prospective germ line stem cells may form compared to the normal one depending on the relation of the altered cleavage planes to the polar cytoplasm. An additional experiment carried out on *Ascaris* egg by Stevens (1909) showed that UV irradiation of the polar eggs cytoplasm results in failure of any primordial germ cells to develop.

Centrifugation experiments on *Ascaris* eggs were performed by Guerrier (1967), with results basically in agreement with Boveri's conception of the organization of this egg. Boveri's studies do not address completely the issue of germ line determination, since they do not test the ultimate function of this lineage, *viz*, the provision of fertile gametes. However, they demonstrate that the key first step of this process is indeed specifically induced by the polar egg cytoplasm. The relation between the embryonic germ line stem cells and the definitive germ cells ultimately located in the gonads has been traced clearly in *C. elegans*, by using fluorescent antibodies that react specifically with granular inclusions of the germ cell cytoplasm. These inclusions, which have been called P granules, are dispersed throughout the cytoplasm at the time of fertilization; but, they coalesce in the posterior polar region of the egg during pronuclear migration, and are thence segregated to the P1 cell. This process is probably mediated by cytoskeletal microfilaments, as it can be blocked by cytochalasin D treatment.

As cleavage progresses the P granules are segregated successively to P2, P3, and ultimately 94 cells. The P granules revealed by the fluorescent antibody are later present exclusively in the germ line progenitor cells Z2 and Z3, the products of P4 and in differentiated larvae they exist only in the gonadal germ cells. Electron-dense bodies that are probably identical with the P granules have also been visualized in the polar cytoplasm of *C. elegans* eggs, and in the embryonic germ line blastomeres. In the uncleaved fertilized egg these structures are located in the peripheral regions of the polar cytoplasm, and by mid cleavage they have assumed a perinuclear location in the primordial germ cells.

The immunofluorescent P granules are similarly disposed, and they remain perinuclear throughout oogenesis, when they again disperse into the distal cytoplasm of the egg. It is not clear whether the anterior-posterior axis of the *C. elegans* egg is determined as a result of events following fertilization or is a preformation that is set up during oogenesis. Maternal mutations are known in *C. elegans* that alter the early cleavage pattern so that two cells form in place of P1. Wood *et al.* (1983) showed that in one such mutant the germ line specific P granules are distributed to the two cells that share the cytoplasm normally destined only for the P2 blastomere, a result that seems exactly to parallel that observed by Boveri in centrifuged and dispermic *Ascaris*. Distinctive cytoplasmic inclusions that are similar in distribution and ultrastructure to the *C. elegans* P granules and are confined to the germ cell lineage are observed in many forms. They were noticed by classical cytologists in the germ cells of crustaceans, chaetognaths and insects. These structures have now been recognized at the ultrastructural level in germ cells of many other animals, among which are annelids, dipteran insects, ascidians, teleosts, anuran amphibians, and various mammals, in which they are usually referred to as *nuage*.

In animals where germ cell determination is precocious, and the germ cell lineage is get aside early in cleavage, the germinal granules are typically found in the region of egg cytoplasm that is to be included in the primordial germ cells, as are the *C. elegans* P granules. In animals such as the mouse, where germ cell determination does not occur until later in embryogenesis, they are not evident until the definitive germ cells can be identified.

Classical writers such as Hegner (1914) thought that the germinal granules of the polar egg cytoplasm, often referred to as *polar granules*, might themselves function as localized germ cell determinates. However, as the following review of germ cell lineage specification in insects and amphibians shows, it is not yet possible to distinguish between this view, which attributes causal significance to the presence of polar granules, and an alternative. This is that these inclusions may function (in some unknown way) in later stages of germ cell differentiation, but not in the *determination* of the germ cell lineage.

In organisms where germ line segregation is at least initiated by means of localized elements of the egg cytoplasm, as in *Ascaris*, such granules could be performed during oogenesis and distributed to that section of cytoplasm which is to be inherited by the germ line stem cell, much as other maternal components of subsequent use during development appear to be localized (e.g., tubulin, yolk, mitochondria, etc.). At present it indeed seems reasonable to regard germinal granules as cytotypic markers for the germ cell lineage.

Germ Cell Lineage in the Eggs of Drosophila

The differentiation of primordial germ cells was studied in chryosomelid beetles by Hegner (1911, 1914). Germ cell formation can be said to initiate when nuclei arrive at the polar region of the oblong egg, and enter a special region of cytoplasm which appears to function as the germ cell determinant. Hegner succeeded in selectively destroying the polar cytoplasm with a hot needle before the peripheral movement of the nuclei had occurred. The injury induced by the needle is rapidly walled off by the forming blastoderm, and normal development of a differentiated gastrula and ultimately a hatching insect takes place. However, primordial germ cells are absent in embryos descended from the cauterized eggs, and adults developing from them are sterile. No genomic material is directly affected, since no nuclei are in the vicinity of the polar cytoplasm at the time of cauterization. Hegner's experiments show that when the polar cytoplasm is destroyed, the capacity of the embryo to elicit germ cell differentiation is lost. The primordial germ cells in *Drosophila* eggs arise at the posterior pole of the egg in much the same manner.

The *pole cells* the first cells formed in the embryo and they lie outside of the blastoderm wall. Geigy (1931) showed that UV irradiation of vegetal pole cytoplasm prior to the migration of the cleavage nuclei into this region results in otherwise normal, but agametic animals. Many subsequent experiments demonstrated that the same cleavage nuclei that when surrounded by polar cytoplasm give rise to pole cells, instead become somatic blastoderm nuclei if the polar region of the egg is destroyed, so that they are surrounded by other cytoplasm. Similar observations have been made with a large number of other insect species. For example, Brown and Kalthoff (1983) and von Brunn and Kalthoff (1983) demonstrated that UV irradiation of a sharply defined region of the polar cytoplasm in *Smittia* eggs that had not yet initiated nuclear division interferes with the migration of nuclei into the polar cytoplasm, prevents subsequent pole cell formation, and also affects that survival of those pole cells that do form. This case is of particular interest because the action spectrum for UV inhibition of pole cell formation closely approximates the UV absorption spectrum of RNA, except for a slightly higher relative efficiency at wavelengths > 260 nm.

In addition all the effects of UV irradiation on the pole cells are photoreversible, which is usually taken to suggest a nucleic acid target (Jackle and Kalthoff, 1980). Togashi and Okada (1983) reported an action spectrum for UV inhibition of *Drosophila* pole cell formation, with resultant adult sterility, that peaks at 280 nm and is also high at 254 nm. In both this and the *Smittia* study the sensitive target areas were shown to be confined to a thin polar region of the cortical cytoplasm. The UV rays could not have reached the zygote nucleus in the *Smittia* experiment, nor in the case of *Drosophila* any of the nuclei, which at the stage when the irradiation was performed are located deep in the interior of the egg. The sterilizing effect of UV irradiation offers on experimental opportunity to demonstrate that the polar egg cytoplasm is specifically required for the formation of functional primordial germ cells. It has been shown by many workers that injection of unfractionated posterior-pole cytoplasm, but not anterior pole cytoplasm, rescues eggs that would otherwise have been sterilized by UV irradiation. The injected eggs are able to produce pole cells, and frequently developed into adults that contain fertile gametes, in contrast to the agametic adults obtained from the irradiated controls.

It is interesting that the cytoplasmic substances required for rescue of pole cell formation can be physically separated from those necessary to restore adult fertility. Thus while unfractionated posterior pole cytoplasm contains both activities. Ueda and Okada (1982) found that a 28,000 ×g centrifugal pellet fraction promotes pole cell formation only. Experiments of Okada and Togashi (1985) indicate that the component required for pole cell formation is an mRNA that can be extracted from unfertilized eggs. However, though injection of UV irradiated eggs with this RNA fraction restores the formation of pole cells of apparently normal structure, fertile files are never obtained. Convincing evidence that *causal pole cell determinants* are localized in the posterior region of the *Drosophila* egg is provided by the cytoplasmic injection experiments. In these studies posterior pole cytoplasm was injected into the anterior or ventral regions of recipient eggs.

Pole cells displaying cytological structures unique to the germ cell lineage were thereby induced to form at the ectopic locations where the cytoplasm had been injected. Among the ultrastructural features used to identify pole cells are intranuclear organelles that consist of an electron dense cortex and in some species an electron lucid core, called *nuclear bodies*. Illmensee *et al.* (1976) also found that posterior pole cytoplasm from stage 13-14 ovarian oocytes induces the formation of ectopic pole cells when injected into the anterior ends of recipient eggs, but cytoplasm extracted from younger oocytes (stages 10-12) does not display this activity. The polar cytoplasm of these oocytes contains some morphologically identifiable polar granules, but evidently lacks some necessary constituents.

The special significance of this demonstration is that the localized determinants for pole cell formation are shown to be present prior to fertilization and ovulation. They are thus in the classical sense a *prelocalization*, of a morphogenetic substance that is synthesized and anchored to the polar region of the egg during ovarian oogenesis. An elegant feature of the cytoplasmic transfer experiments of illmensee *et al.* was the use of donor and recipient strains bearing different genetic marking. For example, the anterior pole cell induction experiment was carried out in flies of *mwh e* (*multiple wing hair*, *ebony*) genotype, and to test the ability of the ectopic pole cells that were formed to serve as germ line progenitors these were transplanted to the posterior regions of *y w sn* (*yellow, white, singed*) hosts. A very small percent of the resulting adults display germ line mosaicism, as expected if the induced pole cells gave rise to functional gametes.

A similar result was reported for pole cells induced to form ectopically in ventral regions of the embryo. This review briefly summarizes much evidence that the posterior polar cytoplasm can specifically induce *pole cell formation*, and that it contains substances *necessary* for germ cell formation. However, additional demonstrations of the ability of ectopically induced pole cells to give rise to functional gametes is clearly required to secure the proposal that the polar cytoplasm actually determines the *germ cell lineage*. The demonstration of Ueda and Okada (1982) that injection of certain cytoplasmic fractions into UV irradiated eggs restores pole cell formation but not adult fertility shows that this is not simply a semantic issue. A difficulty is that in *Drosophila* not all pole cells give rise to germ cells, though the ultimate fate of those that do not find their way to the larval gonad remains controversial. The germinal or polar granules of *Drosophila* eggs have been characterized in great details.

At oviposition these granules consist of dense, membrane-free particles surrounded by cluds of ribosomes (Mahowald, 1968). In various species of *Drosophila* these particles fuse or fragment during early embryogenesis, and the ribosomes and polyribosomes initially associated with them are no longer observed after the pole cell stage of embryogenesis. By the time the germ cells are located in the larval gonad, the polar granules have apparently given rise to characteristic fibrillar structures applied to the nuclear membrane. This progression of forms is similar to what is observed in both amphibian and mammalian oogonia. The fibrillar structures remain throughout the oogonial stage, but are absent in oocytes. Polar granules can again be observed in *Drosophila* oocytes during vitellogenesis. Though there are already localized in stage 10-12 oocytes, as noted above the polar cytoplasm of these oocytes is not yet competent to induce pole cell formation. This observation at least distinguishes the cytological entities recognized as ovarian polar granules from the functional properties of the mature posterior pole cytoplasm.

The *Drosophila* polar granules have been partially purified from pole cells, and there is evidence from two-dimensional analyses of their proteins for a single major species of about 95 kilodaltons. This protein is synthesized during oogenesis, prior to stage 10. The effects of several maternal mutations that result in sterility have recently been interpreted as consistent with a mechanistic role for the polar granules in germ cell determination. Among the examples of interest are a *grandchildless* (gs) mutation of *Drosophila subobscura* which in homozygotes prevents pole cell formation and thus results in sterility.

Ultrastructural examination shows that among other things this genetic lesion affects the deposition of polar granules during oogenesis. Some gs eggs lack any polar granules, while others possess small numbers of polar *granules* that are often incorrectly localized. Unfortunately, this is not the only cytological lesion in gs eggs, as both anterior and posterior poles appear abnormally organized, and during early cleavage, nuclei never descend into the posterior pole cytoplasm. A gs mutation of *D. melanogaster*, gs(1)N26, also prevents nuclear migration into the posterior cytoplasm but has not effect on polar granule disposition. On the other hand, various alleles of another *D. melanogaster* grandchildless mutant, *tudor*, provide a good correlation between the amount of polar granule material and the ability to produce pole cells. Again, however, this is not the unique effect of the mutation, some alleles of which also disturb embryonic segmentation. Perhaps this mutation interferes with a (cytoskeletal?) process by which several different determinants are localized during early development, and the abnormal polar granule deposition in mutant eggs could be merely a visible index of the disturbance, rather than the immediate cause of sterility. A number of other recessive maternal mutations of the *grandchildless* phenotype are also known, but most of these also produce somatic embryonic defects.

Germ Cell Lineage in Anuran Amphibians

In frogs the germ cell lineage originates early in cleavage from blastomeres forming at the vegetal pole of the egg. In normal development the vegetal pole cytoplasm or "germ plasm" is distributed to all of the first four blastomeres, and each has the capacity to generate a functional germ cell lineage. The primordial germ cells reside in the endoderm during early development, where they undergo several divisions, and after gastrulation they migrate to the germinal ridges where they are located by stage 25 in *Rana pipiens* tadpoles, or by stage 44-47 in *Xenopus laevis* tadpoles. At this point they are in a state of mitotic arrest. Although if undisturbed these cells subsequently give rise to the functional germ cells of the organism, they are not yet irreversibly determined to do so, even by this late stage.

Thus Wylie *et al.* (1985) showed that fluorescein-labeled primordial germ cells from stage 45 *Xenopus* tadpoles give rise to a variety of different ectodermal, endodermal, and mesodermal cell types if implanted into the blastocoel of a late blastula stage embryo. Many workers showed that UV irradiation of the vegetal region of the cytoplasm of frog eggs results in a sharp decrease in the number of primordial germ cells, and ultimately, in some animals, in agametic gonads and sterility. These classical experiments suggested that peripheral components of this cytoplasmic region, which is incorporated in the initial precursors of the germ cell lineage, might be required for specification of this lineage.

The significance of vegetal pole cytoplasm for germ line differentiation was confirmed in many subsequent experiments in which this region of the cytoplasm of uncleaved eggs or of 2- or 4-cell embryos was severely damaged by ultraviolet irradiation or by microsurgical means. The number of primordial germ cells observed in the early tadpole (about 35 in normal stage 47 *Xenopus* tadpoles has traditionally been used to assay the effect of exposing eggs and early embryos to UV irradiation. Thus the germinal ridges of stage 47 tadpoles derived from vegetaly irradiated *Xenopus* eggs, zygotes, or 2-cell embryos were often found to be devoid of germ cells. Irradiation at the 8-cell stage, however, has no specific effect on the number of tadpole primordial germ cells. Similarly results were obtained earlier by Smith (1966), *Rana pipiens*, and control experiments showed that parallel irradiation of the animal pole cytoplasm produces no visible defects, and certainly none in the germ cell line.

Furthermore, Smith (1966) demonstrated that the damage to cytoplasmic targets resulting from ultraviolet irradiation of *Rana* eggs can be compensated by injected into the vegetal pole of cytoplasm from the vegetal pole of an unirradiated egg. In a significant fraction of cases the recipient eggs developed into stage 25 tadpoles in which the germinal ridges bore primordial germ cells, while controls receiving no vegetal cytoplasm or cytoplasm from the animal pole lacked germ cells. Rescue of UV irradiated *Rana chensinensis* and *Xenopus* eggs can also be accomplished by injection of vegetal pole cytoplasm. A crude centrifugal fraction of vegetal but not animal pole cytoplasm from non-irradiated eggs was shown to be effective in restoring germ cells to the tadpole germinal ridge. In addition, injection of these fractions into non-irradiated eggs is reported to induce the formation of up to twice the normal number of primordial germ cells. It has recently been seen that primordial germ cells eventually reappear in the germinal ridges of tadpoles raised from some vegetally irradiated eggs, albeit on a much delayed schedule. Thus while the germinal ridges of stage 25 *Rana pipiens* tadpoles grown from UV irradiated eggs lack primordial germ cell, two weeks later these are present, and they ultimately give rise to differentiated gametes, Ikenishi and Kotani (1979) showed that in *Xenopus* embryos raised from vegetally irradiated eggs, fewer primordial germ cell divisions occur, and after the tailbud stage many of the primordial germ cells initially formed disappear.

Furthermore, the surviving germ cells remain in the central endoderm, at tailbud stage, while in control embryos they have already migrated to the lateral and dorsal regions. These observations occasion a somewhat different interpretation than classically proposed of the effect of cytoplasmic UV irradiation. Since surgical removal or disturbance of the vegetal pole cytoplasm does eventually result in sterility or partial sterility there may indeed be a cytoplasmic determinant for the anuran germ cell lineage localized at the vegetal pole of the egg. However, the UV *sensitive component* seems to control only the replicative or migratory behaviour of the primordial germ cells rather than their fundamental specification. We are reminded of the parallel observations in *Drosophila* where, as we have seen, the determination of the germ cell lineage is a process that has several components, and pole cell formation can be separated experimentally from the specification of a functional germ cell lineage. The role of polar granules in the anuran germ cell lineage specification process is also indistinct. These bodies, also known as the "germinal plasm", are initially found in the outer regions of the vegetal pole cytoplasm, and in ultrastructure they closely resemble the polar granules of insect eggs. They are present in unfertilized eggs, having been localized just beneath the vegetal pole cortex during maturation. They may be anchored to the egg cortex, as if the eggs are inverted, they do not redistribute gravitationally as do the yolk platelets.

During cleavage these organelles are incorporated in the primordial germ cell precursors, and they are still present, with associated ribosomes, in these cells at the blastula and gastrula stages. By the time the primordial germ cells are localized in the germinal ridges they no longer contain identifiable polar granules. Instead a fibrous component is found applied to the nuclear membrane. Though transition stages have not been convincingly described, these could derive from the earlier polar granules, an interpretation that is supported by the analogous progression of form undergone by insect polar granules.

In anuran amphibians the fibrillar perinuclear structures persist through most of oogenesis, though they are absent in mature oocytes. Typical polar granules reappear at about the time of germinal vesicle breakdown. Several correlative items of evidence, in addition to the natural history of these germ cell specific organelles, have been adduced in support of the idea that polar granules are involved in germ cell specification. Thus loss of sensitivity to UV irradiation by the 8-cell stage occurs just when the polar granules move inward from the cortex of the egg so that they no longer lie within the shallow penetration range of the ultraviolet irradiation. Furthermore, when *Xenopus* eggs are centrifuged so as to move the polar granules inward precociously, irradiation. Furthermore, when *Xenopus* eggs are centrifuged so as to move the polar granules inward precociously, irradiation at the 2-cell stage fails to affect the number of primordial germ cells in the germinal ridges of stage 47 tadpoles. This result is significant, since only large, dense particles would be likely to have been affected by the low centrifugation forces applied in this experiment, 150 g for 60 seconds.

On the other hand, there is on visible effect of UV irradiation on polar granule ultrastructure, and even if these particles were the essential UV targets, their function in germ cell differentiation could be confined to effects on replication and migration. Polar granules are not evident in the vegetal pole cytoplasm of cleavage stage urodele eggs, and germ cells in these eggs apparently arise from dispressed regions of the marginal zone of the animal ectodermal cap rather than from the vegetal endoderm, as in anurans. Though they may develop from specific cells of the marginal zone, the differentiation of primordial germ cells in urodeles probably requires inductive influences from adjacent caudal mesoderm. The earliest reported appearance of structures resembling polar granules in urodele eggs is at gastrula, though the definitive primordial germ cells of urodele larvae ultimately resemble those of anurans in their ultrastructural characteristics. Thus, the germ cell lineages in urodeles would appear to arise by a different route than in anurans, though in both groups route than in anurans, though in both groups the primordial germ cells ultimately possess the same cytoplasmic organelles. The least artificial view may again be that amphibian polar granules are cytoplasmic structures required later in germ cell differentiation, and that they appear together with whatever localized germ cell determinants may exist in the egg cytoplasm only when the germ cell lineage is set aside at the very beginning of cleavage, as in the anurans.

Cytoplasmic Localization in Molluscan and Annelid Eggs

A characteristic feature of early cleavage in many molluscan and annelid forms is the transient extrusion of a lobe of vegetal pole cytoplasm, which following cytokinesis is resorbed into one of the embryonic macromeres. In the typical case the first cleavage *polar lobe* is resorbed into the CD macromere, the second cleavage polar lobe into the D macromere, the third cleavage polar lobe into the ID macromere, and so forth. As will be recalled it is the D quadrant of molluscan and annelid eggs that gives rise to the mesentoblasts and is responsible for the dorsoventral organization of the embryo.

The polar lobe is often attached to the remainder of the embryo by only a thin strand of cytoplasm. Its removal is easily accomplished without immediate injury to the embryo, which continues to divide approximately on schedule. However, the ultimate developmental consequences of the removal of this anucleate element of egg cytoplasm are both dramatic and specific. In this section we review the experimental effort that has been made over the last 90 years to establish a causal relation between the polar lobe cytoplasm and the establishment of the dorsoventral axis of the embryo, and to understand the nature of the localized determinants that it apparently contains.

(i) Morphogenetic Significance of the Polar Lobe Cytoplasm

Crampton found that deletion of the first cleavage polar lobe of *Ilyanassa* eggs causes the asymmetric cleavage pattern and the delayed division schedule that normally mark the D quadrant to disappear. Cleavage of embryos from which the polar lobe had been removed is radially rather than bilaterally symmetric, and Crampton was unable to recognize in these embryos the 4d stem cell from which the mesentoblasts derive. Wilson (1904b) followed up these preliminary results with an extraordinary study of the effect of polar lobe removal on embryogenesis in the scaphopod mollusc *Dentalium*. These experiments demonstrated that deletion of the first cleavage polar lobe irreversibly blocks the appearance of all the special D quadrant cell lineages. He reported that the mesentoblasts and the mesodermal germ bands never form, and that other larval structures are missing after lobe removal, such as the apical tuft, and the larval shell and foot.

Wilson (1904b) compared the capacities of isolated blastomeres with those manifested by lobeless embryos, and showed that the latter develop in exactly the same way (or fail to develop in exactly same way) as do embryos deriving from AB blastomeres or from single A, B and C blastomeres, none of which produce the mesodermal primordium. From these observations it was possible to infer the existence in the polar lobe of cytoplasmic determinants for the D quadrant cell lineages that build the dorsoventral structures of the embryo. Polar lobe removal is now known to result in the same characteristic complex of developmental defects in a number of molluscan and annelid species.

Among the organisms in which this phenomenon has been studied are the gastropod molluscs *Bithynia* and *Ilyanassa* the *lamellibranch* mollusc *Mytilus* the annelid *Sabellaria* as well as *Dentalium*, the subject of recent investigations Verdonk (1968). Many workers that have confirmed and extended the 1904 conclusions of Wilson. This is not an invariable phenomenon in eggs bearing polar lobes, however. Thus removal of the small polar lobe of the polychaete annelid *Chaetopterus* has only minor developmental effects, and instead vegetal region cytoplasm included in the polar lobe is functionally analogous to the polar lobe cytoplasm in the other species mentioned. Normal development is shown in which Fig. in the D quadrant cells cannot be distinguished.

The structures of the advanced larvae descendant from lobeless embryos are compared to those of normal *Ilyanassa* larvae. Lobeless larvae fall to develop heart, intestine, statocyst, operculum, velum, external shell, eyes and foot. On the other hand, lobeless larvae possess active muscle, nerve ganglia and nerve endings, stomach, some velar tissue with cilia, digestive gland, mantle gland tissue, and pigment cells. Removal of the first cleavage polar lobe cytoplasm thus does not block all cell differentiation, only certain differentiation, though ultimately it produces a cascade of delayed organizational defects. The catastrophic though specific effects of first polar lobe removal illustrated in these figures suggests that diverse morphogenetic determinants are located in the vegetal pole cytoplasm of the CD blastomere by the 2-cell stage. The most straightforward presumption would be that during later cleavage these determinants are physically transferred to the stem cells from which lobe dependent lineages arise, e.g., 4d. The latter cell is the direct ancestor of both the primary mesentoblasts (Me^1 and Me^2) and the primary entoblasts (E^1 and E^2). Thus only embryos in which 4d is present produce heart and intestine at the veliger larval stage. In *Ilyanassa* deletion of 1d produces no observable effect, but deletion of 2d causes absence or impairment of the shell, while 3d forms the primordium for the left half of the larval foot (Clement, 1976). The inference that morphogenetic determinants for these tissues are transferred directly into the d quadrant micromeres during cleavage was tested in *Ilyanassa* and in *Dentalium* by determining the effect of ablation of the D macromere at successive cleavages.

Deletion of the D macromere after the 1d micromere has been given off tests whether the morphogenetic factors of the polar lobe still reside in the macromere or have been shunted into the micromere, and similarly for D macromere deletion after 2d, 3d, or 4d have a arisen. Clement (1962) found that in *Ilyanassa* removal of the 4D macromere after the formation of 4d has no effect on later differentiation, and the resulting embryo is normal except for its small size. This experiment also helps to eliminate the possibilities that the effects of polar lobe removal are due to general injuries, starvation for nutrients, or to disruption of an embryo-wide "morphogenetic gradient." On the other hand, removal of the D macromere before the 2d micromere has been formed results in as severe an inhibition of morphogenesis as removal of the first cleavage polar lobe or of the whole D quadrant at the 4-cell stage. If the 3D macromere is deleted, i.e., after 1d, 2d and 3d have been formed, the resulting larvae display velum, eyes, foot, and some shell, all of which are polar lobe dependent structures, but they still lack the mesodermal primordium.

The morphogenetically significant polar lobe contents therefore appear to be shunted into 3d and 4d micromeres. Essentially similar results are reported for *Dentalium*. Here again removal of the D macromere when only 1d has been produced is equivalent in its effects to deletion of the polar lobe, while consequences of decreasing severity occur if the macromere is ablated after 2d, or both 2d and 3d micromeres have appeared. *Dentalium vulgare* embryos from which 4d is removed also develop into normal larvae.

The general conclusion from both sets of experiments is that the morphogenetic elements localized in the polar lobe are progressively distributed during cleavage to the specific micromeres where they will function. In annelid and molluscan larvae this organ consists of a plate of differentiated cells, several of which bear long cilia, overlying a plexus of neurosecretory cells that is later associated with the cerebral commissure. *Dentalium* embryos from which the first polar lobe has been removed form no apica tuft, but differentiation of the apical organ proceeds normally in embryos from which the second polar lobe has been deleted.

The apical organ forms from progeny of the 1c and 1d micromeres (1974). Microsurgical experiments indicate that the apical organ determinants are localized in the upper region of the first polar lobe and they are later shunted into the upper regions of the C and D blastomeres. Thus, the whole lower end of the lobe and 50-60% of the volume of cytoplasm that it contains can be ablated without loss of apical tuft formation. Determination of the cell lineage that produces these tuft cilia, and probably of the neuronal components of the apical organ as well, occurs after resorption of the lobe into the CD macromere, evidently as a consequence of the inclusion of elements of the upper lobe cortex or cytoplasm in the 1c and 1d micromeres. In *Ilyanassa* the polar lobe cytoplasm *represses* the ability of micromere derivatives other than those normally destined for the apical plate to produce cilia, and this aspect of apical organ differentiation has also been analyzed in the polychaete annelid *Sabellaria*. In the later organism the first cleavage polar lobe bears the determinants for the apical tuft and for the posttrochal chaetae.

As in *Dentalium*, *Sabellaria* embryos from which the second cleavage polar lobe has been removed produce apical tufts. Render (1883) found that any combination of isolated macromeres that includes the C macromere. i.e., ABC, BC, or C alone, will form an apical tuft. Cultured D macromeres never give rise to an apical tuft, but, if the *second* cleavage polar lobe is removed and the C and D macromeres are then cultured, *both* are now able to produce apical tufts. It follows that the second cleavage polar lobe contains a repressor of apical tuft formation, while the first contains only the determinants that induce apical organ differentiation. The elements that induce apical tuft formation are left in the CD macromere after resorption of the first polar lobe, whence they are distributed to both the C and D macromeres, and eventually to the apical precursor micromeres.

The elements that repress apical tuft formation are included in the smaller second polar lobe, and thus are confined to the D macromere after resorption, so that only the C quadrant micromeres are able to produce apical tufts. In accord with this interpretation Render (1983) showed that suppression of second polar lobe formation by treatment with a dilute SDS solution prevents apical tuft formation in both C and D isolates, since the repressive cytoplasmic factors are now included in both macromeres. For formation of posttrochal chaetae, on the other hand, the D cell must be included in the isolate. Determinants for the posttrochal chaetae also remain in the second polar lobe, and are returned to the D macromere alone.

Determination of the Dorsoventral Axis in Equally and Unequally Cleaving Molluscan Eggs

In molluscan eggs they do not produce polar lobes and in which the first four blastomeres are of equal size the D quadrant is the source of the mesentoblast and entoblast cell lineaces, just as in unequally cleaving, polar lobe forming species. Comprehensive analysis of the process of dorsoventral axial fixation has proved illuminating, particularly since both developmental forms are to be found within a single taxonomic class, the gastropods. One possibility is that the dorsoventral axis of the embryo is actually predetermined in eggs of both types, and that polar lobe formation is merely a prominent manifestation of this preformation, or an element of the mechanism by which it is translated into a three dimensional embryo. Cytochalasin was applied after the beginning of first cleavage to suppress polar lobe formation, and then washed out.

Two equal sized blastomeres are formed by the treated eggs, both of which proceed to elaborate a second polar lobe. After second cleavage the embryo can be said to consist of two C and two D macromeres, where D is the macromere into which the lobe is subsequently resorbed, and C is its sister macromere. There are no A or B type macromeres. Development of these embryos results in a duplication of the normal D quadrant cleavage pattern. They give rise to larvae containing double the normal number of polar lobe dependent structures, e.g., two dorsal shell glands, four statocysts, and four mantle cavities, but lack any A or B quadrant derivatives such as the stomodaeum. This experiment shows that the *disposition of the polar lobe cytoplasm is sufficient to determine the dorsoventral axis*, since two such axes form if this cytoplasm is distributed to both macromeres at first cleavage. Choice of the blastomere into which the first cleavage polar lobe flows in the normal egg may be a matter a chance, i.e., the exact position of the cleavage plane with reference to the thin cytoplasmic stalk by which the lobe is connected at the trefoil stage.

The future dorsal end of the egg is in any case clearly *not irreversibly predetermined*. On the other hand, the *vegetal* location of the first polar lobe is a genuine preformation, since the annular constriction by which it arises always occurs in a plane perpendicular to the animal-vegetal axis. Cytological evidence reviewed below shows that the animal-vegetal axis is already clearly evident prior to fertilization. In equally cleaving gastropod eggs the macromeres remain from sometime both morphologically and morphogenetically equivalent. It has been shown for three different species that either of the first two blastomeres can give rise to complete larvae a result that contrasts directly with the restricted potential of the AB blastomere in unequally cleaving molluscan eggs. In equally cleaving gastropod eggs the macromere that gives rise to the primary mesentoblast, 4d, cannot be distinguished until after 5th cleavage. This macromere, 3d, is then identified by the central position that it assumes in the embryo, its non-synchronous cleavage, and the asymmetric disposition of its progeny.

Deletion experiments have shown that prior to 5th cleavage all of the four macromeres retain the capacity to serve as the "D macromere," and to produce the primary mesentoblasts. In normal embryos the 3D macromere is usually, but not necessarily. One of the two "vegetal cross furrow" macromeres. The event that specifies which will become the D macromere, and hence that determines the dorsoventral axis, is apparently an inductive interaction with the first quartet micromeres across the blastocoelic cavity, quadrant function is assumed by that macromere first achieving the greatest number of contacts with micromeres during the interval between 4th and 5th cleavage. This has been established for *Lymnaea stagnalis* and *Patella vulgata* embryos by deletion of first quartet micromeres.

Removal of a sufficient number of these micromeres precludes the differentiation of any D quadrant macromere, and in partial deletions the macromere that assumes the role of 3D can be determined by the position of the remaining micromeres. If the intercellular interaction between macromeres and micromeres is prevented by cytochalasin treatment, the micromere cap remains completely or partially radially symmetric. These experiments show furthermore that in *Lymnaea palustris* all four quadrants of the micromere cap remain undetermined and capable of forming dorsal, lateral, or ventral head structures until after the crucial cell contacts occurs, i.e., during the pause in cleavage following the 24-cell stage. It follows from these and the other results reviewed that the dorsoventral axis of neither the micromere nor macromere blastomere tiers is primordially established, both arising inductively by means of the interactions between them.

A variety of other cell contacts form elsewhere in the embryo during the same interval in which these micromere-macromere interactions occur. For example, in *Patella* embryos functional intercellular coupling is also established at this point, as indicated by the diffusion of iontophoretically injected lucifer yellow dye from blastomere to blastomere. However, contacts of this kind appear not to be required dorsoventral determination, since they are absent from the 3D-micromere junctions, and conversely, they still occur in embryos from which the first quartet micromeres have been deleted. Establishment of the dorsoventral axis in equally cleaving gastropod eggs is thus accomplished by an *intercellular* interaction, while the same end is achieved in unequally cleaving gastropod eggs through an *intracellular* mechanism. The contrast is nicely illustrated by the effect of first quartet micromere deletion in embryos that form polar lobes. Though severe defects in larval head formation result, the whole first quartet can be removed from *Bithynia tentaculata* or *Ilyanassa* embryos. Without affecting determination of the dorsoventral axis and of the specific cell lineages responsible for its construction.

The role of the polar lobe is illuminated by the comparison. It is the device by which determinative cytoplasmic agents or structures initially resident in the egg cytoplasm are spatially disposed to the appropriate blastomeres. This can be seen in detail in the examples of the apical tuft determinants, and the determinants for mesentoblast formation, reviewed above. Whereas both unequally and equally cleaving gastropod embryos rely extensively on inductive interactions for organogenesis, the equally cleaving embryos use this mechanism for dorsoventral axial determination as well while in polar lobe forming eggs axial determination involves the inbuilt location of the constructions by which the polar lobes are formed, and the relative location and timing of the successive cleavage planes. It is interesting that in both mechanisms chance appears to play a decisive role, in equally cleaving eggs in determining which macromere first achieves sufficient micromere contacts, and in unequally cleaving eggs in determining which macromere resorbs the first cleavage polar lobe cytoplasm.

Molecular Constituents of the Polar Lobe Cytoplasm?

An obvious hypothesis that at least in principle has the advantage of testability is that in the polar lobe are sequestered a particular set of regulatory maternal mRNAs. These could code for proteins that directly or indirectly elicit the differential patterns of gene activity required of the cell lineages that are determined through the action of polar lobe cytoplasm. Such lineages sooner or later express different sets of genes than do other embryonic cell lineages, and the basic function of polar lobe constituents must be to set in train events that eventually impose the appropriate states of genomic activity in the nuclei exposed to them. As might be expected, the absence of certain differentiated tissues in advanced larvae grown from lobeless eggs has gross molecular consequences. About the same amount of ribosomal RNA is present in lobeless and normal *Ilyanassa* embryos at three days of development, but after this the lobeless embryo accumulates significantly less total RNA. Similarly DNA content increases faster in normal embryos than in lobeless embryos after four days.

Several alkaline phosphatase and esterase isozyme forms that normally appear in advanced larvae are also lacking in larvae derived from lobeless embryos. The presence of maternal mRNA in polar lobes was established initially by Clement and Tyler (1967), who showed that *Ilyanassa* polar lobes continue to synthesize protein for at least 24 hr after isolation at the trefoil stage. The spectrum of proteins synthesized in the isolated polar lobes was analyzed by 2-dimensional gel electophoresis. Out of the several hundred species visualized there are no newly synthesized polar lobe proteins that are not also synthesized after equivalent incubations in lobeless or normal cleavage stage embryos. Though this result provides no support for the proposal that certain species of maternal mRNA are sequestered exclusively in the polar lobe, nor is it excluded. Proteins whose pI's fall outside of the range of the isoelectric focussing gradient, i.e., any whose pI is >8 or <5, and proteins coded by rare messages, would not be included in these analyzes.

The latter class of course includes the large majority of the diverse species of maternal mRNA present, if measurements carried out on other embryonic material can be considered a guide. An interesting observation is that several proteins are synthesized in isolated polar lobes after 25 hr of incubation *in vitro* that are not synthesized at 4 hr, and that these same changes occur in normal and in lobeless embryos, and after actinomycin treatment of these embryos. This set of molecular changes thus originates at the translational (or post-translational) level, and the polar lobe evidently includes the cytoplasmic elements necessary for their occurrence. The prevalence of a few other newly synthesized protein species increases sharply during the development of normal *Ilyanassa* embryos, but not in isolated polar lobes. Their appearance could be the result of embryonic transcriptional activity, as also implied by the effect of actinomycin, which alters the pattern of protein synthesis during the first 24 hr. However, the same changes occur in lobeless embryos as in normal embryos and thus there is no evidence that the regulation of these protein species is dependent on polar lobe constituents.

The sole observation that assigns to the polar lobe any direct effect on embryonic transcriptional activity is a quantitative one. In mesentoblast or gastrula stages the absolute rate of total RNA synthesis in lobeless embryos is only about 55-60% of that in normal embryos. The difference may be significant, since cell division in lobesame number of nuclei throughout the period studied. During these stages the newly synthesized RNA measured by labeling is mainly of heterogeneous, non-ribosomal nature, and is probably a combination of nRNA and mRNA.

Cytological Observations

Numerous ultrastructural studies have been carried out on the polar lobe cytoplasm. Unusual membrane-bound organelles are present in the cytoplasm of the large polar lobes of *Ilyanassa* and *Dentalium*, as well as mitochondria and other undefined particulate inclusions. However, none of these structures are required for polar lobe function, and they cannot contain the morphogenetic determinants of the lobe, since they can be redistributed to other parts of the egg by low speed centrifugation without disturbing normal development. On the other hand, a special organelle termed the vegetal body that by the same test is functionally important has been demonstrated in the small polar lobe of the gastropod *Bithynia*, and similar structures are present in other species that produce small polar lobes.

The vegetal body consists of a dense aggregation of small vesicles. Though the *Bithynia* polar lobe occupies <1% of the egg volume, it morphogenetic determinants are functionally similar to those of the *Dentalium* and *Ilyanassa* polar lobes. Thus, lobeless *Bithynia* embryos fail to form mesentoblasts and lack eyes, foot, intestine, organized shell, operculum, etc., though they successfully develop digestive gland, ganglia, and some muscle. In *Bithynia*, however, the developmental capacity of the C macromere is the same as that of the D macromere, and lobe-dependent structures form if either the C or D macromeres are present. The progressive distribution of the vegetal body vesicles parallels that of the lobe morphogenetic determinants. The vegetal body is shunted into the CD cell at first cleavage, whereupon the vesicles disperse, and are inherited by both the C and D macromeres. The staining reactions of the vegetal body suggest that its vesicles contain RNA but this remains to be established by biochemical measurement.

Centrifugation at speeds of about 1400 $\times$ g dislocates the vegetal body away from the polar region of the uncleaved egg in about 50% of cases, and these eggs form polar lobes that lack the vegetal body. Van Dam *et al.,* (1982) found that about 50% of centrifuged eggs from which the first polar lobes had been removed developed into *normal embryos*, a result that is never observed after deletion of the lobe from uncentrifuged eggs. Inclusion of the vegetal body in one of the two blastomeres, in the absence of the polar lobe structure, thus suffices for normal development, and it follows that the vegetal body is the site of the lobe determinants. Seen in this light the function of the polar lobe appears essentially to be the distribution of the special package of vesicles to one or the other of the first two blastomeres. In eggs with large polar lobes organelles similar to the vegetal body are not present, and as we have seen centrifugal forces that totally redistribute the particulate cytoplasmic elements fail to dislodge the morphogenetic determinants. Thus it is likely that these determinants are anchored in the cortical structures of large polar lobes, and that they exist initially as regional differentiations of the lobe plasma membrane and cortex. For example, the apical tuft determinants in the first cleavage *Dentalium* polar lobe are apparently fixed in the upper region of the lobe.

The extrusion and resorption of the lobe can be considered as mechanisms for the spatial disposition of the cytoplasmic determinants to the appropriate blastomeres. These movements are accompanied by a complex of cytoskeletal changes. Scanning electron micrographs have provided dramatic instances of regional cortical differentiations at the polar lobe. The vegetal pole is seen to be clearly marked by distinct surface features prior to the onset of cleavage. Furthermore, at least in the egg of *Crepidula* the plane of first cleavage is also predetermined, since the longitudinal orientations of the polar microvillar patch bears a fixed relation to the future plane of cleavage, which is retained during the subsequent cleavages. The initial specification of the vegetal pole in polar lobe bearing eggs may occur far back in oogenesis. Animal-vegetal differences in distribution of microvilli and other cytological features can be observed in the uncleaved egg of the gastropod *Nassarius reticulata*, and could derive originally form the polarized morphology of the oocyte-follicle cell complex. Thus during the growth phases of oogenesis the area of contact is confined to the base of the oocyte, and this becomes the future vegetal pole. An interesting inference is that the animal-vegetal polarity of the egg is imprinted by intercellular surface interactions occurring in the ovary.

Prelocalization of vegetal pole structures extends as well to internal structures such as the vegetal body, which is already localized at the future vegetal pole in growing *Bithynia* oocytes.

Maternal components are also responsible for the asymmetric "spiral" cleavage of molluscan eggs, as shown by genetic studies on the maternal inheritance of left and right handed body forms in the gastropod *Lymnaea*. The orientation of the adult body form in gastropods in determined by the direction in which the micromeres are given off during early cleavage (see the discussion of spiral cleavage in Wilson, Freeman and Lundelius (1982) showed that cytoplasm from right handed or dextral eggs injected into genetically sinistral eggs can transfer the property of dextrality. The direction of cleavage is thus not determined by the orientation of the ovarian oocyte with respect to the follicle cells.

The cytoplasmic elements that confer this property are synthesized during oogenesis, and provide an excellent additional example of the performation of components specifying spatial organization in the molluscan embryo. The studies of polar lobe function reviewed in this section provide an observer of early development with a comprehensive and detailed description of the biological aspects of a cytoplasmic localization process. It seems inescapable that a characteristic set dorsoventral determinants is carried in the polar lobe cytoplasm, and that these are responsible for specification of many of the early cell lineages of the embryo. The spatial distribution of these determinants is an epigenetic, developmental process mediated by the disposition of the polar lobe constituents. However, the structures of substances that later exert morphogenetic effects evidently include or consist largely of preformed maternal components. Their origins in oogenesis, their nature and their molecular mode of action, which directly or indirectly ultimately affects the embryo genomes, remain unknown.

The Development of Localization

The distinction was drawn earlier between spatial assignments in the egg that are prelocalizations, and those that are constructed ontogenically following fertilization. Of course, as Driesch and Wilson pointed out, prelocalized features of the mature egg are the product of prior ontogenic processes that have taken place during oogenesis. The animal-vegetal axes of ascidian and some molluscan eggs, for example, are established during oogenesis, while the anterior-posterior or dorsoventral axes that provide the bilateral organization of the embryo are established by the developmental redistribution of maternal cytoplasmic elements. While common in marine eggs, this scheme of development is by no means universal.

In some animals both axes are prelocalized in the unfertilized egg and in others neither. The examples considered in the following provide some insight into the possible nature of both prelocalized and ontogenically localized determinants, and the processes by which they may be deposited in given regions of the egg.

Prelocalization in the Cortex of Dipteran Eggs

By the cellular blastoderm stage the cortical region of the *Drosophila* egg is already is mosaic of determined cell types. Thus when allowed to develop in intracoelomic culture, cells from anterior and posterior regions of the blastoderm give rise only to the expected anterior or posterior adult structures. In addition, Schubiger (1976) demonstrated that partial embryos formed by ligation at the blastoderm stage can produce all the anterior, thoracic, or caudal adult parts that would normally derive from them. Blastoderm fate maps were based initially on histological reconstructions of the development of organ anlagen and then on analyses of genetic mosaics induced during the first few divisions of the zygote nuclei, in which the developmental disposition of clones of cells bearing visible morphological markers can be followed. By the latter method have been determined the approximate location and number of the primordial cells for adult cuticular structures, as well as for some larval organs. A more direct approach to mapping the morphogenetic values of the blastoderm cells is microblation, by pricking the egg, local cautery or UV irradiation. An UV laser microbeam was utilized by Lohs-Schardin *et al.* to map the primordia for the thoracic and abdominal epidermis of the first instar larva and for adult structures. Cephalic larval structures were mapped by microsurgical blastoderm cell deletions by Underwood *et al.* (1980).

In the experiments of Lohs-Schardin *et al.* (1979a) the UV microbeam was 20 µm in diameter, and affected a patch of about 15 cells (the embryo is about 500 µm × 175 µm and at the stage of irradiation in these experiments it contains about 5000 cortical cells). Specific defects in the larval epidermis resulted from the localized irradiation, and these experiments demonstrate that about one-third of the blastoderm cells comprise the progenitors of the entire larval epidermis. Extensive fate map data have also been obtained by injection of horseradish peroxidase into individual blastoderm cells and other cytological observations. The overall result of these studies is to confirm that long before the appearance visible manifestations of differentiation (except for the precociously appearing pole cells) the blastoderm has acquired a complex pattern of specification in which are encoded both embryonic axes as well as may specific regional assignments. The following evidence shows that in contrast to any of the eggs so far discussed, both embryonic axes are prelocalized in the unfertilized dipteran egg cytoplasm, just as are the determinants for pole cell formation.

Origin of the Axial Determinants in Drosophila

The bilateral rather than radial symmetry of the unfertilized dipteran egg is evident from its external morphology. Thus the future dorsal side is less convex; the future anterior pole of the embryo is marked by the micropyle. The opening in the chorion through which the sperm enters: the two chorionic filaments are located bilaterally at the future dorsal posterior region; and the germinal cytoplasm is already localized at the future posterior pole. A number of maternal mutations are known that disturb primary axial determination in the embryo. There are so far identified 10 different maternally functioning genes, mutations in which alter the prospective fates of ventral and lateral cells of the blastoderm to that of the prospective dorsal epidermis. A well-known mutation of this type is *dorsal* (dl). There are two *dorsal* phenotypes.

Embryos from homozygous *dl/dl* females express the "recessive phenotype," which is characterized by a lack of ventral, mesodermal, and neural structures. Though at the blastoderm stage the embryo appeals normal, invagination of the presumptive mesoderm fails, and the embryo becomes a yolk-filled epidermal tube that lacks any internal organs. A "dominant phenotype" is temperature sensitive. The eggs of *dl/+* females develop normally at 22°C but arrest at 29°C due to the absence of muscles and portions of the ventral epidermis, tissues arising normally from the ventral region of the blastoderm. Nusselein-Volhard *et al.* (1980) concluded that ventral cells of these embryos carry out assignments normally relegated to the more lateral cells, in particular the formation of epidermis instead of mesodermal primordia. These and other observations show that embryonic specification of ventral cell types requires the maternal products of the wild-type *dorsal* gene, and also of the other maternally acting genes that when mutated produce *dorsal* phenotypes.

The *dorsal* gene has been cloned and found to produce a single 2.8 kb poly(A) RNA species, which is present in eggs up to the blastoderm stage, after which it disappears. Studies with germ line mosaics demonstrate that the activity of the *dorsal* gene is required only in germ line cells and *in situ* hybridization using a *dl* probe shows that the gene is active in the nurse cell nuclei at stages 5-11 of oogenesis. After stage 11 dl^+ RNAs are transferred to the oocyte, as are the products of many other genes. Of the ten maternally acting loci known to be required for specification of the normal dorsoventral egg axis, the effects of eight, *viz.*, *dorsal*, *tube pipe*, *snake*, *easter*, *Toll ree*, *spatzle*, and *pelle*, have been shown to be at least partially rescued be injection of wild-type egg cytoplasm and all except *dorsal* can also be rescued to some extent by injection of poly(A) RNA extracted from very early wild-type embryos.

The rescuing activity for *pelle* mutants disappears from the poly(A) RNA fraction at an earlier stage of cleavage then from the cytoplasm, suggesting that translation of the maternal mRNA is by this stage complete while for certain other of these mutants, the rescuing activity is instead preferentially located in the poly(A) RNA fraction of the cytoplasm. Thus translation of *dorsal* mRNA may occur particularly early in development (and/or in oogenesis). The *dorsal* cytoplasmic component required for rescue is by the blastoderm stage spatially localized, since cytoplasm removed from the ventral regions of donor embryos is the most effective in rescue experiments. The mutant effects of the *toll* locus that can be rescued by RNA injection are those caused by recessive mutations, which produce phenotypes similar to those of the other dorsalizing maternal loci.

Dominant *Toll* mutations (*Toll*) have the opposite effects, resulting in the appearance of patches of ventral denticles over the entire dorsoventral circumference, and the absence rather than overextension of dorsal structures (Anderson *et al.*, 1985a). Injection of wild-type cytoplasm into *Toll* mutants restores their ability to produce dorsal structures, and the site of injection defines the ventral region of the structures formed. Thus, whereas lack of *Toll* function results in *dorsalization*, as does lack of function of the other nine loci (all recessive), these observations are consistent with the observation that the dominant $Toll^D$ gene results in *ventralization*, due to production of excess *Toll* product. Double mutants combining $Toll^D$ and recessive mutant alleles of *gastrulation*, *defective*, *snake*, *easter*, etc., lack *both* ventral and dorsal pattern elements but develop lateral structures, while the *Toll*^{D} and *dorsal* combination is totally dorsalized. In any case the demonstration that dorsoventral polarity can be restored to mutant embryos by injection of wild-type egg cytoplasm of poly(A) RNA shows explicity that *maternal transcripts synthesized in the meroistic egg chamber contain genetic information required for normal axial specification in the embryo.* A large number of maternally acting genes are also required specifically for establishment of the anterior-posterior morphogenetic organization of the *Drosophila* egg.

Expression of mutant alleles of seven such genes in cells of the female germ line carried in mosaic mothers has been shown in each case to suffice for complete production of the mutant embryonic pheno. The developmental abnormalities caused by these seven recessive maternal lethals fall into three morphological classes. Mutations in *tyorso* and *trunk* result in failure to develop structures of the extreme anterior and posterior ends of the embryo, and in addition they cause development of structures normally arising more towards the midregion to occur in the vicinity of the anterior and posterior poles. Thus, for example, *torso* embryos typically terminate at the posterior end with cuticular patterns characteristic of abdominal segment A7. Yet, these mutation do not prevent cellularization of the blastoderm, and their effects apparently result from alterations of blastoderm cell assignments. These effects are large scale, disturbing the global fate map of the egg. Thus maternal mutations at *valois*, *vasa*, *staufen* and *tudor* loci that prevent the formation of polar granules and pole cells also cause deletions of abdominal segments, and in the case of *staufen*, head structures; and mutations at the *exuperantia* locus cause replacement of head structures by an inverted posterior end.

It is not clear, however, whether loci such as these are *directly* responsible for the spatial patterns in which maternal morphogens might be distributed. Thus, for example there is some evidence that another gene that when mutated may produce a torso-like phenotype, *viz*, *lethal (l) pole hole,* may be involved primarily in some common cellular function such as proliferation or cell formation and similarly, many *valois* eggs display severe defects in cellularization in the blastoderm. Among maternally acting mutations that affect establishment of the anterior-posterior axis are genetic defects that interfere with determination of *either* anterior or posterior embryonic structures which implies (but does not in itself prove) the existence of separate maternal factors necessary for cephalic and for caudal cell specification. A well-known example is provided by *bicaudal* (bic), a recessive maternal point mutation. Two other loci, which are semidominant, *BicC* and *BicD,* produce similar defects.

Morphological defects produced by mutations at any of these three loci consist either of deletions of various amounts of anterior structure, with otherwise normal retention of embryo polarity, or of replacement of the deleted anterior structures with duplicated posterior structures in reverse orientation. Possibly this mirror image symmetry is the result of a secondary reorganization along the anterior-posterior axis induced in embryos suffering the most extensive deletions of anterior segments. Determination of pole cells is not affected in *bicaudal* embryos, and these remain situated normally at one end. The *bicaudal* mutations affect both the specification of particular cell lineages, for example arising malpighian tubules to form from cells that in normal embryos would give rise to cephalic structures, and the general pattern, for example the number of posterior segments in the embryo.

The variation in the phenotypes of double abdomen embryos is interesting, in that the most anterior structures formed, suggesting that the *bicaudal* mutations affect the distribution of a morphogen that extends over large regions of the egg, rather than being sharply localized to one area of the cytoplasmic cortex, for example. The temperature sensitivity of the *BicD* lesions have been used to demonstrate that, as expected, function of the gene is required only during oogenesis and it is interesting that oogenesis in females homozygous for *BicC* mutations is blocked at stage 8. It seems likely that the distribution of products of maternally acting genes that effect anterior-posterior embryonic axis specification would reflect the polarity of the egg. We have already noted evidence for dorsoventral localization of maternal products of genes affecting that axis.

Mlodzik *et al.* (1985) isolated a gene by its sequence homology with a *Ubx* homeobox and though the morphological significance of this gene, named *caudal* is unknown, the distribution of its maternal transcripts provides a possible example. These transcripts are produced in the nurse cells during oogenesis, and at fertilization are present in the egg. *In situ* hybridization shows *caudal* is unknown, the distribution of its maternal transcripts provides a possible example. These transcripts are produce in the nurse cell during oogenesis, and at fertilization are present in the egg. *In situ* hybridization shows *caudal* RNAs to be more or less evenly distributed during cleavage but at the syncytial blastoderm stage when they are located in the cortex, they have largely disappeared from the anterior end of the egg up to about 20% of total egg length.

The concentration of these transcripts increases monotonically for the next 50% of egg length, and they are present at their maximum level for the posterior 30% of egg length. At cellular blastoderm stage their distribution sharpens, possibly by means of zygotic transcription, and they accumulate in a single sharp band not far from the posterior pole, where are located precursors of the proctodaeum and regions of abdominal segments 9 and 10. Such observations focus attention on the *origins of the primordial polarity of the egg,* which provides an initially oriented matrix for the establishment of differential concentrations of maternal gene products. An interesting insight into the processes by which polarity might initially be imposed derives from morphological observations on another maternal mutation, *decephalic*, which produces an abnormality converse to that arising from *bicaudal* mutations. As the name implies, *dicephalic* embryos of extreme phenotype develop head furrows at both poles, and their cuticular structures suggest a mirror image duplication of the cephalic segments, while differentiation of posterior structures is suppressed. In contrast to bicaudal and *dorsal*, the *dicephalic* mutation visibly affects the construction of the nurse cell-oocyte complex.

In *dicephalic* ovarioles the 15 nurse cells form two clusters positioned at opposite ends of the oocyte, in place of the normal asymmetric arrangement in which the nurse cells are all at the future anterior pole. The chorion of *dicephalic* eggs contains two micropyles, one at each end. The implication is that the anterior determinants are localized near the sites of the nurse cell oocyte junctions or perhaps are loaded into the oocyte may be translated into the spatial organization of the egg, and the preformation of the anterior-posterior axis may descend from intercellular interaction that took place during oogenesis. We have encountered a mechanism of this nature earlier, *viz,* the determination of the vegetal pole of the *Nassarius* oocyte at the unique site of its contact with follicle cells.

Characterization of the Prelocalized Anterior Determinants in Smittia

In the unfertilized egg the UV sensitive elements of the anterior determinants are symmetrically distributed within a polar cone of cytoplasm about 20 µm in diameter. Since little of the incident UV energy penetrates beyond 5-10 µm of the surface, *these elements must lie in or near the cortex of the egg.* After 5-7 hr, when in *Smittia* the nuclei begin their migration toward the egg surface the UV sensitive targets disperse over the entire anterior half of the egg. Double abdomen embryos can then be produced by a low flux of widespread anterior irradiation, a treatment that is not effective earlier. The action spectrum of double abdomen induction by UV light displays a major peak at 285 nm, and a minor peak at about 257 nm. An interpretation consistent with this result is that the anterior determinants include a ribonucleoprotein target. Additional evidence that also suggests a functional RNA component includes the observation of Kandler-Singer and Kalthoff (1976) that anterior determination can be blocked by RNase admitted through a small puncture made in the anterior end of the egg. Up to 40% of eggs so treated display the double abdomen syndrome. Punctures made in other regions of the egg, denatured RNase, or other enzymes do not have this effect.

Furthermore, the UV induction of double abdomen is photoreversible. As much as 60% increase in the frequency of normal embryos occurs if the irradiated eggs are subsequently exposed to long wavelength UV or blue visible light. UV irradiation sufficient to induce double abdomen embryos causes extensive formation of pyrimidine dimers in the ribosomal RNA of these embryos, and photoreactivation treatments that rescue UV irradiated embryos also stimulate the repair of most of these dimers. The same would presumably be true of other RNA species within the UV penetration zone. In any case the conclusion that localized maternal RNAs are required for axial specification in *Smittia* is consistent with that drawn earlier for *Drosophila* from the demonstration that maternal mutants such as *snake* can be rescued by injected wild-type RNA. Potentially useful protein markers for both anterior and posterior determination have recently been identified in *Smittia* eggs.

An *anterior indicator protein* of about 35 kd is synthesized in the anterior but not posterior blastoderm cells of normal embryos. This protein is produced in cells from both halves of the blastoderm in double cephalon embryos, and again exclusively in the anterior end after photoreversal of irradiated eggs. A *posterior indicator protein* of about 50 kd displays a complementary distribution. This protein is synthesized in both ends of the blastoderm in UV induced double abdomen embryos, and its appearance is restricted to the posterior end in irradiated embryos rescued by photoreversion. Synthesis of both indicator proteins occurs long in advance of any visible anterior-posterior differentiation. As the regional synthesis of these proteins is predicted perfectly by experimental interventions that affect the prelocalized cytoplasmic determinants, it follows that molecular regulation of these blastoderm cell functions occurs during the earliest stages of the axial specification process.

Disposition of Maternal Cytoplasmic Determinants During Early Cleavage

Localization patterns in Crenophore Eggs

At the opposite extreme from the eggs of dipterans and other insects in which both embryonic axes are prelocalized are the eggs of ctenophores and hydrozoans, in which until development begins neither axis is specified. Cytoplasmic determinants are nonetheless present in ctenophore eggs, as was shown convincingly by classical experimentalists. These early experiments as well as more recent blastomeres isolation studies were reviewed by Reverberi (1971a). At the 8-cell stage the embryo contains two pairs of external cells denoted "E" cells, located on either side of the four inner cells, which are denoted "M" cells. The derivatives of E cells and M cells have different morphogenetic fates which are expressed accurately in isolated blastomere experiments. In normal embryos the micromeres given off by the E macromeres produce fused rows of large swimming cilia, or comb plates, and the micromeres derived from M macromeres form other structures, including mouth apical organ, and most spectacularly, photocytes. These cells contain a calcium activated photoprotein which enables them to produce flashes of light. The embryo gives rise to bilaterally symmetrical larva organize. Freeman (1977) showed than in three species of ctenophore the embryonic oral-aboral axis is determined by the site of the first cleavage furrow. Though in normal eggs this usually coincides with the location of the polar bodies, which provide a stable orientation marker on the egg surface that is not invariably the case and in centrifuged eggs it is often not the case.

The cleavage furrow apparently forms near the zygote nucleus, which is easily shifted by low centrifugal forces. Normal development ensues in centrifuged eggs whatever the plane of first cleavage, and the consequent orientation of the oral-aboral axis, with respect to the polar bodies. The future oral pole develops wherever the cleavage initiates, and the plane of this cleavage is the future sagittal plane of the larva. The second cleavage occurs at light angles to the first and its plane becomes the future tentacular plane of the larva. Thus neither axis is prelocalized in the egg, and the first and second cleavages are the causal events in the orientation of both. The comb plate and photocyte determinants are localized at different times, and in the unfertilized egg are resident in different components of the cytoplasm.

The comb plate determinants are associated with the cortex. Segregation of these factors occurs between the 2-and 8-cell stages, when they become localized in the M and E macromeres respectively. Experiments of Freeman (1976a, b, 1977) imply that localization in this egg occurs as a *consequence* of cleavage Prior to the 2-cell stage the morphogenetic potential of comb plate formation is not spatially restricted, and it begins to be localized to the aboral pole after first cleavage. If the cleavage schedule is altered by transient exposure to cytochalasin or 2,4-dinitrophenol, or the sequence of cleavage planes is distributed by application of pressure, the result is altered localization of the ability of the embryonic cells to form comb plate cilia or photocytes when cultured. The nature of the localization pattern that develops in the treated eggs was found to depend on the character of the cleavage abnormality that had been induced. Furthermore, cleavage triggers a timing mechanism, and normal localization requires that the correct plane of cytokinesis occur at the appropriate time.

The peripheral cytokinesis that in normal embryos gives rise to the E macromeres at third cleavage can be induced to occur at second cleavage in two different ways. If second cleavage is reversibly blocked for one cycle and the inhibitor is later removed the planes of the following cytokinesis are located similarly to those of third cleavage in normal embryos, though the treated embryo has only four cells. Such embryos display normal localization in the E macromeres of the potential to give rise to comb plates since these macromeres are segregated out at the appropriate time after first cleavage. E macromeres can also be made to form one cycle too early by causing second cleavage to take place under pressure. In these embryos though they are of identical form with those produced by cleavage delay, comb plate forming potential is found to be present in both E and M macromere derivatives.

Ontogeny of Localization in Nemertean Eggs

The eggs of nemertean worms of the genus Cerebratulus were also favourite subjects of classical research on localization. Wilson (1903) Zeleny (1904), and Yatsu (1904) showed that establishment of the definitive localization patterns in the Cerebratulus egg begins with the meiotic divisions set in train by fertilization, and it is complete until the 8-cell stage. The morphogenetic factors that specify the apical tuft and the gut, among other structures, have by then segregated to their respective progenitor blastomeres. Thereafter, the blastomeres behave in an irreversibly determined fashion. However, in the unfertilized egg the morphogenetic factors are globally distributed, although the animal-vegetal axis is clearly predetermined. Thus, for example, Yatsu (1910) showed that normal larvae may be obtained either from fertilized animal or vegetal half eggs.

A developmental map for the apical tuft and gut determinants of the Cerebratulus lacteus egg was constructed by Freeman

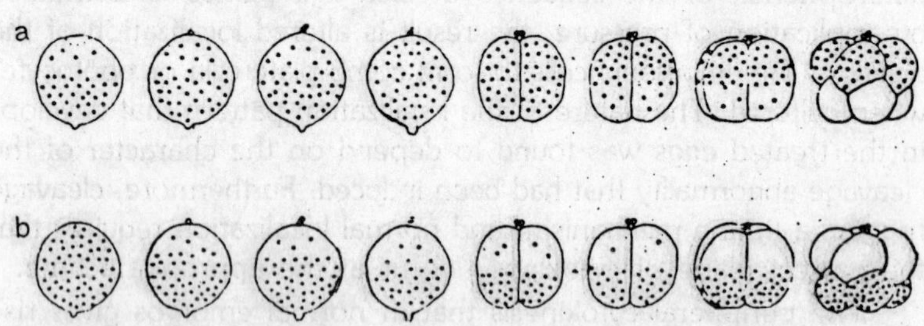

Fig. 7.2. Progressive localization in the eggs of Cerebratulus lacteus. Diagrammatic representation of the distribution factors specifying apical tuft formation (a) and gut formation (b) from the unfertilized egg stage to the 8-cell stage. The period when the meiotic maturation divisions occur is indicated by the appearance of the polar bodies.

(1978), by determining the morphogenetic potential of egg fragments isolated at various stages after fertilization. These factors independently move in opposite directions during the meiotic maturation stages. They achieve their final distribution just before the transverse third cleavage. In this egg, in contrast to the molluscan examples examined earlier, cytochalasin treatment does not interfere with progressive mechanism of localization. Nor does alteration of the cleavage planes by pressure. On the other hand a drug that suppresses aster formation, ethyl carbamate, arrests the localization process for as long as it is present. When the drug is washed out new asters form, and localization resumes.

Freeman (1978) also showed that removal of the meiotic asters with a microneedle has the same effect as ethyl carbamate, while precocious induction of asters with hypertonic sea water accelerates the localization of the gut and apical tuft determinants. Though it is clear from these results that asters formation is required for localization in the cerebratulus egg, the function of these organelles in regard to localization remains unknown. There are clearly many diverse mechanisms by which maternal cytoplasmic determinants are specially distributed. These include intercellular interactions during oogenesis, the disposition of maternal cytoplasm in polar lobes, and the mediation of localization by cytokinesis in early cleavage, or by global cytoplasmic movements triggered at maturation or fertilization. An underlying element of homology in some of these various processes may be the engagement of the cytoskeletal apparatus of the egg, as long ago argued by Conklin (1905a). The crucial role of asters in the localization processes of *Cerebratulus* eggs may be interpretable in these terms. In the following we consider some recent evidence on the participation of cytoskeletal structure in the ontogeny of localization patterns.

Localization Ascidian and Annelid Eggs

We have touched on a number of cases in which cytoplasmic *movements* occurring after fertilization are involved in localization, and yet the determinants appear to be *anchored* in place, in that they are resistant to centrifugal forces that stratify the egg cytoplasm. The cytoskeleton of the egg provides both the possibility of contractile movement and the opportunity of localized anchorage for particles, organelles, and macromolecule. Current evidence specifically implicates the cortical cytoskeleton of ascidian eggs, both in the initial bipolar contraction of the surface elements of the cytoplasm that occurs immediately after fertilization in the localization of the posterior muscle cell determinants. As will be recalled the yellow myoplasm that marks the positions of these determinants is localized by means of cytoplasmic movements that occur prior to cleavage. Conklin (1931) showed that low speed centrifugation fails to interfere with the localization patterns of ascidian eggs. As described in the legend to that figure the pigmented myoplasm remains in place after detergent extraction that removes three-fourths of the egg protein and almost all of the bulk RNA and lipid.

The pigment granules are seen beneath a mesh-like structure termed the plasma membrane lamina, which is composed primarily of membrane proteins, and is supported by action microfilaments. Beneath this lies a complex filamentous lattice that may consist largely of the intermediate filament proteins desmin and vimentin, and to which the pigment granules are attached. After fertilization the plasma membrane lamina, which initially covers much of the egg, contracts over the surface carrying its inclusions, and thus forms the localized yellow crescent. The actin microfilaments of the cortical cytoskeletal complex could contribute to the structural integrity of the cortical myoplasm domain, and also provide components of the contractile engine that drives the spatial segregation of the domain.

By analogy with what is known of echinoderm eggs cytoskeletal myosin could be present, and might provide the motive force. Jeffery (1982) showed that in eggs of the ascidian *Boltenia villosa* the location of the myoplasm crescent can be specified by presentation of a Ca^{2+} inonophore contraction, just as might the penetrating sperm in the untreated egg. A potentially interesting aspect of cytoskeletal localization in ascidian eggs is that a major fraction of the maternal poly(A) RNA is retained after detergent extraction, though most of the ribosomal RNA is not. The possibility thus exists that within spatially localized cytoskeletal domains could be sequestered *specific* maternal poly(A) RNA sequences. However, a two-dimensional examination of the translation products of mRNA extracted from isolated yellow crescents failed to reveal any such mRNA localization. Among the mRNAs present in the yellow crescent myoplasm are some coding for actin but it is important to note that these maternal actin message code for cytoskeletal rather than the muscle actin isoform.

Since only the most prevalent mRNAs of the yellow crescent have been examined, perhaps the main significance is the observation that there are some mRNA tightly associated with it. Cytoskeletal components other than the actin microfilaments are responsible for binding these mRNAs since the cortical actin lattice can be disrupted without causing their release. The possibility remains to be excluded that lower prevalence maternal mRNAs could be specifically in the myoplasm. Maternal poly(A) RNA is also anchored to large cytoplasmic structures in the egg of *Chaetopterus pergamentaceus*.

A typical annelid localization pattern is established in this egg by means of a dramatic series of cytoplasmic rearrangements occurring during the meiotic maturation stages that immediately precede first cleavage. Lillie (1906) showed that regional differentiation occurs in *Chaetopterus* eggs even if cytokinesis is repressed by treatment with KCl, just as it does in cytochalasin treated ascidian zygotes. Cytoplasmic reorganization in both KCl activated and fertilized *Chaetopterus* eggs is blocked by colchicine, which affects microtubule organization, but not by cytochalasin. Stratification of the internal cytoplasm by centrifugation also fails to disturb the localization patterns of *chaetopterus* eggs, and centrifuged eggs develop normally. *In situ* hybridization with H-poly (U) demonstrates that the egg cortex, where the key morphogenetic movements of the cytoplasm take place, is also the site of maternal poly(A) RNA accumulation.

The cortical localization of the maternal poly(A) RNA persists in first cleavage, and extends to the polar lobe formed at that time. After low speed centrifugation most of the maternal poly(A) RNA is sedimented to the centrifugal pole of the egg. This observation shows that correct localization does not require the normal cortical disposition of the bulk of the maternal poly(A) RNA. Nonetheless, the result is interesting because it indicates that in this egg heterogenous RNAs are associated with particulate elements, which for certain sequences could as well be anchored in structural components of the cytoskeleton, just as are the yellow pigment granules and the maternal mRNAs of the ascidian myoplasm. We have encountered in this review several kinds of evidence suggesting that among the morphogenetic determinants of the egg cytoplasm are maternal RNAs, which may be prelocalized in the egg, or progressively localized after fertilization. Examples for which this has been proposed include the UV-sensitive targets involved in germ cell determination in anuran amphibians and dipteran insects; the anterior determinants of the *Smittia* egg; the early embryo poly(A) RNAs that on injection may result in rescue of maternal axial specification mutants in *Drosophila*; and the polar lobe determinants of the *Bithynia* vegetal body. A theory of localization based on regionally confined maternal regulatory RNAs was originally proposed by Davidson and Britten (1971).

The attractiveness of the proposal that the egg stores regulatory RNAs that are or will be localized is based on two features. First, there is the argument in principle that regulatory information transcribed during oogenesis is ultimately required to induce new, differential patterns of genomic function in embryonic cell lineages, information that could reside either in maternal mRNAs coding for regulatory polypeptides, or in regulatory RNA sequences functioning in some other manner. Second, there are observations reviewed in this section that at least in some eggs maternal poly(A) RNAs are associated with regionally localized cytoskeletal elements. This provides in principle the outline of a mechanism for the regional distribution of morphogens. To proceed further will be a difficult task, as it would require experimental demonstration that specific maternal RNA or protein determinants are both necessary and sufficient for cell lineage specification in the embryo.

Regulation, and Localization in the Eggs of Sea Urchins, Amphibians and Mammals

Cytoplasmic factors of maternal origin that ultimately affect the specification of embryonic cells appear to function in the early development of all the creatures so far considered, though the means by which such factors are spatially localized are many and various. The comparative arguments for localization as a general and basic mechanism for early development are thus strong. Nonetheless, as for our predecessors, the problem of explaining regulative behaviour remains difficult. The localization theory has seemed to some observers incomplete, in that it does not obviously predict that subelements of the embryo, when isolated or placed in combination with other subelements, will in some cases produce structures other than those they are normally destined to form, including complete, normally constructed larvae.

Regulative behaviour has been regarded as a major challenge ever since the first reproducible observations of this kind were described by Driesch (1891). Several different factors are now understood to contribute to regulative developmental processes in experimentally perturbed embryos, none of which in any way exclude specification, in the *normal* course of events, by means of cytoplasmic localization. Thus, Wilson pointed out that even on the strict assumption of cell lineage determination by localized morphogenetic agents, certain blastomeres would be expected to behave in a *totipotent* manner if isolated and cultured, while others should give rise to partial embryos according only to the geometry of the localization patterns with respect to the planes of cleavage. We have also seen that localization patterns are in some organisms established ontogenically, at a relatively leisurely pace, so that the morphogenetic potential displayed by a given subembryonic fragment depends on the stage at which it is tested.

In addition, a strong relation exists between the extent to which cell lineages are determined through inductive intercellular interactions during early embryogenesis, and the capacity for regulative behaviour that subelements of the embryos will manifest. As an example already considered, in equally cleaving gastropod eggs in which the dorsoventral axis is determined *inductively*, any of the four macromeres can give rise to this axis, while in unequally cleaving gastropod eggs only the D macromere retains the necessary maternal cytoplasmic factors. Another variable factors is that the structural elements responsible for the spatial disposition of localized determinants are sometimes apparently very delicate, so that in early stage localization is *physically labile*, and is easily reversed by experimental manipulation. The eggs of certain deuterostome groups, including sea urchins, amphibians and mammals, provide illuminating examples both of lability and of early inductive interactions between blastomeres.

Regulation in the Sea Urchin Egg

Localization of determinants along the animal-vegetal axis of the unfertilized sea urchin egg was demonstrated by Boveri (1901a, b), in a study carried out on eggs of *Paracentrotus lividis*. In this species a fortuitous subequatorial band of pigment granules arranged in a plane orthogonal to the animal-vegetal axis indicates the future orientation of the embryo. These granules also indicate location of cytoplasm that is later include in the cells of the archenteron. By shaking the eggs into pieces and fertilizing the fragments. Boveri showed that only fragments containing the pigmented cytoplasm are able to gastrulate and to carry out gut formation.

Horstadius (1928) confirmed that the pigment layer indicates the eventual axial orientation of the embryo and showed that animal half egg fragments develop in a manner similar to the animal halves of 16-cell cleavage-stage embryos, while the partial morphogenesis carried out by vegetal egg fragments corresponds to that observed in 8- or 16-cell vegetal half embryos. Horstadius (1937b) extended these findings to *Arbacia punctulata*, orienting the eggs individually soon after fertilization, and cutting them in half at the pronuclear fusion stage. The animal halves cleaved equally, producing no micromeres, and gave rise to spherical structures with enlarged apical tufts. The vegetal halves, however, formed micromeres and gastrulated, with archenteron and skeleton formation ensuing. It follows that cytoplasmic factors required for micromere, skeleton, and archenteron formation are localized in the vegetal region of the egg cytoplasm even before fertilization.

The orientation and distribution of these materials remain unchanged as the egg cytoplasm begins to be divided up among the blastomeres. Like lobeless *Dentalium* and *Ilyanassa* eggs, the animal egg fragments in these experiments possess complete genomes, but fail to differentiate several important cell lineages. Boveri (1901a) noticed two additional external features that identify the prelocalized animal pole in banded *Paracentrotus* eggs. These are the minute channel known as the *jelly canal*, and the position of the polar bodies.

Schroeder (1980a, b) showed that the jelly canal, visualized by immersion of unfertilized eggs or oocytes in ink, can be used to define the location of the animal pole in unfertilized sea urchin eggs that lack an oriented band of pigment granules. On elevation of the fertilization membrane the ink mark left at the site of the canal remains visible, this mark occurs *opposite* the micromeres at the 16-cell stage. The animal-vegetal prelocalization is thus a general characteristic of sea urchin eggs, and is not confined to those with banded pigment layers. Contrary to Boveri and other authors, both classical and modern, the jelly canal is not a micropyle. Sperm can enter and fertilized the sea urchin egg at any point on its surface with respect to the animal pole. Irrespective of the point of sperm entry, the orthogonal first and second planes of cleavage intersect along the preformed animal-vegetal axis. Figure demonstrates that the polar orientation of the egg is a prelocalization that extends back into oogenesis, since the jelly canal is present even in immature oocytes.

A subequatorial hand of pigment granules similar to the present in unfertilized *Paracentrotus* eggs appears *during early cleavage* in *Arbacia* eggs. Observations of Schroeder (1980a) suggest that the pigment band forms by means of an upward contraction of the cortical cytoplasm that carries the pigment granules away from the vegetal pole. The implication is that a similar event has already occurred in *Paracentrotus* eggs by the time they become available for fertilization, and in this species the pigment granules have been shown to be structurally associated with the cortical cytoskeleton in both unfertilized and fertilized eggs. A prelocalized animal-vegetal polarity is also manifested in various cytoskeletal features of starfish oocytes, including the distribution of actin-filled external spikes and of certain acidic vacuoles, both emeshed in the contractile cortex; the distribution of pigment granules; and the position of the premeiotic aster (Schroeder, 1985). Cytological studies of meiotic activation in starfish oocytes and of fertilization in sea urchin eggs that lies outside the scope of this discussion have revealed a complex cytoskeletal architecture in the echinoderm egg.

Networks of microtubules and microfilaments are present in the cortex, and cortical myosin, actin, and fascin have all been identified in the starfish oocyte. The imperviousness of the axial organization to stratifying centrifugal force exerted in any direction with respect to the egg axis, imply that the oriented elements important for animal vegetal prelocalization reside within or are anchored to a regionally differentiated, preformed, cortical cytoskeleton. Mabuchi and Okuno (1977) showed that in the starfish egg cytokinesis, which involves a cytoplasmic contraction that is normal rather than tangential to the surface, is inhibited by microinjected anti-myosin antibodies.

A reasonable interpretation is that the tangential cortical contractile processes occurring during localization in echinoderm, ascidian, amphibian during localization in echinoderm, ascidian, amphibian and many other eggs are also driven by contractile actin-myosin complexes of the cytoskeleton, and that these structures are associated with the cortex as a whole by actin microfilaments that in many eggs can be dissociated by cytochalasin treatment. The oral-aboral axis of the sea urchin embryos is not evident at a gross level until the gastrula stage, when the ectodermal wall on that side thickens and shortens, the archenteron develops, an asymmetric tilt to the oral side, the oral rods of the skeleton appear, and the secondary mesenchyme cells contract the inner wall of the blastocoel. In undisturbed embryos the oral-aboral axis is apparently specified as early as the 8-cell stage of cleavage however, although even at late cleavage stages meridional half embryos and embryos in which the animal half has been rotationally reoriented with respect to the vegetal half are all able to reproduce new oral-aboral axes and thence to develop normally.

Localization in Amphibian Eggs

Amphibian eggs have been studied since ancient times, and from the 19th century on this familiar material has been the subject of experimental attempts to understand the relation between the visible structure of the egg and the axial co-ordinants of the embryo. The animal-vegetal axis is a preformed feature, which evidently originates far back in oogenesis. Thus, in immature amphibian oocytes the germinal vesicle is always located toward the future animal pole, and during vitellogenesis the large yolk platelets are concentrated mainly toward the vegetal pole. In species whose eggs are coloured the pigment granules are usually confined to the animal hemisphere, probably an adaptation that provides protection from both above and below, since due to the dense vegetal yolk platelets the eggs normally float animal pole upward. Certain macromolecular constituents also seem to be distributed along the animal-vegetal axis even in oocytes and in uncleaved zygotes. For instance, Capco and Jeffery (1982) found that the poly(A) RNA in advanced vitellogenic *Xenopus* oocytes is concentrated along the vegetal pole cortex and in a perinuclear region on the vegetal side of the germinal vesicle.

Shortly after fertilization poly(A)RNA concentrations are higher toward the animal pole. Out of several hundred protein species resolvable on two-dimensional gels a few are particularly localized in the vegetal cytoplasm of matured *Ambystoma* oocytes and a variety of protein species have been reported to be distributed in a gradient along the animal-vegetal axis in newly fertilized *Xenopus* eggs, though most are not. A detailed study was reported by King and Barklis (1985) in which poly(A)RNA was isolated from animal, middle and vegetal third of eggs section perpendicular to the animal-vegetal axis and translated in a cell-free system.

The proteins synthesized were displayed by two-dimensional electrophoresis, and by this means five mRNA species of the approximately 600 whose products were resolved are found to be confined to the vegetal region of the unfertilized egg, while others are preferentially (i.e., 10 to 100-fold) concentrated there. One species was identified that is significantly more prevalent in the animal region. Several clearly localized maternal mRNAs have also been recovered from a cDNA clone library constructed from unfertilized *Xenopus* egg poly(A)RNA, including one clone representing a transcript localized at the vegetal pole, and three representing animal pole messages. All of these transcripts belong to the low abundance class of maternal messages of which there are $> 10^4$ diverse species stored in the egg.

Though primordially imposed can be at least partially reversed by causing the eggs to develop in an inverted orientation. Eggs treated in this manner arrest after gastrulation, but transplantation experiments reveal that reversal of some axial specifications has taken place, so that, for example, neural structures now derive from explanted cells of the original vegetal hemisphere. Unlike the animal-vegetal axis, the dorsoventral axis of the amphibian egg is an ontogenic construction, that is specified only after fertilization. In appropriately pigmented eggs the location of the future dorsal pole is forecast by the appearance of a zone of cortical cytoplasm that is depleted of pigment granules, traditionally known as the *grey crescent*. Classical observers, beginning with Newport (1854) noted that the plane of first cleavage usually becomes the sagittal plane of the embryo, and that in most (though not all) cases this plane bisects grey crescent.

The later normally forms *opposite* to the point of sperm entrance. Since the sperm can enter, or will fertilize if experimentally deposited at any spot on the animal hemisphere, the egg is initially radially symmetric, and before fertilization all equatorial loci thus possess the capacity to participate in the formation of either ventral or dorsal structures. At gastrulation the lower edge of the grey crescent becomes the dorsal lip of the blastopore, which is the site of formation of the dorsal structures of the neurula. The inductive process by which these structures arise was revealed in a series of studies begun by Spemann and associates. Dorsal structures such as notochord, somites and neural tube form as a result of a sequential series of inductive interactions.

Attention was initially focused on the induction of dorsal axial structures during gastrular invagination carried out by cells of the "Spemann organizer", which is located above the dorsal lip of the blastopore. So powerful is effect of the organizer and so plastic the condition of the cells prior to invagination that organizer cells transplanted to the prospective ventral side of the early gastrula induce there a secondary dorsal axis. A consistent and until recently widely accepted interpretation was that the cells of the Spemann organizer are initially specified by maternal cytoplasmic factors localized after fertilization in the grey crescent area of the egg. This theory found direct support in experiments of Curtis (1962), who reported that prior to the embryo suffices to produce a second dorsal axis, just as does ventral transplantation of cells of the dorsal organizer region at the gastrula stage. Unfortunately these results appear to have been the consequence of experiments artefacts and a wholly different understanding of the role of localization in the specification of dorsoventral axis of amphibian eggs has emerged from recent cell lineage, transplantation and cytological observations.

Origin of the Dorsoventral Axis

The fate map of the pregastrular amphibian embryo shows that the cell located in the dorsal marginal zone ultimately give rise to mesodermal structures of the dorsal axis. It is some of these cells which in the Spemann experiments display the capacity to induce dorsal structures. The problem we now the capacity to induce dorsal structures. The problem we now address is whether the initial determination of these cells as mesoderm progenitors with dorsal inductive capabilities in fact occurs as a self-differentiation, such as might follow from their inheritance of cytoplasmic factors localized on the dorsal side. Explantation experiments carried out in *Triturus* embryos and in *Xenopus* embryos suggest that the dorsal marginal cells of the animal blastomere cap have become fully determined by the early blastula stage. Thus, if these cells are excised after this stage and cultured, they produce various mesodermal structures, including myotomes, notochord, mesothelium, etc., but if isolated earlier they give rise only to ciliated epithelium. From these and similar studies, it can be concluded that in *Xenopus* determination of dorsal mesoderm precursors takes place between the 64-cell and the 512-cell stage.

In *Rana* and *Ambystoma* as well, *the capacity to induce a second axis,* i.e., behave as a Spemann organizer, appears to have been fixed in dorsal marginal cells in the course of the blastula stage. However, as these tests challenge the state of determination of the dorsal organizer precursor cells, they do not reveal directly the mode of their initial specification. It is now evident that both the self-differentiation of early dorsal equatorial blastomeres, and the inductive differentiation of cells in this region by vegetal pole blastomeres on the dorsal side lying *below* the original gray crescent are involved inductive determination of mesoderm in animal cap cells by vegetal region cells was demonstrated in the urodele *Ambystoma* and in *Xenopus*, in experiments in which the normal axial mesodermal precursors were excised at the blastula stage and the remaining animal hemisphere ectoderm, when combined with the vegetal hemisphere, was shown to be able to recreate the axial mesoderm anew.

The dorsoventral orientation of the induced mesoderm is controlled by that of the vegetal hemisphere in these recombinants, and the induced dorsal mesoderm is then capable of behaving as a Spemann organizer, Gimlich and Gerhart (1984) demonstrated explicity that the ability to induce dorsal mesoderm is originally resident in a few *vegetal* blastomeres at the 64-cell stage. In their experiments the original dorsoventral axial orientation of the egg was destroyed by UV irradiation of the vegetal hemisphere, and the irradiated embryos were shown to be rescued by transplantation of one to three blastomeres from the quadrant underlying the prospective dorsal marginal region. Injection of fluorescent label demonstrates that the transplanted vegetal cells are themselves ancestral to no mesodermal derivatives, and their progeny are later found incorporated only in the gut.

On the other hand, similar experiments, in which two dorsal equatorial blastomeres drawn from the third cell tier of a normal 32-cell embryos are transplanted into equivalent positions in UV-irradiated hosts, show that these blastomeres possess the capacity to rescue dorsal axis formation by *self-differentiation*. Thus injection of fluorescent lineage tracers demonstrates that the progeny of these transplanted cells contribute directly to the notochordal and other mesodermal structures of the organizer region. The inductive activity of the vegetal blastomeres is shown by the experiments of Gimlich and Gerhart (1984) to be sufficient to account for the specification of the future organizer region. Nonetheless, the evidence reviewed clearly implies that there exist maternal cytoplasmic factors that specify the cells inheriting them to give rise to dorsal mesoderm, including that of the Spemann organizer, and that these factors localized in the dorsal equatorial blastomeres of the early cleavage state embryo.

The inductive mechanisms also operating no doubt assist in imposing the appropriate spatial patterns of dorsal specification. We have seen that the vegetal pole cytoplasm is also the location of determinants for the germ cell, at least in anuran amphibians. Though the animal hemisphere blastomeres display innate tendencies to differentiate as epidermis they seem in general to remain plastic and undetermined into the early blastula stage, pending their reception of appropriate signals from self differentiating blastomeres below. In the sea urchin embryo as well the study vegetal pole blastomeres appear to be more highly determined *ab initio*, and the fates of the animal cap blastomeres to be inductively determined particularly with respect to the second axis. In considering the mechanism underlying this mode of development it is important to note that in the amphibian embryo the initial inductive interaction required for the determination of the dorsal marginal cells of the animal hemisphere takes place *in the complete absence of genomic transcription.* As reviewed earlier, transcription does not resume in amphibian until the midblastula stage.

The Mechanism of Localization

We now return to the initial process by which the dorsal determinants are localized on the grey crescent side of the egg. The primary dorsoventral asymmetry is caused by the movement of the sperm nucleus, and a displacement of internal cytoplasmic constituents to one side that may be due to formation of the large sperm aster. This is followed in normal development, by a strong cortical contraction toward the future ventral side, occurring in the plane of the future dorsoventral axis of symmetry that results in the rotation of the cortex with respect to the deeper cytoplasmic contents.

Cortical microtubules and microfilaments are evidently involved in these cytoplasmic movements though the contractile process itself is nor sensitive to injected cytochalasin. The consequence of the rotational contraction if the appearance of the grey crescent at the future dorsal side. This is a zone of cortical cytoplasm that has been depleted of pigment granules. As in the ascidian egg the pigment granules are engaged in a network of cytoskeletal actin microfilaments. The same contractile movements result in elevation of vegetal region yolk along the future ventral side, and in an asymmetric distribution of poly(A) RNA, which appears to be related the redistribution of the yolk platelet mass. However, there is no evidence that the internal redistribution of yolk is of functional significance. The mechanism by which the dorsal determinants are distributed is sensitive to application pressure, to transient exposure to cold (1°C for 4 min) and to UV irradiation.

Malacinski *et al.* (1974) and Chun and Malacinski (1975) found the UV-sensitive neuralizing factors localized mainly in the future dorsal region of the vegetal egg cortex in both urodele and anuran eggs. UV irradiated has a purely cytoplasmic effect, since nucleic from irradiated eggs are able to support development when injected into enucleated eggs as well as do control nuclei, and the irradiation penetrates only 5-10 μm or less into the egg cortex. However, ultimately UV irradiation clearly affects subsequent differential gene function in the embryonic cells, and thus alters the distribution of protein products that are confined to dorsal or ventral tissues. Cold, hydrostatic pressure, and UV irradiation all prevent axial determination if applied to the zygote between about 0.4 and 0.8 of the interval between fertilization and first cleavage. Thus the subsequent differentiation of axial dorsal structures is suppressed, to a variable extent, by these treatment, and when most effective they result in embryos that consist only of an ectodermal tube containing endoderm and some mesenchyme cells. All three treatments probably affect the cytoskeleton of the egg, and interfere with the redisposition of cytoplasmic constituents occurring during the period of sensitivity that precedes first cleavage.

Microtubules are stabilized by D_2O, and Scharf and Gerhart (1983) found that if immersed in D_2O, cold treated eggs display much better axial development. The effect of UV irradiation on the cytoskeleton is not known. However, to some extent D_2O also rescues axial formation in irradiated eggs. Unlike some examples considered earlier, the action spectrum for the UV inhibition of axial differentiation in amphibian eggs peaks at 280 nm. indicating a protein rather than a nucleic acid target. An additional argument that the UV effect could be due to damage to processes mediated by microtubules is that UV irradiation also blocks grey crescent formation as well and this is known to be sensitive to treatment with colchicine as well. The localization pattern established in the sensitive interval before first cleavage is remarkably labile.

It has been known for a century that the dorsoventral axis can be shifted merely by rotating the egg and holding it in a turned position after grey crescent formation Schultze (1894) showed that double axes from in anuran eggs that are inverted just before and during first cleavage. This observation was extended to the 2-cell stage by Penners and Schleip (1982a, b,) and was analyzed as a consequence of the gravitationally driven movements of the heavy internal yolk mass with reference to fixed cortical components. Ancel and Vintemberger (1984) showed that in parthenogenically activated *Rana*, eggs, which lack the aster normally induced by the sperm, the location of the grey crescent and the future dorsal axis can be perfectly controlled by a brief rotation. This suggests that gravitational rearrangements of the inner cytoplasmic constituents may *substitute* for the cytoskeletally mediated rearrangements that occur in normally fertilized eggs. To examine this inference, Gerhart *et al*. (1981) developed a procedure by which *Xenopus* eggs could be tipped at any given angel off vertical, and maintained in the altered position as long as desired. This was accomplished by immersing the eggs in Ficoll, thus dehydrating the privitelline space and destroying its natural function as a lubricating bearing within which the egg freely rotates.

Regardless of the original point of sperm entry or the position of the grey crescent, a 90° rotation suffices to determine a new dorsoventral axis. The blastopore can be made to form at, rather than across from, the point of sperm entry. The eggs are most sensitive to gravitational rearrangement before grey crescent formation but by application of low centrifugal force the dorsoventral axis can also be shifted later in the precleavage interval. Centrifugation at two successive times in different orientations or centrifugation followed by gravitational reorganization induces the formation of doubly axiated larvae. This result is consistent with the view mentioned above, that potential dorsal determinants are initially distributed radially around the egg, and that in normal development they become active on the future dorsal side following contact with, or additions form, the relocated deep cytoplasm.

The major translocation occurring in normal eggs is a rotation in the plane of the future dorsoventral axis, that results in a 30° displacement of the subcortical cytoplasm relative to the egg surface. The destructive effects of cold or pressure treatment, and of UV irradiation can be overcome merely by tipping the egg. Egg irradiated so as to completely inhibit dorsal development are rescued by being held for 30 min 90° off axis. It follows that in tipped eggs the gravitationally driven reorganization that occurs in unirradiated fertilized eggs. In both cases a necessary interaction along the future dorsoventral axis must occur between the deep cytoplasm and the cortical components. There is no evidence that such interactions are involved in the anuran egg, and rotation does not rescue primordial germ axial determination. Nor are germ cell determinants redistributed in inverted eggs. As for the eggs of several other species that we have considered, there is evidence that amphibian eggs also contain localized maternal morphogens that appear to promote certain pathways of self-differentiation in the blastomeres that inherit them. Thus, for example blastomeres of the subequatorial zone tend to produce differentiated muscle cells; blastomeres of the animal pole region differentiate as epidermis. The studies considered here show that dorsoventral axial determination depends on *structural* rearrangements, mediated in normally developing eggs by cytoskeletal cortical movements.

There is no evidence that the cortical rearrangements which specify axial orientation are to be equated with the direct localization of specific maternal morphogens. Thus a series of ecentrifugation experiments of Black and *Gerhart* (1984) show the exertion of gravitational force at various angles with respect to the animal-vegetal axis may provoke the appearance of an embryonic dorsal pole at unexpected locations, the immediate cytoskeletal structure of which should not have been perturbed asymmetrically in the treated eggs. For example, eggs centrifuged down along the animal-vegetal axis produce dorsal poles in equatorial regions located randomly with respect to the sperm entry point. Rather than directly redisposing a specific maternal "dorsal morphogen," treatments that affect the cortical cytoskeleton of the egg should perhaps be thought of as altering a spatially oriented *matrix*, in which such morphogens may at some later time regionally associate. Ultimately there indeed is localized in the vegetal blastomeres of the 32-cell embryo the capacity to induce the overlying cells to give rise to the "organizer" tissue on which later axial development depends.

Cell Lineage Determination in Mammalian Eggs

Mammalian embryos differ from all others considered in this review in both their slow absolute rate of early development and in the relative temporal order in which various developmental events occur. Thus first cleavage in the mouse embryo occurs only toward the end of the second day after ovulation, there are only 10-30 cells on day 3; and 100 cells are not attained until day 4. The rabbit egg, which is about 8 times the volume of the mouse egg, cleavages somewhat more rapidly and at two days it contains 16 cells; at 3 days 128 cells, and at 4 days over 1000 cells. A major distinction of mammalian embryos is the early developmental stage at which they switch from the use of maternal mRNA to mRNA synthesized in the embryo nuclei. In the mouse embryo, which has been studied the most intensively, mental mRNA has largely disappeared as early as the two-cell stage.

The synthesis of new proteins from embryo transcripts becomes a prominent feature at this time. Furthermore, the early mammalian embryo develops from the beginning in a nutrient environment so that relative to other embryos, net growth, including rRNA synthesis, starts very early. The storage of relatively huge amounts of maternal cytoplasmic constituents such as ribosomes and yolk is for the same reason unnecessary. Most of the differentiated cell lineages giving rise to the tissues and organs of the fetus appear after implantation of the blastocyst in the uterine wall, which in the mouse occurs at day 6. The first overt differentiation of cell types in the embryo can be distinguished at blastocyst formation late in the 5th cleavage stage, with the separation of trophectoderm cells from the inner cell mass (ICM).

The trophectoderm is a functionally differentiated cell lineage, as discussed in detail below, which after implantation gives rise only to extraembryonic membranes. The ICM includes the progenitors of all cell lineages, participating in the construction of the embryo proper as well as of some additional extraembryonic structures. The first specific cell layer to derive from the ICM appears with the delamination of the primitive endoderm. This takes place late on the fourth day of development in the mouse. The slow space of determination within the ICM has been demonstrated by transplanting genetically marked primitive endoderm and other ICM cells into dissimilar host blastocysts. Gardner (1982, 1984) demonstrated by this means that primitive endoderm cells are already definitively committed since single cells of this layer from 5 day embryos participate only in the development of endodermal tissues in the postimplantation fetus. However, they give rise both to the visceral yolk sac endoderm, and to the parietal endoderm cells remain plastic with respect to their ability to be included in either parietal or visceral structures. This decision apparently depends on contacts with other cell layers.

Individual cells other than those of the primitive endoderm of the 5th day ICM (usually described by the misnomer "primitive ectoderm") may give rise to all other types of fetal tissue including the definitive embryonic gut, and mesodermal and neural structures. We have already seen that germ cell determination in the mouse embryo also occurs in the "primitive ectoderm," and not only until after the 7th day of development. Probably many of the primitive ectoderm cells have the capacity to give rise to germ cells. It is most unlikely that the late determination of cell lineages within the ICM is any way influenced by localized maternal cytoplasmic components.

Though it is difficult to exclude this categorically, in view of the extreme plasticity observed in chimeric recombinations of mammalian blastomeres. Determination in the post-implantation mammalian embryo instead is probably mediated by intracellular interactions, but little is yet known of the detailed mechanisms. Discussion of the role of cytoplasmic localization in mammalian embryogenesis has thus focused on the separation of the trophectoderm from ICM cell at the early blastocyst stage, a process which has been studied in unusually great detail. The caveat should be kept in mind, however, that the separation of a determined layer of *extraembryonic* cells from the totally uncommitted embryonic progenitor cells of the ICM is not necessarily a phenomenon that is analogous to the determination of muscle or germ line or skeleton-forming cell lineages *within* the embryo.

Blastomere Totipotency

A fundamental observation is that the cells of the early mammalian embryo appear completely totipotent when tested by culture of isolated blastomeres or of partial embryos and in chimeric blastomeres recombination experiments. Individual blastomeres from 2-, 4-, and 8- cell rodent and rabbit embryos can develop into complete blastocysts. Blastomeres from 8-cell mouse embryos fused with late morulae are able to participate in normal development and fused embryos up to the late morula stage also develop normally. Furthermore although cells from the inside of the morula normally give rise to the trophectoderm, inside cells isolated from 32-cells morulae or from the ICM's of blastocysts remain totipotent. Not until the late blastocyst stage do the ICM cell lose the ability to regenerate trophectoderm. The outside cells of 16-cell morulae also retain the capacity to contribute to the ICM, along the same lines Ziomek *et al.* (1982a) showed that experimentally produced aggregates of pure inside cells, pure outside cells, of mixed blastomeres populations from 16-cell morulae will all give rise to blastocysts that implant and develop normally.

In normal embryos cells ancestral to both trophectoderm and ICM are present at later as the 32-cell stage. Marked clones of cells descended from single outer blastomeres that had been injected with horseradish peroxidase at the 16-cell morula stage. These clones can be seen to include progeny in both trophectoderm and ICM. After the 32-cell to the ICM, and this process seems mainly to be complete by the 16-cell stage. Soon after its formation the mouse blastocyst contains about 14 prospective ICM cells, and there are about 20 prospective trophectoderm cells. The totipotency of the blastomeres of rodent embryos at least through cleavage and morula stages implies a large reliance on inductive intercellular processes during normal early development. The plasticity manifested by these cells under the artificial conditions of the experiments reviewed also indicates the difficulty of determining whether cytoplasmic localization might be involved in the differentiation of the trophectoderm in undistributed embryos. Thus even if particular spatial relations predicting cell fate were to exist in the normal embryo, these assignments might be very easily reversed, and hence not easily detected.

Origin of the Trophectoderm Cell Lineage in Normal and Experimentally Perturbed Embryos

By the 6th cleavage there are about 45 trophectoderm cells, and these can be functionally distinguished from the ICM cells by a number of cytological, physiological and molecular criteria. The trophectoderm cells are linked to one another by characteristic junctional complexes, and they are distinguished in ultrastructure from ICM cells by a number of both surface and integral cytoplasmic features. They perform the special function of pumping fluid into the interior of the blastocoel. On presentation to the uterine wall they elicit an implantation response, and in their absence implantation cannot occur.

Ultimately they give rise to the extraembryonic ectoplacental cone, to a population of endopolyploid giant cells that divide no further and probably to chorionic. Two-dimensional analyses of the proteins synthesized by the trophectoderm cell of preimplantation embryos have revealed a number of species not included in the set of proteins synthesized by ICM cell at the same stages. After implantation some of these proteins continue to be synthesized in the extraembryonic layers descended from the trophectoderm. A monoclonal antibody against intermediate filament proteins of trophectoderm cells has been described and a cDNA clone representing the message for this trophoblast-specific protein has also been recovered. Since cells on the outside of the morula usually give rise to the trophectoderm, it was proposed that *outside position* is the causal factor in the determination of this cell lineage, and conversely, that *inside position leads* to the alternative developmental pathway, participation in the ICM. This idea was supported by experiments of Hillman *et al.* (1972), in which the fate of marked cells inserted in chemeric blastomere combinations was shown to depend in the expected way on their position.

Though as we have seen each blastomere of the 4-cell embryo is totipotent, Hillman *et al.* (1972), showed that if placed on the outside of a 4-to 6-cell embryo the progeny of the test blastomere is often found only in the trophectoderm, and after implantation in the yolk sac. If surrounded by other blastomeres, on the other hand its progeny is usually recovered in the ICM and in the post-implantation fetus. Blastomeres from the inner regions of the 16-cell embryos can also be steered toward either trophectoderm of ICM differentiation by being placed on the outside or inside; respectively of arrays of 15 other blastomeres of the same age. In these last experiments the test blastomeres and its clonal progeny were labeled with a fluorescent stain. However, Ziomek and Johnson observed that by this stage test outer blastomeres display a predisposition to give rise to trophectoderm, though not an irreversible one. Ultimately the ICM is covered by a trophectoderm layer, and this may be the event that represses further delamination of trophectoblasts, by permanently eliminating an "outside" environment.

Several other kinds of experiments contribute to the impression that external or internal position exerts a powerful influence. Thus, for example, alkaline phosphatase, which is an inner cell mass maker, is reported to appear in chimeric aggregates only when these include blastomeres totally surrounded by other blastomeres. Stern-flattened mouse embryos under a glass plate and showed that in these embryos trophoblast cells appear even in internal regions where the blastomeres about the glass interface rather than other cells. The observations may be interpreted in two distinct ways. The external cells could become committed to the trophectoderm pathway as a direct result of their position on the outside of the embryo their natural process of commitment could be mediated by their inheritance of cytoplasm localized radially at the periphery of the embryo.

The latter interpretation is allowed because observations on intact embryos have shown that in the course of cleavage, morulae and blastocysts formation, relatively little radial or migratory displacement of cell occurs. Thus cells marked with fluorescent or enzymatic stains, or by injection of oil droplets, give rise to progeny that remain clonally clustered detailed study of the cell lineage of the cleavage stage mouse embryo and of the geometry of early intercellular contact has been carried out by Graham and his colleagues. At third cleavage one of each pair of daughter cells normally lies deep within the embryo while the other occupies a peripheral position. By injecting oil drops into inside and outside blastomeres (Graham and Deussen 1978) showed that as expected the inner cells of these pairs contribute preferentially to the ICM, while oil drops injected into the outside cells are normally recovered in trophectoderm cells. The important point is that this radial correlation with blastomeres fate reflects an earlier intracellular cytoplasmic distribution.

Thus oil droplets injected in the peripheral cytoplasm of 2-cell stage blastomeres also tend to be distributed to later trophectoderm cells, while those injected centrally ultimately appear in both trophectoderm and ICM. Thus in normal development there appears to be a relatively fixed radial relation between the peripheral cytoplasm at early cleavage, the peripheral cell layer of the 8-cell embryo, and the outer trophectoderm progenitors of the 16-and 32-cell morula. There is unfortunately insufficient evidence to indicate whether relevant regional localization pre-exists in the uncleaved mammalian egg or the fertilized ovum. A closer focus on the events occurring at the 8-cell stage reveals a radial intracellular *polarization* of the blastomere cytology, which is mediated by intercellular contact. At 4th cleavage the peripheral cytoplasmic regions are physically segregated to the outer cells of the 16-cell morula. The early 8-cell stage blastomere is spherical, and both surface and internal features are arranged in a homogenous way. By the end of the 8-cell stage a major cytoplasmic reorganization has occurred. The nuclei move toward the basal, inner surface of the cell, where contact with an adjoining cell is established.

At the apical or external pole endocytotic vesicles accumulate. These have been displayed in cytological preparations by enzymatic staining of ingested horse-radish peroxidase. Cytoplasmic actin also accumulates at the polar regions, as visualized with fluorescent antibodies. The most extensively studied changes occur near the surfaces. At the basal side gap junctions and tight junctions form; arrays of microfilaments and microtubules are erected parallel to the lateral walls of the blastomere; and the apical surfaces are distinguished by localized arrays of short microvilli. These can be easily detected at light microscope magnifications by the density of binding sites for various fluorescent ligands that they provide. The overall consequence of this polarization is a change in form of the embryo from a loose aggregate of spherical cells to a tightly compacted, radially organized embryo, in which intercell contacts have been maximized within, while the outer surface has become covered with microvilli. When 16-cell morulae are disaggregated into couplets of sister blastomeres and examined the *outer cells* display the same polar cytological features as do the *outer regions* of the 8-cell blastomeres.

The inner cells, which are derived from the basal and lateral portions of the 8-cell blastomeres remain non-polar. Outer cell features, such as the external microvillar surface, are also retained by the outside cells of the 32-cell morula and the trophectoderm cells of the early blastocyst. The fate of the progeny of individual polar cells at the 8-cells stage depends on the plane of cleavage. If this bisects the pole, two new polar cells are formed, both of which become trophectoderm, while if it is normal to the polar axis a trophectoderm and an ICM progenitor cell are formed. These observations imply directly that the peripheral cytoplasm of the polarized 8-cell stage blastomere acts as a determinant for trophectoderm differentiation, while the cells inheriting the internal cytoplasm are destined to become ICM cells, i.e., unless they are placed in circumstances where they are induced to polarize anew.

The early determination trophectoderm and ICM precursors at once affects their biosynthetic activities. Thus Handyside and Johnson (1978) showed that marker polypeptides synthesized only in the trophectoderm cells of the definitive blastocyst are not synthesized in inner cells of the 25-30 cell morula, while polypeptides characteristic of the later ICM are. Polarization is an inductive process which requires intercellular contact. Furthermore, continued cell contact is needed to maintain the states of polarization initially imposed. When cultured *in vitro*, polar 16-cell blastomeres display trophectoderm-like behaviour and they wrap around non-polar cells, while apolar cells culture together tend to polarize, and to produce aggregates including both external trophectoderm-like cells and inner, ICM-like cells. Polarization occurs *in vitro* when a blastomere from an 8-cell embryo that has not yet undergone compaction is brought into asymmetric contact with other blastomeres.

Polarization is induced in the 8-cell stage blastomere by contact with the surface of a 2-cell, 4-cell, 8-cell or 16-cell stage blastomere but not with an uncleaved zygote or an unfertilized egg. As we have seen, the 2-cell stage, when this inducing ability first appears, is also the point at which the biosynthetic processes of the embryo become dependent on its own transcription. Induction of polarity does not require the formation of intercellular junctional complexes, since in normal embryos it precedes the formation of these structures, since 8-cell blastomeres polarize when placed in contact with 2-cell blastomeres, but form no junctional complexes; and since ionic media, antesera, and drugs that block junction formation do not prevent polarization.

Perhaps the most convincing demonstration that contacts with competent inducing cells account for polarization is that the axis of polarity can be controlled by the geometry of the contact. The apical pole always forms at a point opposite the contact, and in the case of multiple contacts, at a point equidistant from them. Within the intact 8-cell embryo, this property would naturally result in a radially organized embryo in which the apical faces of the blastomeres are oriented outwards, since the greatest number of intercellular contacts is in the inside. The ability of the early cells of the rodent embryo to undergo polarization as a result of cell contact lies at the root of their totipotent behaviour in isolated blastomere as chimeric recombination experiments. When isolated, the blastomeres divide, and then induce polarization in each other, with the eventual segregation of cells that inherit the apical cytoplasm of the new external poles and give rise to trophectoderm precursors.

In chimeric aggregates cells placed on the inside do not polarize, because they are symmetrically surrounded by contacts on all sides. However, on the outside they polarize anew, whatever their original orientation, and thus give rise to trophectoderm cells. Since outside cells tend to form extended contacts with underlying cells and to wrap around them, their dimensions expand in the plane normal to the axis of polarity. The plane of cytokinesis in these cells this is usually radial, and two new polarized outside cells are generally formed at each division in the undisturbed morula or in complex aggregates. However, isolated outside cells may appear totipotent since the plane of cleavage now frequently bisects the axis of polarity, producing both new outside and a new non-polar inside cell.

In short the experiments that display the totipotency of mouse embryo blastomere perturb these cells in two fundamental ways. They cause repolarization, and they result in reorientation of cleavage planes. Neither effect is inconsistent with the view that in *undisturbed embryos* the determination of the trophectoderm cell lineage begins with the segregation of radially localized cytoplasm into the outer cells, though their specification clearly remains reversible at least through the 32-cell stage. The ability of the early cells of the rodent embryo to undergo polarization as a result of cell contact lies at the root of their totipotent behaviour in isolated blastomere as chimeric recombination experiments. When isolated, blastomeres divide, and then induce polarization in each other, with the eventual segregation of cells that inherit the apical cytoplasm of the new external poles and give rise to trophectoderm precursors.

In chimeric aggregates cells placed on the inside do not polarize, because they are symmetrically surrounded by contacts on all sides. However, on the outside they polarize anew, whatever their original orientation, and thus give rise to trophectoderm cells. Since outside cells tend to form extended contacts with underlying cells and to wrap around them, their dimensions expand in the plane normal to the axis of polarity. The plane of cytokinesis in these cells are generally formed at each division in the undisturbed morula or in complex aggregates. However, isolated outside cells may appear totipotent since the plane of cleavage now frequently bisects the axis of polarity, producing both new outside and a new, nonpolar inside cell. In short, the experiments that display the totipotency of mouse embryo blastomeres perturb these cells in two fundamental ways. They cause repolarization and they result in reorientation of cleavage planes. Neither effect is inconsistent with the view that in *undisturbed embryos* the determination of the trophectoderm cell lineage begins with the segregation of radially localized cytoplasm into the outer cells, though their specification clearly remains reversible at least through the 32-cell stage.

8

Pattern Formation and Regeneration

Positional Information

This volume attests a renewal of deep interest in the problem of pattern formation in developmental biology. In large measure, the resurgence of enthusiasm coincides with Wolpert's reformulation of this fundamental problem in terms of the concept of positional information. Prior to Wolpert's introduction of the positional information concept, the dominant theory available postulated the existence of developmental fields possessing "prepatterns", non-uniform spatial distributions of biochemical substances in a tissue, whose local peaks would induce the formation of pattern elements such as specific digits, sensillae, or bristles. In contrast to Stern, Wolpert proposed the more abstract idea that cells within a developmental field possess positional information about their locations with respect to the boundaries of the field, through access to an underlying positional co-ordinate system. The behaviour of each cell in the field was assumed to be due to two independent processes. The cell *assesses* its position, then *interprets* its positional information according to the type of cell it is, and forms a specific structural element in the overall pattern.

The chief difference between Wolpert's positional information and Stern's prepattern, is that positional information is free of assumptions about the existence of specific morphogen peaks underlying the subsequent differentiation of specific pattern elements. This freedom in one sense makes the theory of positional information less predictive, yet allows for two important possibilities; first, that the positional information system in *all* the developmental fields of one organism are identical, and, more radically, that the positional information system in all organisms or all epimorphic fields is identical. The general success of the positional information concept led to a search for the co-ordinate system which supplies positional information.

At present, polar, spherical, and generalized Cartesian, models have been proposed. Differences among, these alternatives are far from trivial. Although it is always possible mathematically to transform from one to another co-ordinate system the "forces" or tissue properties which must be postulated to explain the observed features of pattern regulation differ sharply in the different models. While the morphogens have yet to be found, one task in this area of biology consists of efforts to discover the co-ordinate system and "forces" or requisite cell properties that most simply account for the observations, with the hope that the proper formulation will provide both macroscopic laws at the tissue level, and aid in discovery of the underlying molecular variables.

The Phenomena of Pattern Formation

To assess the relative success of the alternatives that have been proposed, it is necessary to review briefly at least some of the major phenomena of pattern generation and regeneration.

Intercalary Regeneration of Intervening Structures

If an amphibian limb capable of regeneration is transected proximal and distal to the elbow, and the distal wrist fragment grafted to the proximal shoulder stump, cell proliferation forms in the wound area, followed eventually by regeneration of the missing elbow region. This basic phenomenon is fairly ubiquitous and fundamental. Juxtaposition of normally nonadjacent tissues from a single developmental field is generally followed by regeneration, in proper spatial order, of the structures normally lying between the juxtaposed tissue edges. This notion of "betweenness" is necessarily central to any theory of pattern formation. The simplest physical model to account for "*betweenness*" in intercalary regeneration postulates the existence of one or more chemical concentration gradients spanning the tissue, whose concentration levels specify the positional information of cells at each point in the domain. As shown in which a proximal/distal gradient along an amphibian limb is envisioned surgical removal of the elbow and grafting of wrist to shoulder creates a discontinuity in the gradient, (S), at the graft junction. If one imagines that gradient concentrations remain fixed in the "old" tissue fragments, while diffusion occurs in the new cells of the wound blastema, then simple diffusive averaging of the concentration discontinuity at the graft junction smooths over the discontinuity, recreating all the intervening gradient values in proper spatial order.

Sequential Formation of Positional Axes in Development

In several systems, positional axes appear to be established sequentially during development. In classical experiments, Harrison removed the right forelimb bud of *Ambystoma* and grafted in its stead the left forelimb bud. Such grafts must invert either the anterior-posterior limb axis while keeping donor and host dorsal and ventral axes aligned; or invert dorsal-ventral donor and host axes, while keeping anterior and posterior axes aligned.

Harrison found that if very early left limb buds were grafted to the right, they developed into normal right limbs. If late left limb buds were grafted, they formed normal left limbs with that axis inverted with respect to the host that was inverted at surgery. But if left to right grafts were made at an intermediate stage, the outcome depended upon which axis was inverted at the graft junction. If the anterior-posterior axis remained normally aligned and the dorsal-ventral axis was inverted, the donor left limb bud formed a right limb, while if the dorsal-ventral axis remained aligned and the anterior-posterior axis was inverted at the graft junction, the donor left limb bud formed a left limb which remained inverted at the host donor junction. Harrison interpreted his results to imply that donor anterior-posterior axis becomes autonomously self-sustaining prior to the dorsal-ventral axis. Similar data suggest that amphibian eye axes, as well as limb axes are established sequentially, although the status of data on the eye is in dispute.

Distal Transformation

If an amphibian limb or tail is transected at the elbow, the proximal stump can form a regeneration blastema and regenerate the distal limb. If the digits of the transected distal fragment are implanted into a host flank to establish an adequate blood supply to the distal fragment, the cut surface at the elbow, which initially faced proximally, forms a regeneration blastema and regenerates second distal wrist and hand structures that are mirror symmetrical to the initial implanted distal limb. In these experiments, both the proximal stump and the implanted distal limb fragment regenerate the *same* set of distal limb structures from the cut surfaces at the elbow, identified as a regenerate in the proximal stump and duplicate on the implanted distal limb fragment. The fact that both fragments form distal limb has been called the "*rule of distal transformation*". Similar results have been found in many insect legs and in the imaginal discs of *Drosophila*.

Supernumerary Limbs

Among the most striking observations in pattern regulation is the induction of extra or supernumerary limbs following grafting operations. After both the anterior-posterior and dorsal-ventral amphibian limb axes are fixed, transplantation of a left distal limb to a right proximal stump which reverses the anterior posterior axis of donor relative to host but leaves the dorsal-ventral axis aligned, typically results in formation of two supernumerary limbs at the anterior and posterior margins of the donor host junction. If instead the anterior-posterior axes of host and graft are aligned, but dorsal-ventral axes are inverted the two supernumeraries emerge from the dorsal and ventral margins of the host donor junction. These supernumerary limbs generally have the handedness of the proximal stump.

Similar results have been obtained in transplantation of cockroach limbs. Rotation of a left distal limb by 180^0 and regrafting to its own proximal stump results in a more variable range of results. After such rotations, the donor may partially rotate back toward its normal alignment; sometimes zero, one, two, or more supernumerary limbs are formed at the gifts site, with the same or opposite handedness.

Duplication and Regeneration by Complementary Fragments of a Tissue

Distal transformation by both proximal and distal limb fragments of amphibians is one example of duplication and regeneration by complementary fragments of a developmental field. The phenomenon is common, however, and has been studied in greatest detail in the imaginal discs of *Drosophila melanogaster*.

Drosophila Imaginal Discs

Drosophila melanogaster is a holometabolous insect with egg, larva, pupa, and adult stages. During metamorphosis, the larval ectoderm lyses, and the ectoderm of the adult is formed by the terminal differentiation of special larval organs called *imaginal discs*. In the late third instar larva, each imaginal disc is a two-dimensional sheet of cells forming the surface of a hollow sphere. The columnar cells on one hemisphere form the imaginal disc proper, while the thin squamous cells on the hemisphere form the peripodal membrane which is lost during metamorphosis.

Imaginal discs are found as bilaterally symmetric pairs, each destined to form specific left and right regions of the adult ectoderm: the left and right first leg discs form the two prothoracic legs. The two wing-thorax discs form the left and right mesothoraces and wings, etc. By injecting specific fragments of each disc into host larvae for metamorphosis, it has been possible to construct a fate map of each type of imaginal disc. For example, the fate map of the wing-thorax (hereafter wing) disc, shows that the upper and lower margins of the disc along its longitudinal axis form ventral and dorsal thoracic structures.

During metamorphosis, the wing disc folds along a bent arc running from the anterior to posterior disc margin, and apposes ventral and dorsal thorax areas, ventral and dorsal wing hinge areas, and ventral and dorsal wing blade areas, creating a "bag" that everts through the peripodal membrane. The center of the disc maps to the distal wing tip, while an arcing line from anterior to posterior disc edge through the disc center corresponds to the wing margin. Grafting experiments like those in amphibian or cockroach limbs are not yet feasible in *Drosophila*.

However, closely analogous experiments can be performed by cutting the wing disc into known fragments, and injecting each fragment into the abdomen of a fertilized adult female. In that environment, the disc fragment heals its cut edge, apposing tissue regions which are normally nonadjacent and new cells form in the wound area. Over a week in culture, the mass of a disc fragment typically doubles. After culture in adult abdomens, fragments may be recovered and injected into host larvae for metamorphosis, then recovered from the emerged adult host. By comparison of the patterns of hairs, sensillae and bristles which form when a known discs subfragment is injected directly into larvae for immediate metamorphosis, it is possible to characterize the pattern regulation which occurs in the cultured fragment.

The following are the dominant results:

1. Anterior-posterior and ventral-asymmetries exist in the capacity of narrow anterior margin and ventral thorax margin structures have been found by a number of workers to regenerate extensively, while narrow posterior and dorsal margin fragments do not. The results in other imaginal discs are fundamentally similar although it should be stressed that an interior "high point" has been demonstrated in other discs.
2. If an interior "distal" circular region containing the "high point" is cut and cultured, it duplicates, the duplicate forming a mirror symmetric hemisphere to the original high point region. If the outer "proximal" annulus remaining after the high point region is removed is cultured, the outer annulus regenerates the central "distal" high point region.
3. It is critical that complementary fragments do *not* strictly duplicate and regenerate. Under different experimental conditions and with varying frequencies, normally duplicating fragments can regenerate most of the pattern elements in the complementary fragment.
4. If the wing disc is cut in two fragments by a straight cut, the roughly smaller fragment duplicates some or all of its structures. In favourable cases, a mirror symmetrical duplicate is generated whose symmetry line lies near the cut on the wing disc; the larger complementary fragment regenerates structures normally formed by its smaller complement. Thus complementary fragments often exhibit complementary behaviour, one duplicating, one regenerating. Consequently, it is possible to draw an arrow across each single straight cut on the disc, pointing from the fragment which regenerates into that which duplicates. As such arrows point radially outward from a small region in the interior of the disc. The polarity of regeneration and duplication reverses around this interior point, termed the "high point."
5. If two narrow normally duplicating crescent fragments from opposite edges of the disc are mixed, they regenerate the intervening pattern elements spanning across the disc.
6. Mutants causing pattern duplication in different imaginal discs have been analyzed by a number of workers (e.g., Jurgens and Gateff and will be discussed further.

7. If the disc is cut into arbitrary 3/4 and 1/4 "pie" sectors, the 3/4 fragment regenerates, the 1/4 fragment duplicates.

Models of Positional Co-ordinate Systems

The first major advance in predictive use of the positional information hypothesis lay in the formulation of the polar co-ordinate, or "clockface" model for pattern regulation in epimorphic fields by French *et al*. and Bryant. The initial model was based on some of the results described in amphibian limbs, cockroach limbs, and imaginal discs, and is well illustrated by application to the wing disc of *Drosophila*. The existence of a "high point" in the central (distal) region of the disc about which the direction of regeneration reverses suggested that cells might measure their distance in the tissue from this special point. This raised the possibility that the position of cells is specified in a polar co-ordinate system with the high point as its origin.

Since the wing disc is a two-dimensional surface, an azimuthal angle must be measured. Where an angle specified by a single scalar variable, that variable would necessarily be discontinuous along some radial line in the tissue; for example, along the radial line $\phi 2\pi=0$. But Bryant found that any 1/4 pie-wedge fragment of the wing disc duplicates, while its 3/4 complement regenerates. If an azimuthal discontinuity were present, the • Wedge fragment containing it should behave differently from the rest, and regenerate. Failure to find evidence of an angular discontinuity implies that, if cells measure an angle, they do so seamlessly. Therefore the model proposes that cells measure radial distance from a distal "high point" origin, and an angle seamlessly modulo 2π. In order to account for the bulk of the data on epimorphic regeneration and duplication, the polar co-ordinate model initially proposed two rules of intercalary regeneration, and a third special rule for distal transformation.

Rule 1. If cells having different radial values are apposed, cell proliferation will be stimulated, and the missing intervening radial values will be restored back to a resting radial gradient, then proliferation will cease.

Rule 2a. If cells having different angular values are apposed, cell proliferation will be stimulated and the missing angular values will be intercalated back to a resting angular gradient.

Rule 2b. Since two angular arcs around a 2π circle of values join any two juxtaposed angular values, a choice rule is needed.

The simplest postulates that angular intercalation occurs along the *shorter* arc.

Special rule 3, the *complete circle rule*. If a complete circle of angular values at a proximal radial level is exposed, distal regeneration occurs. The special nature of Rule 3 will be discussed later.

The two intercalation Rules 1 and 2a, b, suffice to explain major features of epimorphic pattern regulation. For example, the classical example of grafting an amphibian hand to shoulder with regeneration of the missing elbow region is explained by Rule 1. Proximal radial values in the shoulder apposed to distal radial values in the hand lead to intercalation of the intervening radial values specifying elbow. Duplication and regeneration by complementary fragments of the wing disc is explained by Rules 2a, b. A single, straight cut on the wing disc yielding a narrow anterior fragment and a large posterior fragment.

During culture, the narrow anterior fragment folds over, apposing the cut edge such that the ventral and dorsal thoracic regions heal together, ventral and dorsal wing blade regions also heal together. This healing juxtaposes cells with similar radial values, but discordant angular values. This discontinuity stimulates cell proliferation, and smoothing of the angular discontinuity along the shorter angular arc. Since this shorter angular arc is the arc present the original narrow anterior fragment, the positional values in the new cells form a mirror symmetrical duplicate of those in the original anterior fragment, and the fragment duplicates. The positional values present along the cut margin of the large posterior fragment are identical to those along the cut margin of the narrow anterior fragment. If the large fragment wound heals in a similar way, the same pairs of positional values must be apposed in the posterior fragment as in the anterior fragment.

Therefore, the posterior fragment must intercalate, along the shorter angular arcs, the same intervening positional values as did the anterior fragment. Therefore, the posterior fragment regenerates. The polar model demonstrates a more general result. Whatever the co-ordinate system specifying position in a developmental field may be, the positional values along the two margins of a single cut are identical. If the two fragments heal in similar ways, both will appose essentially identical pairs of positional values. Therefore, if subsequent pattern regulations is governed by diffusivelike averaging of positional disparities, both complementary fragments must reform the *same* set of structures. If one fragment duplicates, the second must regenerate. *The prediction of complementary behaviour in complementary fragments is a coordinate free property* which follows from the postulates of graded positional value, and averaging of disparities. Since the polar model is symmetric about the "high point," a narrow posterior fragment will duplicate, its complement will regenerate. Therefore, the direction of regeneration will reverse about the high point. Further, any 1/4 wedge section will heal its two cut margins and duplicate, while its 3/4 complement will regenerate.

Intercalary Regeneration, Betweenness and Convex Sets

A central feature of the postulates of intercalary smoothing is that only those positional values lying *between* the apposed values can be reformed. This leads to a critical restriction in the predictive consequences of any given co-ordinate system, since it implies that diffusivelike smoothing can *only* recreate positional values lying in the *convex set* bounded by the positional values in the apposed tissue edges. This restriction, in turn, implies that different co-ordinate systems may demand different special cellular behaviours beyond simple diffusivelike smoothing to account for the data. The concept of a convex set, and the limitations it imposes can be brought out in the polar model.

Radial positions can be visualized without loss of generality as a radially symmetric gradient whose peak is at the distal "high point." I show a wing disc from which the distal high point region has been removed, thus removing the radial gradient "peak." In the remaining outer proximal annulus, only lower values of the radial gradient are present. Therefore, no juxtaposition of tissue edges in the proximal annulus can lead to "diffusivelike" filling in of the missing high-point radial peak. In a polar model, if the region containing the origin is removed, that region does not lie "between" the positional values in the proximal annulus. That is, the region around the origin is not in the *convex set* of all those positional values derivable by averaging *any* pairs of positional values present along the cut margin of the proximal annulus. The implication of this feature of any polar co-ordinate model is that distal regeneration by a proximal fragment cannot be obtained by diffusivelike averaging of positional discontinuities, and some special rule is needed.

In the initial formulation of the polar co-ordinate model, special Rule 3, the complete circle rule was proposed. According to this rule, exposure of a complete circle of angular values at a proximal level leads to regeneration of distal radial values. With the assumption of this rule, the model accounts for the capacity of a truncated amphibian limb to undergo distal transformation, and regenerate distal structures. The same postulate accounts for distal regeneration by a proximal wing disc annulus. Equally, Rule 3 accounts for duplication of the cultured wing disc high point region from the exposed circle of angular values at a proximal radial level, and similarly explains duplication from the proximal cut surface by a newt hand whose digits are implanted into a host flank. Finally, Rule 3 accounts for the striking observation that grafting a left hand to a right stump yields two supernumerary limbs at the position of maximal discord of angular values.

Such a graft creates two complete circles of angular values at the radial level of the graft. These undergo distal transformation and yield two supernumeraries, with the handedness of the host. The successes of the polar model are considerable, and have led to testing several features of the model. I turn now to an evaluation of weaknesses in this mode. First, it should be stressed that special Rule 3, or any modified form of Rule 3, is formally equivalent to postulating a mechanism beyond diffusivelike averaging of positional values, to regenerate a missing radial gradient "peak." For example, in the current version of the polar co-ordinate model, its authors propose that short arc intercalation itself generates distal regeneration. While the postulate of a special mechanism to regenerate distal radial values is not a flaw, its status should be made explicit.

As shown next, distal regeneration does not require special rules in other co-ordinative systems. If special mechanisms exist to recreate missing gradient values, then those mechanisms are of central importance. They are likely to play a role in the initial *establishment* of positional gradients as well as subsequent pattern regulation. The forms of special mechanisms which are suggested depend on the choice of co-ordinate system. Thus different co-ordinate systems suggest that different cellular properties beyond diffusive averaging are required to account for the data on pattern formation. Second, as noted by Winfree, in the vicinity of the origin, the gradient of the angular variable in the tissue becomes infinitely steep, since the origin is a singularity in the angular map on the plane.

If cell proliferation is proposed to occur proportional to the gradient in radial or angular values, then proliferation near the origin should be unbounded. It is not. As Winfree points out, this implies a lack of independence of radial and angular gradients near the origin. Third, anterior-posterior and dorsal-ventral developmental axes appear to be established sequentially. This is not deducible, hence not explained, by the proposal that position is measured in polar co-ordinates. For example, if a radial axis is established first, the amphibian limb bud should be angularly symmetric in transplantation between the time when the first axis and second axis are formed, in contravention to Harrison's classical observation.

If angular values are established first, then reversal of anterior-posterior or dorsal-ventral areas should yield similar, not dissimilar results, at each stage, but do not. Fourth, a considerable body of work testing the complete circle rule has shown it to be false in imaginal tissue, cockroach limbs, and amphibian limbs (18,26,28). It now appears that distal regeneration is roughly proportional to the proximal angular arc present. Confirmation of the complete circle rule would have been striking evidence in favour of a polar model.

Its disconfirmation does not rule out such a model, but alternative co-ordinate systems have the deductive consequence that distal regeneration should be proportional to the proximal arc; hence the data disconfirming the complete circle rule leave a polar model with an *ad hoc* postulate as Rule 3, while alternatives deduce, hence explain, this phenomenon. Fifth, accumulating evidence shows that normally duplicating imaginal disc fragments can with variable frequency regenerate some, most, or perhaps all of the complementary domain. Such results are now known in the wing, haltere, leg, and genital discs. No convincing comparable data are yet available on amphibian or cockroach limbs or other epimorphic systems, although *divergent regeneration* from symmetrical half limbs in amphibians might be due to regeneration by normally duplicating fragments.

The polar co-ordinate model is constructed such that intercalation causes complementary fragments strictly to duplicate and regenerate. The model is unable to account for regeneration by normally duplicating disc margin fragments by simple positional averaging. In fact, no proposal co-ordinate system, coupled with the assumptions of simple diffusive averaging of apposed positional values, can account for extensive regeneration by both the normally duplicating and normally regenerating fragments of a tissue. All co-ordinate systems share the property that both complementary fragments have identical positional values along their cut margins, hence if both fragments heal in similar ways, apposing similar pairs of positional values, both fragments must reform the same set of structures lying in the convex set bounded by the apposed positional values.

Evidence that each complementary disc fragment can regenerate large domains of the other fragment is evidence that mechanisms beyond simple diffusive smoothing must operate during imaginal disc pattern regulation. The polar co-ordinate model can account for this phenomenon by weakening the short arc intercalation rule to allow long arc intercalation. Formally, this is equivalent to an active mechanism to regenerate angular values not in the convex set reached by diffusive smoothing. Sixth, mutants in *Drosophila* are available which can produce completely duplicated legs. In order for a disc fragment to produce a complete duplicate appendage by apposing cut surfaces of the fragment and positional smoothing, the apposed surfaces would have to form a convex set containing the *entire* tissue image. Examination of the polar co-ordinate model shows no single fragment has this property. Within the frame of the polar model, a *partial* disc fragment can produce a mirror symmetric duplicate, but no fragment of a normal disc can produce both a normal leg, and a complete duplicate appendage.

Transverse Gradients, or Modified Cartesian Co-ordinate System

Recently several workers have independently suggested a modified form of a Cartesian co-ordinate model to account for the data on epimorphic pattern regulation. Suggestions by Winfree and myself are nearly identical. The *Drosophila* wing disc with roughly orthogonal monotonic anterior-posterior and ventral-dorsal gradients of two chemicals, X and Y. Lines of constant concentration are bowed outward on the disc, symmetrically about the "high point." The "image" of this cross gradient system in a "tissue specificity space," i.e., the XY morphogen space. In this "image," lines of constant X or Y chemical concentration, are straight lines, so a convex, bowed line of constant X is a straight line. Similarly, a straight cut on the actual wing disc corresponds to a *concave* line. The model assumes that a cut wing disc fragment heals its cut margin, apposing nonadjacent positional values, and that simple diffusion smooths discontinuities in X and Y, and fills in the convex set bounded by the apposed XY pairs along the cut margin. As the bowing of lines of constant X and Y concentration on the disc implies that diffusive smoothing of X and Y discontinuities in new cells of a large posterior fragment of a single straight cut will fill the shaded convex set and *regenerate* anteriorly to the anterior-most value present along its cut margin.

The complementary anterior fragment apposes the same pairs of discordant values, and duplicates to the same anteriormost value. Symmetrical convex bowing of X and Y concentrations about the high point insures that the direction of regeneration reverses about the high point. A large anterior fragment from a straight cut will regenerate posteriorly, its posterior complement will duplicate. Similarly, any 1/4 pie-wedge fragment which apposes its two cut margins will duplicate, its 3/4 complement will regenerate. Therefore, a transverse gradient Cartesian model can yield reversal of the direction of regeneration about the high point, without the assumption that the high point is a locus from which cells measure position. Distal regeneration is a consequence of simple diffusive smoothing in a transverse gradient Cartesian model. As shown, deletion to the high point region leaves a proximal annulus.

Wound healing apposes tissue around the circular cut margin, thereby creating discontinuities in the X and Y gradients. Since the deleted high point region lie in the convex set reached by diffusive smoothing from the proximal annulus, the distal high point region is regenerated. Similarly, the high point region itself purse-string closes its circular wound margin, apposing the same pairs of positional values as did the proximal annulus, and therefore duplicates the X and Y high point values in the new cells of the wound area. The same principle predicts that any proximal arc of tissue will undergo limited distal regeneration of structures lying in the convex set of the positional values along the cut margin. Therefore, *limited* distal regeneration proportional to the proximal is present.

It has recently been demonstrated in imaginal discs and amputated symmetric half limbs of amphibians and cockroaches, and is a deductive consequence of a Cartesian model. As a transverse gradient model explains the striking observations that grafts of distal left limbs to proximal right stumps can generate two supernumerary limbs with the handedness of the host. With minor distortion from symmetry, the Cartesian model also has the consequence that 180° graft rotations of the left hand onto the left stump can yield variable numbers of supernumeraries, including two supernumerary limbs of opposite handedness. Finally, unlike the polar model, a transverse gradient Cartesian model is in direct accord with the widespread data suggesting that anterior-posterior axes are established prior to dorsal-ventral axes. Any transverse gradient model suffers certain defects analogous to those of the polar model in which the "high point" region in the polar model lies outside the convex set of the proximal annulus. Its regeneration requires special mechanisms.

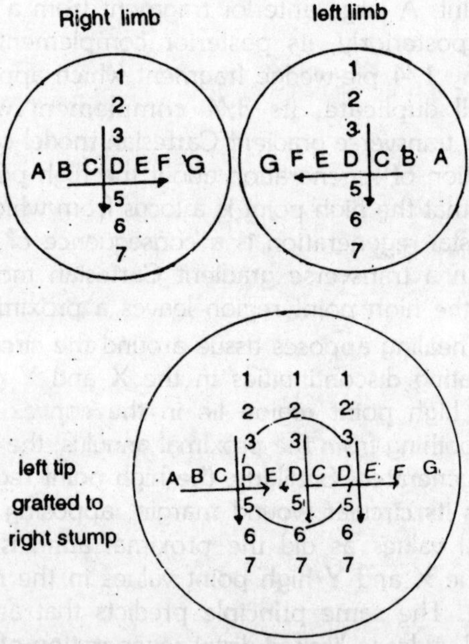

Fig. 8.1. Schematic picture of anterior-posterior and dorsal-ventral Cartesian X, Y co-ordinates for right limbs, projected onto the plate. The position at the mid values of the two variable, 4, D, corresponds to the distal limb tip. Grafting a left distal limb to a right proximal stump leads to smoothing of discontinuities in X and Y and the formation of two supernumerary distal limb tips with the handedness of the host proximal stump, at the positions of maximal disparity of host and graft axes.

In a transverse gradient Cartesian model, the extreme disc margins lie outside the convex sets of their complements. Thus, as the large posterior fragment only regenerates partially. It does not reach the anterior disc margin. Similarly, the narrow anterior fragment duplicates partially. Special mechanisms to recreate X and Y gradient maxima and minima are required to account for the observed capacity of disc fragments to regenerate or duplicate completely. Like the polar model, no Cartesian model can explain the capacity of normally duplicating fragments to regenerate large domains of the complementary disc fragment by diffusive smoothing. Nor can a Cartesian model account by diffusive smoothing alone, for mutants that form complete limbs and complete duplicate limbs, since no subregion of a Cartesian co-ordinated tissue can have the entire tissue image in its convex set. The assumption of a transverse gradient model leads to the suggestion that cells in a tissue have special mechanisms capable of reforming X and Y gradient maxima and minima lying outside the convex sets of cultured fragments. Explicit mechanisms are described next.

Spherical Co-ordinate Model

Russell proposed a spherical co-ordinate model using three orthogonal X, Y and Z gradients to form a Tissue Specificity Space. Position in a tissue is specified by a *solid angle* θ, ϕ corresponding to latitude and longitude angles. Each positional value is a unique ray at constant X: Y: Z ratio emanating from the origin. That is, each ray is a *line* of equivalent positional value in XYZ space. The longitudinal angle ϕ, is defined by the ratios of X and Y in the equatorial plane, and the latitude angle θ, is defined by the X: Z, or Y: Z ratio.

The image in XYZ space of a two dimensional tissue is a two dimensional surface pierced by a set of solid angle rays emanating from the XYZ origin, which itself does not normally lie in the physical tissue. The spherical model can account for almost all the available data using only the concept of diffusive smoothing. This is most easily visualized in the wing disc by remembering that the disc is topologically a spherical surface, the disc proper, backed by the peripodal membrane. Let this tissue spherical surface be embedded in XEZ space such that each solid angle ray pierces the closed two dimensional surface of the image once, specifying the positional ϕ, θ value at that point. Therefore, the image

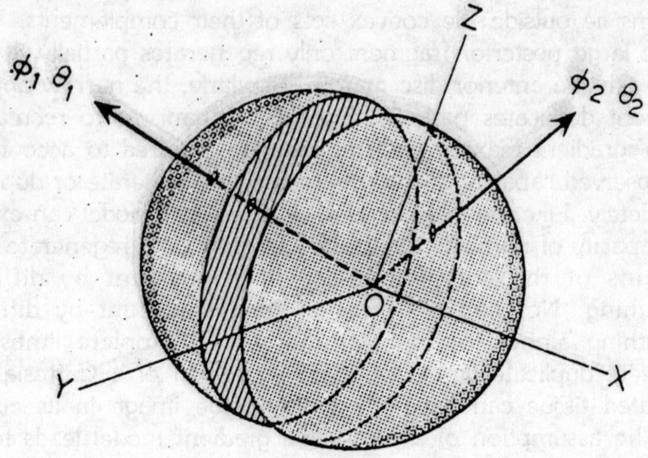

Fig. 8.2. Russell's (1978) spherical co-ordinate model. X, Y and Z are orthogonal gradients. Position is specified by ratio of X:Y:Z with respect to origin, 0, each ratio is a ray at a unique solid angle ϕ θ. The diagram shows a disc as a spherical surface (dotted region), with an anterior fragment cut off. Wound healing in the anterior fragment creates an additional new XYZ surface (parallel line). ϕ, θ Rays pierce both surfaces, hence the anterior fragment duplicates. The posterior fragment forms the same new XYZ surface and regenerates.

surrounds the origin, and each ray pierces a unique point on the tissue. If a narrow anterior fragment of the disc is cut, the margin purse string wound heals, and smoothing of X, Y and Z discontinuities in the new cells forms a *second* surface in XYZ space whose edges join those of the original anterior fragment image. Rays from the origin pierce both the new surface and the original anterior hence the narrow fragment duplicates completely.

The large posterior fragment wound heals similarly, and forms the same new image surface in XYZ space in the new cells of the wound area. These are pierced by the same rays the pierce the anterior fragment, hence the posterior fragment regenerates completely. The model is spherically symmetric. Therefore not only do narrow anterior, posterior, dorsal or ventral fragments duplicate while their large complements regenerate, but a distal fragment containing the high point region will purse string close, creating a second image surface in the new cells in the wound area. This second surface will be pierced by the same rays that pierce the high point region, therefore it will duplicate the high point. Equally the proximal annulus will regenerate distally.

Finally, this model directly explains distal regeneration proportional to the proximal arc cultured, and the incidence of supernumerary limbs. Despite its strengths, the spherical model suffers weaknesses. First, the spherical model, like the polar model, has a singularity at the north pole high point, where all longitudinal angular values meet. The hypothesis that positional gradients regulate proliferation confronts infinitely steep longitudinal angular gradient at the "high point." More importantly, while the spherical model is very appealing, it is unable to account for all the data by simple diffusive averaging of gradient discontinuities. Like the polar and Cartesian models, diffusion in the spherical model can yield only duplication and regeneration by complementary fragments. It cannot account for regeneration by normally duplication fragments, nor the existence of mutants which form normal limbs, and complete duplicate limbs. To account for these phenomena, it is necessary to imagine cellular processes capable of recreating missing X,Y or Z maxima and minima not in the convex set of the cultured fragment.

General Critique

The review of the polar, Cartesian, and spherical co-ordinate models leads to several comments.

I. What pattern regulation properties are co-ordinate-free, assuming only the fundamental process of intercalary smoothing of positional discontinuities?

1. The sequential formation of anterior-posterior or dorsal-ventral axes is not co-ordinate-free.
2. Distal transformation using only positional smoothing is not co-ordinate-free for it does not arise in a polar model. Further, the arguments that distal transformation can be co-ordinate-free implicity assumes embedding the tissue in a two variable co-ordinate system, i.e., in an R^2 space, with no singularities. This is equivalent to assuming *transversal gradients* of scalar variables. It is only coordinate-free in that the gradients need not be orthogonal.
3. Like distal transformation the generation of supernumerary limbs using only intercalation is not co-ordinate-free.
4. Duplication and regeneration by complementary fragments is co-ordinate-free in the restricted sense that both fragments

must make the *same* new structures in pattern regulation. However, that any specific fragment should wholly duplicate or wholly regenerate is not co-ordinate-free, as seen in the specific "bowing" of X and Y concentrations in the Cartesian model.

5. Intercalation itself, in one sense is co-ordinate-free occurring in any co-ordinate system with graded scalar variables. Yet it is not entirely free. In the polar model, for example, one might suppose *angular* intercalation occurs faster (or slower) than radial. In the former case, intercalary regeneration would initially spread around annular rings, in the latter, spread radially. In a Cartesian model, assumption of different X and Y intercalation rates leads to yet different directionalities to regeneration.

II. The geometries of positional gradients do not identify the co-ordinate system. It might be supposed that determination of actual morphogen gradients in a tissue would suffice to determine the co-ordinate system used by the cells. That this is not so can be seen by re-examining the spherical co-ordinate model. Here the tissue is imbedded in an orthogonal X,Y,Z tissue specificity space. Ratios of these three chemicals specify rays of a solid angle emanating from the origin. A different use of the same three gradients provides a simple molecular interpretation of the polar model. Let the *concentration* of Z specify distance from the high point and the *ratio* of X and Z the azimuthal angle modulo 2π. Each ray in a plane of constant Z, emanating from the Z axis corresponds to a unique angle at a unique Z radial distance from the high point. Simple diffusive smoothing of X, Y and Z, discontinuities recovers the intercalation rules of the polar model. Finally, position might be specified by the actual *concentrations* of X,Y and Z, not their ratios. Thus, the same three orthogonal gradients might be used by the cells in *diverse* co-ordinate systems, with differing predictions about the regulative properties achievable by positional intercalation alone.

III. There is no persuasive evidence of positional singularities. Although the same gradients might subserve different co-ordinate systems, one hope of distinguishing among them is to ask whether singularities exist as expected in the polar and spherical models. The data reviewed offer no persuasive evidence. Not only can "high point" behaviour be generated in a Cartesian model without singularities, but behavioural high points, about which the direction of duplication and regeneration reverse have not been demonstrated in other discs. Indeed, if there is a "high point" in the leg disc, it lies at or near the edge of the upper medial quadrant, and certainly not at the distal tip.

IV. Since no co-ordinate system can account for regeneration by normally duplicating fragments through intercalation of positional discontinuities alone, further processes must occur in epimorphic pattern regulation. This implies that choice among potential co-ordinate systems cannot be made simply by excluding those, like the polar, that require active processes to explain large fractions of the data.

V. The conceptual framework of positional information itself, and of discussion of alternate co-ordinate systems is inadequate. The general problems are these:

1. Under the postulates of positional information, the pattern which results from *interpretation* of the information is *wholly arbitrary* with respect to the features of the positional field itself. Wolpert's French Flag can switch arbitrarily to a "Wolpert Flag" of Jackson Pollock complexity. No constraints are imposed by the field.
2. Further, even given the assumption of a co-ordinate system, whether polar, spherical, or Cartesian, its relation to the tissue domain is wholly arbitrary and free of the geometry of the tissue itself. For example, in a polar model, we have no grounds whatsoever to think that the high point will be located in the middle of the wing disc and at the upper medial leg disc margin. In short, beyond those properties that flow from smoothing positional discontinuity alone, the hypothesis of positional information is at best weakly predictive.

VI. The observations show that processes beyond smoothing of positional discontinuities in *pre-established* fields are involved in epimorphic pattern regeneration. This suggests that regeneration includes those processes initially involved in pattern generation. While alternative basic premises for considering pattern generation are possible, I suggest the following:

1. Pattern generation is fundamentally self-organizing. While this may not be true in mosaic eggs, the capacity for small duplicating leg disc fragments to regenerate entire legs is but one example of such self-organization.

2. The same example supports a preference for theories which do not postulate special (e.g., source or sink) cells.
3. Intercalation of positional gaps suggests position behaves as though it were spatially averagable. The simplest physical example is diffusion, which suggests that the Laplacian v^2 operator expressing such averaging is likely to be fundamental.

The implication of the assumption that diffusivelike averaging of positional variables occurs lies in the fact that such processes naturally couple the size and shape of a tissue to the profiles of gradients which occur in it. The central mathematical concept is an eigenfunction; intuitively this corresponds to a gradient profile having the property that, under the action of diffusion, the profile keeps its *shape*, but may change in amplitude. For example, in a long tube with sealed ends and initial cosine distribution of diffusible stain with maxima and minima at the ends would maintain the cosine distribution but relax to the uniform distribution. Even for modified forms of diffusive-like averaging of positional information, describable by modified operators, the eigenfunctions of such operators provides a natural link between the tissue geometries and the shapes of gradients of positional variable within it. Thematically, the tissue shape controls the profiles of the positional field, which in turn controls tissue geometries and internal inhomogeneities, providing one framework for a self-organizing theory of pattern generation.

Further, if gradient profiles are naturally related to tissue shape, it also becomes reasonable to search for natural, rather than arbitrary relations between those profiles, and subsequent interpretation events. For example, "thresholds" might follow contours of constant concentration, rather than being placed arbitrarily. Consistent with the hypotheses of self-organization, absence of special cells, and diffusivelike averaging, the task is to construct theories able to generate at least two-dimensional positional information. In the next section, I show that the familiar two variable reaction-diffusion model does just this is a particularly natural way.

Sequential Generation of Two Positional Axes and Two-dimensional Positional Information

It is now well-known that model biochemical systems in which synthesis and degradation of chemical species are coupled, and their diffusive transport occurs, can lead to the formation of spatially inhomogeneous gradients of the constituents in the domain. As a particular model, we postulate a single biochemical system of two components, with concentrations X (*r, t*) and Y (*r, t*) at position *r*, time *t*, which are being synthesized and destroyed at rates F (X, Y) and G (X, Y) and are diffusing throughout a tissue. The equations for this system are:

$$\frac{\delta x}{\delta t} = F(X, Y) + Dx\nabla^2 x$$

$$\frac{\delta x}{\delta t} = G(X, Y) + Dy\nabla^2 y \qquad (1)$$

which are chosen to have a spatially homogenous steady state X_o, Y_o at which $F(X_o, Y_o) = G(X_o, Y_o) = 0$.

Analysis of system (1) begins by linearizing about the specially homogenous steady state by substituting $X(r, t) = X_o + x(r, t)$, $Y(r, t = Y_o + y(r, t)$ and retaining only terms up to first orders in x and y in a Taylor expansion of F (X, Y) and G (X, Y). The resulting linear equations in deviations x and y from the homogenous state are:

$$\frac{dx}{dt} = K_n x + K_{12} y + Dx\,\Delta^2 x \qquad (2)$$

$$\frac{dy}{dt} = K_{21} x + K_{22} y + Dy\,\Delta^2 y$$

The stability of the spatially homogeneous steady state of the linearized equations to spatially inhomogeneous perturbations in the concentration of x or y may be analyzed by evaluating the determinant of the matrix

$$\begin{vmatrix} K_{11} - k_1^2 D_1 - \lambda & K_{12} \\ K_{21} & K_{22} k_1^2 D_2 - \lambda \end{vmatrix} = 0 \qquad (3)$$

yielding the dispersion relation between the two temporal eigenvalues

$$\frac{\lambda_1^{\pm} + {}^1/_2\,[K_{11} + K_{22} - K_2\,(D_1 + D_2)] \pm}{{}^1/_2\,\sqrt{[K_{11} + K_{22} - k^2(D_1 + D_2)^2 - 4[k^4 D_1 D_2]^2 - k^2(D_2 K_{11} + D_1 K_{22}) + K_{21} K_{22} - K_{21} K_{12}]}} \qquad (4)$$

Here $k_1 = k$ is inversely proportional to the wavelength of the perturbation λ_1. For a given wavelength perturbation λ_i, corresponding to a specific value of k_1, if both λ_i + and λ_i have negative real parts, that perturbation decays. In an unbounded domain, all perturbations with wavelengths corresponding to values of k_i for which the dispersion relation has at least one temporal eigenvalue with positive real part, will grow in amplitude in time and create a spatial pattern. With appropriate constraints on the linearized reaction and diffusion constants (19) the dispersion relation $\lambda = f(k)$ is positive for a restricted range of k, $k_1 > k_1 > k_2$, hence for a restricted range of wavelength 1, $l_i < l_i < l_2$. In this case, system (1) acts as an amplifier, selecting and amplifying from thermal noise those wavelength components 1 in the range between l_1 and l_2. If system (1) exists in a bounded spatial domain and no flux boundary conditions are imposed, then a chemical pattern will be established only if it simultaneously has a wavelength in the range $l_1 = l_2$, and also satisfies the boundary conditions that the gradient have zero component normal to the boundaries assuming no flux boundary conditions, the shapes of chemical patterns which can "fit" on to the spatial domain must satisfy the Laplacian diffusion operator, and are an infinite series of specific *eigenfunction* patterns determined by the geometry of the domain. The first several eigenfunctions on a wing disc shape are shown.

Establishment of a First Positional Axis

As the spatial domain in which system acts grows gradually larger, a succession of the discrete eigenfunctions or their superposition will sequentially become established in the domain. For any reaction-diffusion system which acts as an amplifier, perturbations below some minimal wavelength, l_1, will die away. Therefore, if the maximum physical length of the tissue is less than l_1, the spatially homogeneous, state is stable. As the tissue grows in length past l_1, the first eigenfunction of the tissue domain will bifurcate from the homogeneous solution.

A fundamental theorem of bifurcation theory guarantees that the first pattern will be a stable solution of the full non-linear equations. Therefore, as a wing disc shape gradually enlarges, no chemical pattern forms for lengths below l_1, then eigenfunction 1, grows of the homogenous state, and forms a monotonic gradient along axis of the tissue. It has been recognized for sometime that this primary bifurcation provides a means to establish a first axis of positional information in a growing domain and supply one dimension of positional information. The new question 1 wish to address is: Can a single reaction-diffusion system like (1) establish two transverse axes in a growing tissue? At first glance, the answer appears to be "No."

In the linearized equations of or any similar system, each eigenfunction pattern which forms in the tissue is a spatially inhomogeneous distribution of a *fixed ratio* of the underlying chemical variables x and y, measured as deviations from the homogenous steady state X_o Y_o. This property follows from the fact that unstable wavelength l_1 which is amplifying on the tissue, corresponds to a unique scaling factor k_1. Each unique choice of k_1 specifies a particular set of values in matrix having at most one positive temporal eigenvalue l_1; and a unique corresponding eigenvector x_i:y_i, which is a fixed ratio of the linearized deviations of x and y from X_oY_o. This constraint implies that lines of constant concentration of X on the tissue parallel those of Y. Therefore, although two chemical gradients are present in a two-dimensional domain, they supply at most one dimension of positional information. More generally, if *N* chemical species are in system (1) $N > 2$, all N occur as a fixed ratio when a single temporal eigenvalue λ_i, is positive, hence lines of constant concentration of all *N* are parallel to one another and supply only one positional axis. Further, if any single eigenfunction pattern, is amplified alone in the tissue through a single positive eigenvalue, the ratio of all *N* species is again fixed, and at most one positional gradient is established.

Establishment of a Second positional Axis

It has not been generally recognized that the familiar class of two-variable reaction-diffusion systems capable of establishing a first axis can *sequentially* establish transverse axes in growing two-dimensional domains. This, can be exemplified on a wing disc shape. If the range of wavelengths l_1 and l_2 which are amplified by the linearized system (1) is made broad enough, then as the tissue domain enlarges past l_1, the first eigenfunction mode,. forms. As the two-dimensional domain enlarges further, both the first ventral-dorsal monotonic pattern and the second anterior-posterior monotonic eigenfunction pattern, satisfy the boundary conditions and may be amplified together.

If the length and width of the tissue are unequal, each mode corresponds to a different specific wavelength l_i, l_i ($l_i > l_i$), and the corresponding scaling factors k^2 are unequal, k^2i = k^2j.

Therefore, the eigenvector ratio of the chemical variables *x* and *y*, derived from matrix, differs for the two modes. The superposition of the two distinct modes, each amplifying a different ratio of X and Y, yields non-parallel (transverse) gradients of X and Y in the tissue. As a second example, on a growing rectangle whose sides are of unequal length, $L_1 > L_2$, the linearized equations initially amplifying the first eigenfunction creating parallel X and Y monotonic gradients in the L_1 direction, thereby creating a first axis of positional information. Then when the rectangle enlarges further the chemical system amplifies the *superposition* of the first two monotonic eigenfunctions in the L_1 and L_2 directions creating transverse gradients of X and Y, and two positional axes:

$$c_1 \begin{bmatrix} x_1 \\ y_1 \end{bmatrix} e_1^{\lambda} T \bullet \cos \frac{\pi r_2}{L_2} \begin{bmatrix} x_1 \\ x_1 \end{bmatrix} + c_1 \begin{bmatrix} x_2 \\ y_2 \end{bmatrix} e_1^{\lambda} T \bullet \cos \frac{\pi r_2}{L_2} \begin{bmatrix} x_2 \\ y_2 \end{bmatrix} \quad (5)$$

The first major property of this class of models, therefore, is the prediction that two transverse positional axes should form *sequentially* as the tissue grows in size.

The linearized model of is of limited value for several reasons. The amplitude of each of the two superimposed modes grows exponentially without bound. Whichever mode grows faster (λ_1 or λ_2) eventually dominates the superposition, therefore a single eigenvector $[x_1{:}y_1]$ or $[x_2{:}y_2]$ dominates the pattern and the underlying X and Y gradients become parallel. Second, the contributions of the first and second patterns to the superposition are given by arbitrary initial conditions, c_1, c_2. Finally, each eigenfunction pattern can occur in two opposite forms or branches, with maxima replacing minima. Which is established depends entirely on initial conditions. The conjunction of these defects implies that the relative orientation of X and Y gradients on the tissue is not uniquely fixed by tissue geometry in a linear theory. These defects of a linear model are fully met by a fairly broad class of non-linear reaction-diffusion systems in an *asymmetric* two dimensional tissue such as the wing disc. A number of workers have proposed non-linear models which are known to bound the amplitudes of chemical patterns in one spatial dimension and two dimensions.

Generally, the establishment of the first pattern parallels the linear analysis, except that for large amplitudes, the pattern may deviate from the eigenfunction pattern. As the spatial domain enlarges, the second pattern commonly is gradually established through a secondary bifurcation from the primary spatially inhomogeneous monotonic pattern, yielding at each tissue size a fixed mixture or amplitude of distorted first and second modes. Finally, in a spatially asymmetric two-dimensional domain, the cube of the eigenfunction is almost always non- zero. From bifurcation theory this implies that, as the tissue enlarges, only one of the two branches (or forms) of the primary monotonic pattern emerges smoothly from the homogenous state, while the other branch is stable but requires a macroscopic perturbation to be attained. For at least some non-linear systems, only one branch of the subsequent secondary bifurcation emerges smoothly from that first mode. The novel implication is that the entire sequential establishment of a first positional axis, then two transverse positional gradients in two dimensional domains, their shapes and orientations, can be uniquely determined by the changing tissue size and geometry, and the kinetic parameters of the non-linear model. In Fig. 10a, b, I show the results of simulations of one non-linear model given by:

$$\frac{\delta X}{dt} = -AX + \frac{BY^6}{1+Y^6} + D_1\,\Delta^2 X \qquad (6)$$

$$\frac{\delta Y}{\delta t} \pm CX + \frac{D(Y6+6)}{1+Y^6} + D^2\,\Delta^2 X$$

with parameter values specified in the legend. As a wing disc shape enlarges, a first ventral-dorsal monotonic gradient of X and parallel gradient of Y is established and supplies one-dimensional positional information along one positional axis. As the disc enlarges further and the underlying anterior-posterior second mode begins to contribute to the pattern, the X and Y gradients both rotate slowly on the disc, Y faster than X, gradually establishing distinct non-parallel transverse, but not orthogonal, monotonic X and Y gradients spanning the tissue forming a second positional axis. The model sequentially establishes transverse axes and two dimensions of positional values through two gradients whose forms and orientations are initially identical, but gradually come to differ as the tissue enlarges. The positional field is uniquely determined by the shape trajectory of the tissue, and the specific non-linear model.

Novel Predictions

The inherent simplicity and minimality of this two-variable reaction-diffusion mechanism commends it as a serious model for the establishment of transverse positional axes and positional information in developmental fields. It may be that a more complex three variable model coupled with the superposition of three of more modes can yield three transverse gradients which might underly Russell's spherical co-ordinate model or a polar co-ordinate model. The minimal model has a number of predictive consequences.

1. During establishment of the second axis, the positional gradients may shift or rotate on the tissue. Duranceau reports that positional values do appear to rotate on the developing wing disc.
2. Since the monotonic gradients have values above and below the homogeneous steady state values $X_o Y_o$, that pair of values will typically occur at least once at an interior (hence, distal) point in the tissue.
3. The axes are transverse, but oblique, not orthogonal.
4. Positional axes are established sequentially, as seen in several systems
5. The longer axis in a field is established before the shorter.

The model predicts that very small fragments cannot maintain gradients and will decay toward the steady state; therefore small fragments should drift toward the map position in the tissue corresponding to the steady state. Consistent with this prediction, Strub in fact found that dissociated and reaggregated proximal leg of wing imaginal fragments from small transform distally. Predictions of the two variable, non-linear reaction-diffusion model with regard to duplication and regeneration are a combination of properties arising from simple passive diffusive averaging of positional discontinuities, and properties arising from the active dynamic and boundary conditions, and are described more fully elsewhere. It is important to stress that the capacity of simple two-variable reaction-diffusion systems to generate transversal gradients supplying two-dimensional information is not restricted to a special subset of these models, but is rather a robust feature to be expected in essentially all members of this class of models.

The general reason this is so is that these models generically amplify a finite *range* of wavelengths, $L_1 - L_2$. As the tissue in

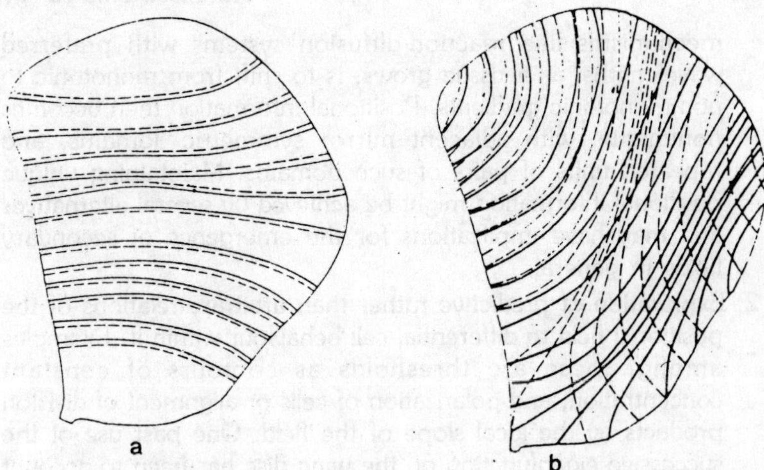

Fig. 8.3. A—Establishment of a first axis. First longitudinal monotonic spatially inhomogeneous steady state of non-linear system (6), corresponding to first eigenfunction. Ones of constant X concentrations (solid lines) parallel lines of constant Y concentrations (dashed lines), Spacing between lines of constant X concentrations, or of Y concentrations are not strictly proportional to gradient slope, but are used to indicate shapes of X and Y gradients. B—Establishment of Second Positional Axis. As wing disc grows, second eigenfunction, makes a contribution to mixed non-linear steady state solution. X and Y gradients rotate away from positions in (A), and lines of constant X concentrations (solid lines) separate from lines of constant Y concentrations (dashed lines) creating transverse gradients and a second positional axis. Parameter values used for the non-linear model (6).

which such a reaction-diffusion system occurs grows larger, or other parameters alter, it must eventually reach a size at which N of the longer wavelengths "fit" on the tissue, while $N + 1$ of the shorter wavelengths simultaneously fit. Since both modes satisfy the boundary conditions, both will be amplified and the resulting (non-linear) superposition of gradients will be locally transversal. A very substantial amount of mathematical work will be required to clarify the character of such patterns. I believe this class of models which can couple the shape of a tissue to the shape of the transverse gradients which arise in the tissue, warrants deeper exploration. Important directions for further elaboration of this class of theories include:

1. Analysis of the implications of non-monotonicity in positional gradients. The typical behaviour of pattern generating

mechanisms like reaction-diffusion systems with preferred wavelengths, as a tissue grows, is to shift from monotonic to non-monotonic gradients. Positional information then becomes nonunique, with adjacent mirror symmetric domains, and repeated units of pairs of such domains. Maintaining unique positional information might be achieved by several alternatives and may have implications for the emergence of secondary fields in general.

2. Exploration of predictive rather than arbitrary relations of the positional field to differential cell behaviour within it. Examples among, these are thresholds as contours of constant concentration, and polarization of cells or alignment of division products by the local slope of the field. One past use of the successive eigenfunction on the wing disc has been to account for the numbers, positions, two-fold symmetries and sequence of wing disc compartmentalization.

9

Regeneration and Morphallaxis

The term morphallaxis was applied originally in the study of animal regeneration, early in the 1900s. It defined a process where structure appeared to be regenerated by remodeling over a wide area within remaining tissues, so that complete pattern was restored without any localized zone of special growth being responsible for the replaced parts. Such regeneration can be seen in adult planarians and urochordates, but the precise roles played in it by special reserve cell populations, cell migration, and the alteration of differentiation and growth remain imperfectly understood. Recently, the term has been extended to initial development to refer to those regulative systems, seen in early embryos of many (perhaps most) types, where cell's contributions to the final pattern of the body are set by their *relative positions* within a cell sheet or a mesenchyme during the period across which determination occurs.

The proportions and completeness of the pattern are thus independent of growth schedules and of the size of individual examples. This contrasts with patterns of cell differentiation built up by directed lineage, or by the local control of growth and cell fate by interaction among neighbours i.e., epimorphic processes. Some far-reaching mechanism of intercommunication or signalling within the embryonic tissue is obviously implied by this positional phenomenon, but although the idea that there is essentially just one form of such mechanism is forceful and attractive, we still lack evidence for this.

Moreover there is, in evolutionary terms, less a priori reason to expect a universality of mechanism for control of multicellular patterns than for, say, the hereditary transmission of genetic information. In this sense, continued use of the term morphallaxis as if it referred to one as-yet-unelucidated mechanism may be disadvantageous. The system is that which initially specifies the body plan in the early vertebrate (specifically, amphibian) embryo. Our own membership of the zoologically very homogenous vertebrate group gives an incentive to attempt to understand the early development even when we currently lack some of the technical advantages evident for other groups, notably insects. Certain vertebrate systems have a classic status as experimental objects, probably because of ease of access for surgical manipulation. This has resulted in an extensive backlog of geometrical knowledge about the fate map of material and the cellular movements in undisturbed development, such knowledge being prerequisite for any approach to analytical rigor in experiments of the type to be described.

The Medio-Lateral Dimension

The Anatomy of the Developing Pattern

The amphibian mesoderm develops by immigration of cells from a surface or sub-surface equatorial zone, the marginal zone, of the blastula. The future dorsal midline cells lead the process but the zone from which tissue is recruited soon comes to be annular, so that a bilaterally symmetrical, cylindrical cell sheet develops with an extended dorsal side. Keller Nakatsuji and co-workers have studied in fine detail the origin, fates, and behaviour of the mesoderm cells during normal gastrulation and pattern formation. By the time the medio-lateral pattern of its future contributions to the body is being established, it is one or a few cells thick, and situated between the neurectodermal layer and the yolky endodermal mass.

Cells are continually added to form the cross sections of more posterior levels of the cylinder (and thus body axis) from an annular zone around the closing blastopore. The four territories comprising the archetypical vertebrate medio-lateral pattern are the mid-dorsal notochord, the paired somite-forming columns, paired pronephric zones and the lateral plate mesoderm with blood-forming islands ventrally. The tissue to be devoted to these structures is established as zones, of particular relative extents, within the cross-section of the mesodermal cylinder just described. The total field between the extremes is in the order of 100 cells and 1.5 mm within the cell sheet. Thus with four stripes and bilateral symmetry as trivial added features, this is an instance of the problem of the regulating "*French flag*" used by Wolpert in his general treatment of the concept of pattern formation.

The initial polarity for the pattern, i.e., the dorsal midline position and site of the first immigrating portion of the future mesodermal cylinder, is set from an "organizer" boundary that derives, in amphibians, from a material localization in the activated egg. But with this exception, the evidence is that ooplasmic localizations, cell lineage, and the cell cycle, and even total cell number or tissue extent across a considerable range are of little consequence for the completeness and proportioning of the pattern attained. Position, by contrast, is determinative of the fate of material until quite late stages before the first visible expression of the determinations.

Outline Of Classical Findings And Concepts

Cell Fate Determined by Position

The term *determination* as used of embryonic tissue must always be defined operationally; that is, in terms of what has been done to the cell to test their state of commitment, or restriction in development potential. Unclearness about the implications of particular observations, in assessing the state of determination of parts of the embryo, has dogged vertebrate developmental biology ever since the period of its classic works. Thus one surprisingly often encounters views to the effect that the body plan of amphibian embryos is built up because the successive cleavage products, throughout the egg, inherit different subprogramms in the form of "ooplasmic" precursor localizations, or because small groups of early cleavage products are set aside (e.g., by binary sequential decisions) as the only possible founders of parts of the pattern, as is seen in insect embryogenesis.

It is appropriate here to summarize the evidence for three assertions, namely (a) that cell potential for most parts of the mesodermal body pattern is liable until late stages of gastrulation, (b) that the property of acting as a pattern-controlling boundary "the organizer" is restricted to a particular group of cells that normally develop the mid-dorsal territory, and (c) that cell-cycle related changes of differentiated state, and/or differential migration and adhesion among determined cells cannot make more than local and late—acting contributions to primary pattern establishment. Despite Driesch's original characterization of the unknown mechanism of morphallactic development as a "harmonious equipotential system" there are few embryo types which long remain in the condition where all large coherent parts, when isolated, will adjust the fates of their cells to form complete, normally proportioned editions of the body plan.

Vertebrate embryos are indeed among the most highly regulative in this way, but from the onset of cleavage in amphibians, isolated pieces (e.g., halves) will only regulate and form substantially normal pattern if they include material from the restricted region, the organizer, whose fate is to contribute to dorsal mesoderm and the head endoderm. As already mentioned, properties that are the precursors for this region are first localized within the egg near the onset of development. If an embryo fragment has the organizer region near one of the edges that must reheal to restore its geometric continuity (i.e., if a presumptive lateral, rather than a dorsal, half embryo is created), the achievement of normal pattern is unreliable from all stages and seldom found from operations as late as late blastula (5000 cell) stages. But this loss of regulative capacity does not mean that an advanced degree of determination has taken place among the individual cells of the embryo.

Large mesodermal isolates cultured from much older (50,000 cell) gastrulae often develop according to presumptive fate only (i.e., no regulation), but even this does not imply determination among their cells at the time of explanation. Such observations tell us only that intercellular communication systems that normally regulate pattern (including any biomechanical as well as biochemical factors) have dynamics such that a spatially complete set of events can no longer be achieved from the starting conditions obtaining within the isolate. No cellular "decisions" need have occurred, and such isolates, placed as part of another embryo that retains the appropriate dynamics (effectively, one that possesses the organizer), may deviate from their original fate in whatever direction is required for the production of normal pattern in the host. Much about the communication system that regulates pattern might still be learned from a systematic exploration of the times at which each sector of the marginal zone other than the organizer loses the capacity to be assimilated into the normal pattern, when relocated to particular sites.

Spemann and co-workers were responsible for a body of work which led to concepts, about the dynamics of such development, that have essentially been inherited by a majority of vertebrate embryology laboratories today. Their empirical discovery of the organizer and of the induction of the nervous system were undoubtedly achievements of crucial importance, but their theoretical legacy may have done much to retard the subsequent progress of understanding. It was found that one specific cell group, when relocated to remote, presumptively ventral parts of a host embryo, was able to cause development of a second edition of the body pattern, centered on itself but sharing the host material with the pattern of original orientation. This group included the

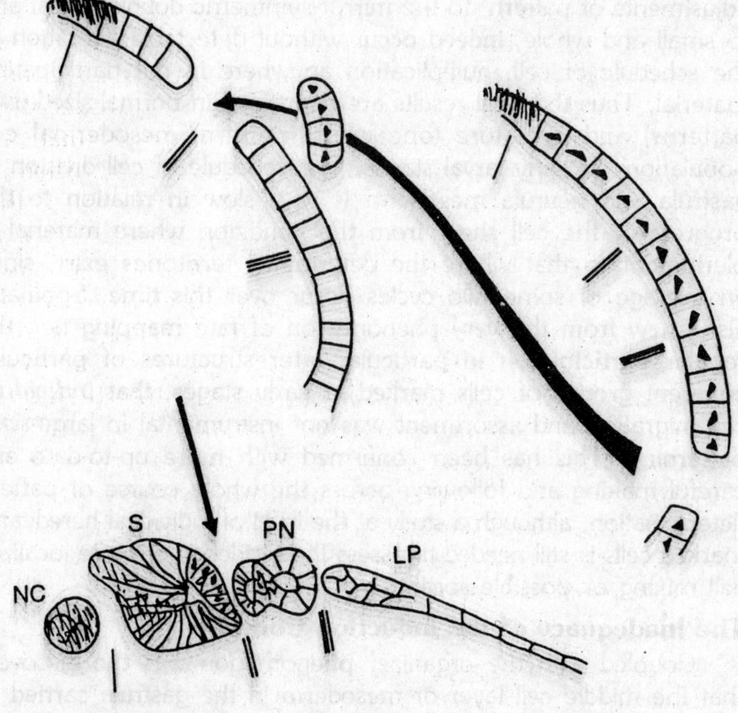

Fig. 9.1. The irrelevance of cell lineage and cell sorting for pattern establishment.

dorsal midline of the zone of mesoderm immigration (dorsal lip) and future head endoderm at early gastrula stages.

The graft developed to make its normal, dorsal, and anterior presumptive contribution in the second pattern, while the extent of host participation was variable but could, under some circumstances, approach a 50% subdivision of the available material with the original pattern. To confine ourselves momentarily to the mesoderm, the real subject of this chapter, it can be deduced from Spemann's writings that he believed that the re-specification of the fates of material, involved in this phenomenon as well as in that of regulation to wholeness by early dorsal or lateral half embryos (see above), was essentially on a geographical or relative positional basis. Any special stimulation of cell division and thus new tissue production played at most a minor role.

Contemporary work has shown that these two major regulative adjustments of pattern, to the mirror-symmetric double dorsal and to small and whole, indeed occur without detectable alteration in the schedule of cell multiplication anywhere in the participating material. Thus the final results are expressed in normal-sized (two patterns) and miniature (one whole pattern) mesodermal cell populations at early larval stages. The schedule of cell division in gastrula and neurula mesoderm is very slow in relation to the progress of the cell sheet from the condition where material is pluripotent to that where the determined territories exist, since an average of some two cycles occur over this time. Spemann also knew from the very phenomenon of fate mapping i.e., the regular participation in particular later structures of particular coherent groups of cells marked at early stages, that *individual* cell migration and assortment was not instrumental in large-scale patterning. This has been confirmed with more up-to-date and careful making and following across the whole course of pattern determination, although a study at the level of individual hereditarily marked cells is still needed to assess the incidence of more localized cell mixing or possible sorting.

The Inadequacy of the Induction Concept

Coupled with the organizer phenomenon was the discovery that the middle cell layer or mesoderm in the gastrula carried its dorsal midline the influences responsible for positioning and regionalization of the neural plate in the overlying neurectoderm. The extra central nervous system accompanying the second body pattern after dorsal lip implantation was thus a natural result of incorporation of an organizer into the mesoderm beneath it, being a territory *induced* within the neurectoderm. As implied by Wolpert, it would have been well for developmental biology if the term, *induction*, had then been restricted to events like this evocation of the nervous system from neurectoderm by contact with mesoderm; that is, to the transfer of information determining a pathway of development between one cell layer and another. As it was, Spemann and his school expanded the concept to refer indiscriminately to all hypothetical signals, emanating from particular determined territories of pattern to evoke other determinations in neighbouring material.

The spatial organization of development came to be seen just as a sequential hierarchy of inductions, with failure to distinguish conceptually between processes believed to occur along the plane of one cell layer or tissue and those occurring between two opposed layers or tissues and co-ordinating the patterns of differentiation between them. This has remained the conceptual framework for most embryologists. Thus most textbook accounts of the grafted amphibian organizer phenomenon fail to distinguish between the direct effect of the graft itself, which is to set a new boundary for the dynamic system organizing the pattern of mesodermal territories, and the accompanying induction of a second central nervous system with this new dorsal midline of pattern as a secondary expression. We are here concerned only with events controlling primary pattern, which arise in the mesodermal layer.

It can be seen intuitively that without some ancillary concept corresponding to negative feedback or self-limitation, models based on hierarchies of induction fail fundamentally to explain the salient features of morphallactic regulation, the size-independence of the completeness and proportions of pattern. Briefly, if cells on finding themselves to be developing as type A manufacture a specific substance which diffuses or is otherwise exported into the surroundings, and causes there the development of cell type B, and so on, there is no way that the system as a whole can take account of the overall space available in setting the spatial scale of the territories devoted to A, B etc. The kinetics of manufacture, diffusion/export and breakdown of signals and of cells responses must be part of the species metabolic equipment, and not naturally susceptible to feedback modification against individual embryo size without a special mechanism.

It is furthermore hard to see what such a feedback mechanism might be. Without it, the intrinsic property of sequential induction is to use up a particular amount of tissue or space for each particular territory induced, until either pattern is completed and the remaining space makes up the final territory, or the process runs out of tissue and pattern remains incomplete. Spemann, and most experimental embryologists writing since, while aware that vertebrate embryos avoid such incompleteness or variable over a wide, experimentally induced range of sizes at pattern formation, have been unable to confront this fact analytically.

The Testable Hypothesis that Positional Information Regulates the Pattern

Several biologists working with invertebrates, earlier this century, had tried to encompass morphallactic regulation in development and regeneration within a clear theoretical framework. In recent years, Wolpert has proposed a clear and, in principle, testable concept in the hypothesis that *positional information* underlies the control of pattern. He further proposed that this principal is universally used, and even that a universal mechanism underlies it in the development of regulative embryos across the animal kingdom. This vigorous and challenging notion has caused a wealth of new work, and undoubtedly stimulated the re-awakening of interest in the fundamental problem of pattern.

It is currently in danger of degeneration or "dilution" however in that wherever pattern develops, then cells are said to have interpreted "positional information" and the two phenomenon are treated as almost synonymous. There is, accordingly, a certain impatience with the concept among experimental scientists. This is unfair, since although cast in abstract terms, the real scientific value of the hypothesis lies in its precise postulates about the form of the mechanisms, and thus its predictions which are potentially refutable in an appropriate test. It is not a *necessary* concept, since patterns could from and be regulated in other ways. The abstract hypothesis is that normal pattern is achieved, despite any earlier disturbances which relocate cells within the field or create abnormally sized fields, because a communication system operates through the entire tissue. This system ensures that some cellular variable, whose local values set cells states to from the pattern territories, is distributed as a gradient with absolute boundary levels at the extremes of pattern and a particular, monotonic profile in the space between. The variable concerned might be a substance concentration or ratio of such concentrations or a range of cell surface states or configurations.

Boundary states are set up by early localizations in the egg structure giving special properties to regions, or by symmetry breaking processes, and the consequent boundaries are the organizer regions seen in experimental embryology. Correct pattern depends on the appropriate interpretation of the gradient values by cells in terms of epigenetic action. There is then a family of more or less plausible hypothetical, but concrete physico-chemical systems whereby such a positional gradient might be set up and regulated, bearing in mind that normal pattern depends upon normal boundary values for the single state and a normal gradient profile between them. The application of the formal idea to the pattern under discussion, and then the principal model gradient mechanisms that have been proposed. These are diffusion from a focal source with degradation elsewhere, and reaction/diffusion models.

The latter propose that autocatalytic processes of short-range diffusion produce a particular state of cellular activation as a local peak near one boundary, but lead to a larger-range diffusion gradient of an inhibitor of their own activity. The profile of either activation or inhibitor level may be used as the positional gradient. Diffusional transfer of substances of appropriate molecular weight to carry certain biological specificity might be consistent with the formation of gradients on the space and time scales (10 hr-1 mm) that are required. The possibility of a series of cell surface states modified by neighbour-interaction, as a positional signal, exists but has not been formally modeled due to our profound ignorance of the machinery involved.

The amphibian mesoderm system offers no evidence from grafting experiments for more than one organizing boundary, the organizer of Spemann. Ventral tissue experts no reciprocal effect to change the fate of its surroundings if grafted, but is assimilated into normal pattern if given time. One area of test for the hypothesis concerns limits on the performance of these model physicochemical systems in producing complete and normal gradient profiles (thus complete patterns) under various abnormal conditions

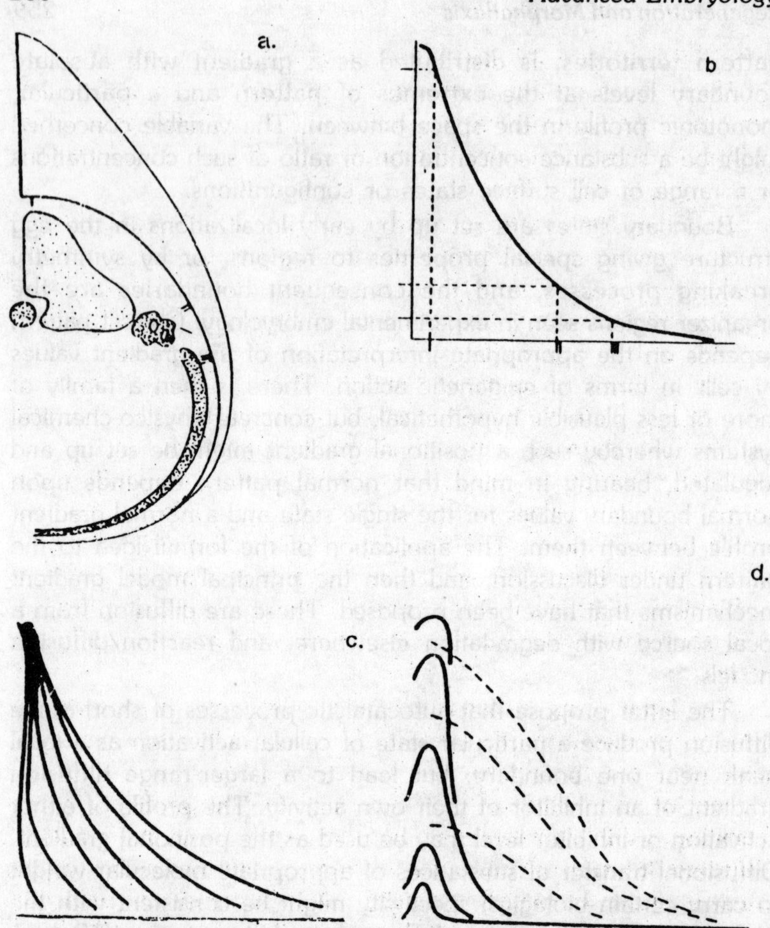

Fig. 9.2. The idea of positional information in media-lateral pattern formation. (a) The formed mesodermal pattern in situ. (b) The formal idea of the gradient extending through the earlier cell sheet. (c) A source of fixed concentration of a morphogen at one boundary. (d) A reaction/diffusion created positional gradient.

such as reduced size and bipolarity. Alteration of reaction kinetics and diffusibilities of the consistent molecules in any particular such system, in relation to individual embryo size, is impermissible because implausible; regulating behaviour must come, if at all, from the system dynamics. Does the performance of embryos under various circumstances match what plausible models can be shown to produce?

The exploration of possible performances by model gradients from computer simulation or analysis is hardly exhaustive enough for quantitative observations about behaviour of real patterns to be critical tests of the validity of positional information. A much stronger test concerns the postulate, distinctive to this theory, that there are no specific interactions between emerging pattern territories, in the control of their relative proportions. Such control is stated to occur automatically from the independent positioning of the divides between the territories according to positional signal levels in the gradient. This suggests quite critical experiments in that specific, determined territories that do not include the organizer can be grafted into cell sheet in which pattern is to be determined in somewhat younger embryos, and the effect upon sizes of the territories subsequently formed in those embryos can be measured.

The Pattern in Miniature Embryos and in Those with Two Organizers

The region of the long axis, in tail-bud stage larvae, where the proportions and overall scale of the pattern can be assayed as soon as histodifferentiation has revealed the character of the component cells. Overall proportions of the total cells assigned to each of the four territories are estimated by counting the nuclei seen in each structure in a regular subset of the series of precisely transverse histological sections cut at standard thickness. Meaningful comparative results are obtained without correcting to absolute cell numbers present in each tissue, because pilot investigations reveal that nuclear size and internuclear distance in the various tissues, in the principal planes of section, do not differ as between normal and experimentally small embryos from within an egg-batch (i.e., sibling embryos). That is, their tissues differ in overall scale (cell number), but no significantly in texture.

The rate of cell cycle between the period of determination of pattern and that of pattern assay is either zero or slow in the various territories and again there is no evidence for differences as between normal and experimental embryos during this interval, which lasts some 24 hrs. or 72 hrs. from the middle of the period of actual determination, in the fast-developing *Xenopus* and the slow-developing *Ambystoma*, respectively. The sampling involves several thousand cells per embryo, and repeat sampling of individual embryos gives very comparable results in terms of percentage of the sample population found in each structure. The control data have revealed that as expected, normal embryos, especially sibling sets, achieve rather constant proportions for the pattern.

More remarkably, proportions are almost indentical in the two species used in the study, which represent the two major amphibian subclasses, and which differ about two-fold in the mean numbers of cells between medial and lateral extremes of pattern in cross-section (*Xenopus laevis*, the clawed frog, c.70 cells *Ambystoma mexicanum*, the axolotl c. 140 cells). The assessment of performance of the regulative system in this work is based on comparisons between synchronous, sibling sham-operated and experimental embryos in matched sets. One completed study has assessed the response of the pattern specification system to the challenge created by two different abnormal circumstances.

In the first test, presumptively dorsal part-embryos were made by removing ventral blastomeres (early cleavage products) from young blastulae and causing them to round up and heal. Morphologically small gastrulae and neurulae are produced which develop to tailbud larvae in which the mesoderm contains as little as 25% of the normal cell number, thus some 40% of normal cell numbers along the medio-lateral dimension. In the second test, late blastula stage embryos were provided with a second organizer (see later section), midventrally, by grafting a dorsal blastoporal lip from an early gastrula into a site made by removing a few cells from the ventral marginal zone.

The new organizer diverts the development of some third or more of the total mesodermal cylinder into a new axial pattern having full cellular continuity with the "host" pattern at their adjoined lateral margins. Thus, a double dorsal, mirror-image duplicated pattern is produced, involving a normal-sized cell population in total cross-section. These operations and their results are described elsewhere in full. In single pattern of reduced dimension down to 40% of normal cell number, regulation achieves proportions for the four territories that are statistically indistinguishable from normal. The normal precursor material for lateral territories (even including pronephros) has probably been entirely removed in the more extreme examples.

"Over-regulation" appears to have occurred. In patterns of the second experimental type, the tissue lines between two organizing boundaries where notochords have differentiated, and normal cell numbers are involved. Regulation produces mesoderms where the total percentages devoted to the four territories, across the two opposed patterns, lie much closer to control values than might be expected. Since the presumptively lateral third of the cell sheet has been reorganized in relation to a new dorsal midline, loss of lateral representations and the joining of two incomplete pattern might be expected. This is largely avoided, although the lateral plate is statistically under-represented.

Most strikingly and unexpectedly, the two patterns *both* contribute to this avoidance of lateral loss in the territory between them, since both show smaller cell populations than normal devoted to pronephros, somite, and notochord. The new organizer, mid-ventral in the original blastula/gastrula, has caused a scaling down of the positions of boundaries between territories in the original, host-controlled pattern. The phenomenon occurs most with respect to pronephros and somite territories, but to a significant extent even in "host" notochords, whose territory we believe to become fixed soonest after the onset of gastrulation. These relationships and their implications for the positional information theory are represented. The curve represents the profile of a signal gradient whose levels specify normal pattern proportions at normal tissue size (shown along the baseline as cell numbers) on the positional information hypothesis for this system.

A profile from a concave family is selected simply because plausible model gradient mechanisms characteristically produce such curves, but any monotonic form would do as well. The proportions for the pattern are known at some 4%, 40%, 10% and 46% for notochord, somite, pronephros, and lateral plate/ blood island respectively, with 10-20% coefficient of *relative* variation in each. We can thus mark, on the ordinate, threshold signal levels which would establish boundaries between territories so as to divide up the tissue space or cell number (baseline) in the appropriate ratios. The response empirically seen in small and in doubly organized mesoderms is shown as version 3 in terms of cell number along the baseline. If true regulation of an interpreted gradient were to underlie this response, therefore, the gradient profiles achieved would bear the relationships to the normal one that are shown in curves, and because of the new positions of boundaries.

No gradient models having one organizer, as a signal source, with diffusion and degradation of signal elsewhere, can replicate such changes of profile in response to simulated equivalents of the biological experiments described. Few plausible models based on the more dynamic reaction/diffusion mechanisms (see later section) achieve the required behaviour either. Profiles are produced which more resemble numbers 1 and 2 of the figure, which when interpreted would produce baseline pattern versions 1 and 2, corresponding to loss of more lateral regions of the pattern, and even an associated expansion of medical territories. A particularly challenging feature of the present biological pattern is that a restricted territory, the pronephros, far from either of its borders, is both rescaled and appropriately positioned in small versions. A steeply non-linear curve corresponding effectively to a peak of signal level near one border, the most usual reaction diffusion component that actually responds to change of field size by repositioning the peak shoulder, would not be adequate here. The need to produce the implied regulation with respect to a profile that would show slope, thus carry positional information, across most of the field, imposes even greater stability and plausibility problems on such models.

In fact, only very narrow ranges of parameters for the simulated systems result in anything like effective behaviour. Certain "*solutions*" to the problem of steepening gradients by feedback from reduced overall scale of the system have been suggested. Thus, it is proposed that by various ancillary signals which are functions of tissue size, the diffusion terms or those describing the kinetics between autocatalytic and inhibitory components of the morphogen system are appropriately adjusted to restore the necessary gradient at equilibrium. All such "*solutions*" could be designed to work, by a human engineer. For the biologist, however, the intellectual choice would seem to lie between continuing to use such models, meanwhile wondering about their plausibility for quite basic and primitive instances of biological regulation, and on the other hand, turning to other classes of model which suggest new experiments. A more detailed discussion of the implications of the results just described, for various positional information models, appears elsewhere.

The Hypothesis that Serial Diversion of a Wavefront of Determination Regulates the Pattern

Biological patterns like the one we are considering have two main features, and an appropriate model must explain both of them. The territories that are determined are correctly ordered in space, and follow the "*rule of continuity*" when one edition of the pattern joins another in tissue, in that territory B is seldom omitted to leave A and C adjoining. Pronephros is occasionally absent in small amphibian patterns, but this may be because its visible expression is subject to a threshold number of determined cells being created. In addition, the relative numbers of cells assigned to the territories are quite highly regulated. There follows the outline of a model which departs from the positional information idea but which would explain the data thus far. It is related to previous ideas of Rose and of Meinhardt and also marks a reversion towards the original conception by Child of physiological gradients and dominance hierarchies controlling development. The model involves a wavefront of cell determination passing across the available tissue, organized by a primitive gradient or by propagated cellular "activation" from the organizer, and a set of specific signals which divert determination from one course towards another according to a logical pattern.

The organizer is assumed to control only an overall polarity within the embryo's tissue, this polarity being expressed as graded rates of physiological progress towards determination by cells, smoothly ranked from medial (dorsal) to presumptive lateral (ventral) extremes. Such control could be via a diffusion gradient of a signal state, declining smoothly with distance from source and setting local developmental rates, or else via early propagation of an activated cellular state, in the form of a wave that starts or accelerates progress towards determination, spreading from its origin at the autonomous organizer region. This spatial communication system need have no further, accurate informational function such as would be required of regulated positional information.

It could therefore be a primitive and plausible diffusion-controlled or propagated signal, lacking regulative capacity against scale variation and maintaining only the direction and continuity of the wavelike sequence of development irrespective of precise position of material in relation to the boundaries of the pattern to be formed. If of gradient form, it could assume profiles 1 and 2 under the conditions corresponding while if of propagated wavelike form, its transit time would be proportional to tissue distance traversed rather than of particular absolute duration in both normal and small embryos. Normal ordering and relative extents of the determined territories are assumed to follow because of the logic governing access to the various determinations by the cells.

An early biosynthetic product of cells which have advanced into each determined state is assumed to diffuse rather rapidly into the rest of the tissue and to prevent, at a threshold concentration, development towards that same state by "*younger*" i.e., slower-developing or as-yet-unactivated cells that receive it. Thus, in normal circumstances all cells that can develop fast enough move initially towards notochord determination. But a diffusable signal, produced specifically by cells that have become committed to the notochord state, builds up in the system as a whole until it attains a level that diverts less advanced cells towards the next state, somite. Such a process might be termed the *serial diversion* of a wavefront of *determination*, to produce pattern. Each specific signal is assumed to be so diffusable or transportable within the cell sheet as a whole that its level or concentration is a negative function of the size of the mesoderm as well as a positive function of the numbers of cells producing it. It thus acts as a sensor of the proportion of the total tissue that consists of cells which have been switched to produce it.

The logic of proportion control is somewhat similar to that of the regulation of a single frontier position by a reaction-diffusion system, and the whole mechanism is like a linked sequence of such events that avoids the regulatory problems associated with positioning all the frontiers by one informational gradient. Due to more rapid building up each specific signal to threshold, the wavefront of determination will be diverted in character more frequently, marking off smaller territories between frontiers, when the cell sheet is smaller than normal or is being simultaneously invaded by another wavefront originating elsewhere as in organizer grafting. If the organizer is in fact the source for a primitive gradient, rather than the origin for propagated activation then the gradient is assumed to produce a wavefront of development simply by having smoothly graded, at some prior time, the rates of maturation in cells across the embryo.

The conceptual separateness of the wavefront of the determination (spatially organized), and then the succession of territory-specific signals that monitor the proportions of each pattern part is crucial to the understanding of this hypothesis. The positional information hypothesis is distinguished from the foregoing one by its strong proposition that there are no specific interactions between territories during pattern control. It has been reported that in double dorsal patterns, even the two notochords and pairs of somite columns, far apart in cellular terms, can affect one another's sizes. If specific feedback signals that exert effects on determination are to account for this, they must be sufficiently mobile so that significant spread throughout the mesoderm occurs within fractions of the time—course of pattern determination. This has consequences that suggest experimental studies.

Firstly, a quantitative study of the time course of dissemination through cell sheets of labelled molecules of various characters and size is long overdue. Such a study could be carried out by following dissemination with time from small grafts of tissue heavily labelled with marked molecules, that had been injected into a blastomere of the original donor embryo. We have currently no idea how this time course would compare with that for free diffusion in aqueous space of the same dimensions, because the effects of restricted communication between cells via gap junctions, and possible effects of active transport or mixing in intracellular space, could modify effective "diffusion" coefficients in opposite ways. Figures derived from such studies could offer plausibility estimates for a variety of models, but no more than this. The biosynthetic arrangements underlying actual signals used in pattern regulation could give their progress through tissue a character intermediate between that of diffusional spread (proportional to distance squared) and that of a propagated state or wave (proportional to distance).

Specific Suppression of Territories by Heterochronic Grafts; a Test of the Serial Diversion Hypothesis

Another study suggested by the serial diversion model involves transplantation of large pieces of already-determined territories, from older embryos, to ectopic locations in gastrulae whose own patterns have yet to be determined. With certain donor-host age relationships, such grafts should exert specific suppressant effects upon the percentage of the host mesoderm finally devoted to the homologous territory. Any significant observation of such a phenomenon would constitute strong evidence against positional information as the effective control mechanism for the pattern. Experiments have been designed to search for such effects and the results of a study in progress are outlined in this section. A type of operation, the heterochronic graft, where a dorso-lateral stripe of neurula mesodermal mantle together with overlying neuroectoderm and posterior zone of recruitment at its blastopore margin, is grafted to a narrow mid-ventral slot in a younger gastrula stage sibling.

Zones of mesodermal recruitment and neurectodermal layer are matched up in graft and host and heal into continuity. The developing mesodermal cylinder now has, in addition to an approximately normal-sized host-derived cell population, a mid-ventral strip of older tissue which proves to have maintained its advanced schedule of development and to be in fact determined, under these conditions, to give rise to somite alone or to somite flanked on one side by pro-nephric differentiations. The notochord territory has been excluded from grafts and none of this structure in fact develops, but a small neural plate or rodlike neural formation usually overlies the ectopic somite, since neural fold was part of the graft. Summarizes in histogram form the results of quantitative pattern analysis (see earlier section) in experimental *Xenopus embryos* of this type, in relation to their sham-operated control siblings.

Total cross-sectional cell number in *host* mesoderm is not significantly different from normal, but the proportion devoted to somite is quite significantly reduced, the effect being very variable but present in most embryos. This somite deficit is largely concentrated in those parts of the axis lying opposite the relatively posterior position of the ventral ectopic somite in the mesoderm. Pronephros proportions are much more variable among experimentals than in controls and, on average, somewhat lower. But analysis of individuals reveals that, in those experimental embryos where pronephric proportions are diminished right below the normal range of variation in controls, a pronounced pro-nephric differentiation has occurred in graft-derived structures ventrally.

Axolotl embryos behave similarly in response to these heterochronic grafts, though there the specific response of somite-proportion is distributed more widely along the host's axis. These results make it quite improbable, in my view, that a positional information principle underlies control of the medio-lateral dimension of the body plan. The alternative theory proposed here

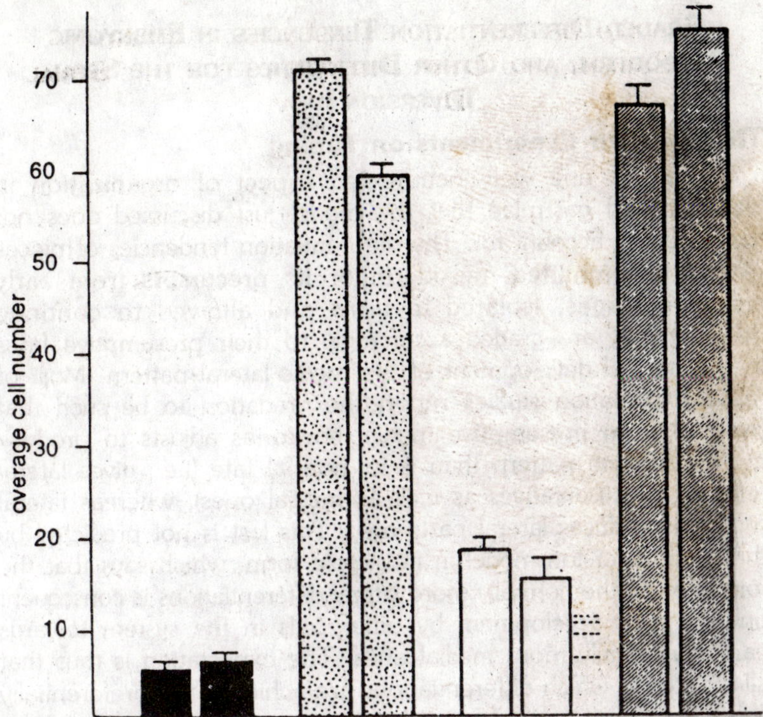

Fig. 9.3. The effect of heterochronic grafting. Data from a population of heterochronically grafted Xenopus embryos with ectopic somite, and their matched sibling controls. Means of absolute cell numbers per pattern territory per section, within embryos, and their standard errors for the population.

would, however, accommodate and even predict them. They would represent feedback effects exerted by specific signals of the sort that normally regulate the pattern, but in this case emanating from advanced representatives of the territories concerned, that have been joined to the cell sheet at one extreme of the pattern. Buildup of signal specific to a particular territory would be accelerated, if older cells of that territory had already primed the system with an appreciable level at the site in the host where the control of proportions (i.e., the positioning of a frontier) is taking place. The result would be abnormally early diversion of the character of the wavefront of determination, and thus an abnormally small proportion of the tissue space devoted to the territory in question.

Graded Differentiation Tendencies in Embryonic Mesoderm, and Other Difficulties for the Serial Diversion Idea

The Need for Experiments on Timing

There is one well-documented aspect of organization in blastulae and gastrulae that the model just discussed does not automatically account for. The differentiation tendencies of pieces of the uncommitted mesoderm or its precursors from early cleavage stages, isolated in *vitro* and allowed to continue development, are graded accordingly to their presumptive fates in the normal development of the medio-lateral pattern. Most of these explanation studies report the gradation to be such that material from presumptive medial territories adjusts to produce more complete pattern than is its normal fate (i.e., gives lateral territory representatives as well as medial ones), whereas lateral material produces lateral parts only. This last is not predicted by the serial diversion model in its simple form, which says that the formation of the normally more lateral differentiations is consequent on the prior development by other cells in the system towards earlier, normally more medial ones. The expectation is thus that all isolates in which differentiations are achieved will preferentially form representatives of medial territories, and that more lateral ones will only be achieved insofar as each maintains or reachieves a spatial gradation in intrinsic rates of maturation within its material, such as in normally set up by communication with the organizer. But differentiation patterns in explants can be produced without notochord. Furthermore, under particular conditions in whole embryos, notochordless body patterns can be found that are otherwise normally organized in the medio-lateral dimension.

There is no reason, in such cases, to postulate the death of cells that would to otherwise have become notochord; such cells have presumably contributed to the somite that now marks the medial extreme of pattern. From limited published work and from informal exchange of information, it appears that the capacity for differentiation as isolated explants develops earlier in the mesodermal precursors of embryos of the frog type than in those of salamander type, and that its dorso-ventrally graded properties when explanted are somewhat less pronounced in the former type. This is reminiscent of another quantitative difference in the progress of pattern control as between the two types. If a salamander-type embryo is divided into a smaller dorsal and a larger ventral fragment when gastrulation activity has already started in the dorsal midline (the organizer), the dorsal fragment can continue development and regulate its fate to form pattern to a considerable degree.

The ventral fragment however, although surviving for a long time, usually differentiates little expect epidermis, yolky endoderm, and after prolonged delay, a primitive lateral platelike mesenchyme with blood islands. In frog embryos subjected to the same procedure, the ventral fragment has already acquired the capacity to gastrulae and to develop mesodermal pattern to a large degree, though the medial extreme of pattern is missing. The general trend is for all these partial patterns deficient in medial territories, however produced, to result from abnormally slow schedules of development and (presumably) determination in the cells. These are associated either with the early treatment of the whole embryo, or with the ambient conditions following grafting operations or the culture of isolates. Proper quantitative study of the time schedules associated with the development of all examples of abnormal patterns will be of prime importance in assimilating all the data into a coherent theory for pattern control in this system. However, the following outline might accommodate both the concept of serial diversion arrived at in a previous section and the phenomenon of patterns without medial parts mentioned above.

The effect of the organizer is assumed to be one of "spreading activation,' propagating into surrounding tissues at sometime before its first expression as the spreading of gastrulation activity to each location. Such activation of progress towards gastrulation and the determined stage may start sooner and progress relatively more rapidly in frog-type than in salamander-type embryos, hence the systematically less graded character of development in frog isolates, and the less delayed and incomplete morphogenesis of the ventral fragments in these embryos. The degree of activation achieved locally may steeply affect the rate of cellular development, ensuring the gradient rate needed to effect a wavefront of determination susceptible to serial diversion. But such an organization would also mean that in cases of subnormal activation or attenuation of the apical (organizer) end of the activation gradient due to specific early treatments, or perhaps loss of activation through early isolation in culture or tissue disruption, development is slowed to varying extents.

The more "medial" determined states may become sequentially unavailable with prolonged passage of physiological time in cells, even in the absence of diverting signals from "older" cells elsewhere. Total isolation from activation may lead always to blood island/lateral plate development after a prolonged period. In short, the normal pattern developed in *vivo* may be the combined outcome of a (nonregulatable) gradient system in degree of activation, itself biasing the "first choice" state of determination at each location, together with the series of specific diverting signals which ensure proportionality and which prohibit the formation of *laterally* incomplete patterns in small and in organizer-grafted embryos. The specific signals would also underlie the results of heterochronic grafting. The gastrulation movements themselves occur according to a pronounced graded time schedule that appears quite well controlled as between individuals. Presumptive blood island cells begins their immigration a set time after the presumptive notochord (organizer) region.

The precise relation between the endogenous controls of rate in these early activities and those controlling onset of determination for pattern is unknown, but it will be of great relevance for the theories so far discussed to time the visible dorso-ventral gastrulation sequence in small and in organizer grafted double dorsal embryos in relation to their normal siblings. Embryos of both type are able to reduce the sizes of territories so as to complete pattern in the spaces available. Regulation of a true positional signal gradient, if successful might be expressed in the gastrula stage as normal schedules of progression in absolute time, spanning the reduced dorso-ventral tissue spaces available. That is, little gastrulae and the two half fields of double gastrulae should each take a *normal* time for immigration activity to proceed across them from its onset mid-dorsally.

By contrast, the primitive sort of activation system, necessary for pattern coherence on the serial diversion model, might be expected to behave like the profiles number 1 or 2. If timing reflected the gradient profile, this would lead to a contrasting phenomenon at gastrulation; the completion of the blastoporal ring by rapid spread of the immigration activity in less than normal time. These simple, relevant but exacting experimental observations remain to be made. The units of pattern also called "territories" do not correspond simply to determinations for development into cell types as recognized by the cytologist or the cell physiologist.

They are, rather, particular geographical units where cells possible development becomes restricted to formation of particular components of the final body, each component then contributing a multiplicity of cell types and structures.

Somite contributes skeletal and muscle tissue, while lateral plate and blood-forming islands contribute an even greater variety of histological compartments in the finally differentiated frog. There are thus limits to an analogy between the concept of pattern formation by serial diversion of determination due to specific signals, and that of control over the functional sizes of different tissue systems in the later body by systemic "chalones" (e.g., erythropoietin, liver-chalone). At the time of pattern control, the embryo has no vascular system, but the cells of the entire mesoderm are probably in communication *via* gap functions allowing passage of molecules up to around 1,000 molecular weight, and *via* the restricted extracellular diffusion space between the sealed faces of ectoderm and endoderm. The postulated signals are thus more allied to traditional morphogens than to hormones or chalones, and could conceivably be their common evolutionary precursors.

The Antero-Posterior Dimension

The Regulation of longitudinal Pattern

Pattern in the antero-posterior or cephalocaudal dimension of the body plan is also subject to a morphallactic regulatory process at early stages, at least as regards anterior regions down to the level of the hind limb rudiments. Experimental embryos of the type discussed in an early section with medio-lateral pattern down to below half-normal scale, also show normal numbers of abnormally short somites (mesodermal segments) between unique markers such as ear vesicle and root of the tailbud. There is reduced cell number in the original long dimension of each somite at its formation, so that a normal complement of these structures can be made with the tissue available. These embryos also show reduced cell number in the long dimensions of the pronephros, the heart and the brain parts, the last presumably a reflection of shorter zones of inductive activity in the mesoderm. In order to normalize longitudinal pattern to small size in this way, embryos must heal after size reduction by middle blastula stages.

Transection of later gastrulae and neurulae in the transverse plane, or transposition of somitogenic tissue along the axis at these times, results in highly mosaic behaviour. That is, pieces of embryos isolated or transposed at such stages differentiate, with respect to numbers of somites developed in each piece and the timing of their segmentation, like pieces of a photographic film exposed to a complex scene, then cut up, then developed. It therefore appears that although the visible processes demarcating successive segments from the somitogenic tissue occur in mesoderm which has been in its definitive position for sometime, the original factors which scale the processes and which set their local timing to give a "wavefront" progressing down the body plan, must be subject to a regulatory process acting before or at gastrulation.

The result of this early process must be some graded cell state along the presumptive axis, which sets the smoothly graded time-sequence for local events as well as the different determinations attained, and which transits the same absolute range of values from anterior to posterior pattern limits in normal and in small embryos. A system which can do this is indeed reminiscent of true positional information, if in fact the cells that are to make the pattern are laid out in coherent sequence in a sheet at the time their "*values*" are given them. But our understanding of the fate map for this pattern dimension, whereby tissue is recruited into the mesoderm in its antero-posterior sequence during gastrulation, is not complete. Extensive cell mixing could conceivably occur with respect to position along this axis as between blastula and gastrula stages, though this seems unlikely.

Positional Signals, Nonequivalence and Timing

Most embryologists feel that we are presently much further from understanding the mechanism of antero-posterior pattern regulation than that of medio-lateral pattern regulation in the vertebrate embryo. The positional information concept still seems helpful for the longitudinal axis, especially if it is broadened to include the idea of positional value being given by a mechanism that measures time spent by cells in the marginal zone before their recruitment into the mesoderm, rather than their physical positions within a spatial gradient of a signai. It is along the antero-posterior axis that the phenomenon referred to by Lewis and Wolpert as non-equivalence is most evident in primary pattern formation. Thus, the medio-lateral dimension of pattern described earlier, with its zones, however, tissue undergoing similar initial behaviour and differentiations, e.g., as somite or lateral plate, can be shown to be encoded with unique properties at each level.

The most striking work demonstrating this has been carried out on bird, rather than on amphibian embryos. The tissues early non-equivalence is revealed by the autonomous nature of the overall development sequences carried out by regions when transplanted elsewhere along the axis, but also, most obviously in the specific physiological "age" in development at which the morphogenesis takes place in tissue from each level. Thus, pre- segmental material from the neck somite region, transplanted to the thoracic or lumbar region, will ultimately produce there vertebrate of "neck" character. Similarly, lateral plate mesenchyme of fore- and hind-limb regions acquires very early its specific character, determinative of the type of limb produced when ectodermal jackets and mesoderms are cross combined. Time, or rate as a continuous variable, seems to be fundamental to the expression and perhaps the genesis of pattern in the longitudinal axis.

If experimentally small amphibian embryos having undergone morphallactic regulation are synchronized precisely with normal siblings at the early neurula stage (i.e., at the anterior onset of somite segmentation), their subsequent progress is parallel in terms of *numbers* of somites segmented and thus proportion of the axial pattern developed, rather than in terms of absolute cell numbers incorporated into somites and other axial structures. The latter arrangement would complete the development, by using up tissue, earlier in the smaller embryos, whereas in reality the same absolute time span is linked to the development of the complete pattern in both cases. Taken overall, the evidence is for a great multiplicity of states coded into tissue to give rise to this dimension of pattern. Such a multiplicity could not readily be regulated by a series of unique size-assessment/switching events such as were postulated for the relatively simple medio-lateral dimension of pattern. Instead, some property occupying a graded range between absolute boundary values is implied, as in positional information theory, although there are problems in considering how such information might be used in controlling numbers of similar structures.

A Note on Universality

There is no priority reason why control of the body pattern in the mesodermal cell layer should not be achieved by different mechanisms in each of its two dimensions. Frankel has suggested that the regulation of cortical geometry in ciliated protozoans has such a dual character, as well as drawing out the parallel phenomena of organization as between this and pattern regulation in the early development of vertebrate embryos. A similar dualism may apply to pattern establishment in the insect blastoderm. In another early episode of vertebrate development, namely the establishment of antero-posterior limb pattern, the data are currently consistent with the idea of a true, interpreted positional gradient underlying the character of the skeletal elements.

There is even evidence in the chick limb-bud to suggest that by contrast with the primary body pattern discussed in this chapter, growth control is correlated with pattern control so that the gradient mechanism for the latter may be simple and primitive, and yet avoid the difficulty of regulation for size because size is itself controlled accurately at an early stage. Should this prove to be the case, the way is open to reconsider yet other concepts for control of the relatively complex pattern that characterizes, say, the limb skeleton. These concepts come under the general heading of "*pre-patterns*".

Unlike positional information, or the serial diversion model outlined here, pre-patterns involve the concept of a spatial distribution in the morphogen levels that is isomorphic with the future pattern, i.e., a set of peaks and troughs of concentration, rather than being a simple gradient that requires a complex genetic response from the cells. A problem with the physico-chemical model systems whereby such pre-patterns might emerge, has always been that the pattern produced is crucially dependent upon the scale of the system, for reasons similar to those whereby dependent. If limb rudiments in fact contrive to keep spatial scale constant during the initial specification of pattern within them, then some (though not all) aspects of these somewhat complex patterns may depend upon pre-patterning processes. The possibility is emerging that, rather than a single universal form of process, there may be a selection from a small repertoire of available processes according to functional considerations connected with particular dimensions of pattern in particular embryos at particular embryos at particular times. The various classes of process touched on in this chapter are all, nevertheless, to be thought of as morphallactic indeed the term is to continue in modern use. In none of them is the development of a particular spatial array of cells with specific potentialities intimately tied, via neighbour interactions, to the process of tissue production by cell division itself.

10

Regenerative Development

The vertebrate limb had long been the center of the debate leading up to the publication of the *Origin of species* and there seemed to be a common plan to the structure of all vertebrate limbs. The term coined to describe such similarities was '*homology*'. Much was made of the homology of vertebrate limbs by Darwin and the supposed explanation of the phenomenon was that the limbs of each species had descended from a common ancestor whose archetypal limb pattern had subsequently undergone modification. Despite the fact that these conclusions were derived from studies of the anatomy of adult limbs, as many others have been since then, the common ancestor theory had implications for embryology. First, it became accepted that all vertebrate limbs passed through this common ancestor stage in their development, without any embryological evidence to support it.

Only recently have detailed studies been performed to discredit this view and it is now clear that the prechondrogenic condensation pattern of the limb bud mirrors, in miniature, the pattern of the adult form. The limb of each species therefore develops uniquely in terms of the prechondrogenic condensation pattern, a conclusion which is reinforced below. Secondly, in evolutionary terms, development is viewed as a highly ordered sequence of hierarchical processes. This means that the earliest events in the development of the limb should be common to all vertebrates, with adaptation only working on the latest stages to provide the variety of adult forms we see. In this hierarchical scheme, one of the earliest processes to occur in limb development must surely be the establishment of the limb axes, without which organized growth could not take place.

All vertebrate limbs have an antero-posterior, a dorso-ventral and a proximo-distal axis, and, according to the implications of the above evolutionary thinking each axis should be organized by the same mechanisms throughout the vertebrates. We do not know even today the precise details of how the axes of developing limbs are organized. The only techniques presently available to us to gain this knowledge are the observation of normal development and experienced interference in various ways. By performing these experiments on several species of embryo we can test the above hypothesis.

Clearly, the implication are that, if all vertebrate limbs organize their axes by the same mechanisms, the effects of experimental interference should be the same. There have up to now been almost no studies of what might be called experimental homology, since most experimenters prefer to concentrate their efforts on the embryos of only one species. The beginnings of this study were reported recently and the work described here continues this effort. Four species of vertebrate limb are compared–the developing limb of *Ambystoma mexicanum* (urodele amphibian), *Rana temporaria*, *Xenopus laevis* (anuran amphibian) and *Gallus domesticus* (*Aves*) as well as data on the regenerating limbs of *Ambystoma mexicanum*. We can therefore look for similarity not only between the developing limbs of amphibians and birds, but also between developing and regenerating limbs.

Normal Development

To observe the pattern of emergence of prechondrogenic elements limb buds at various stages of development can be stained with Alcian green. We find that this pattern reflects faithfully the cartilages in the adult form. There are no similarities between the stages of developing limbs of axolotl, *Rana*, *Xenopus* or *chick*, indeed the elements do not even develop in the same antero-posterior sequence. For example, in axolotl limb buds the digits appear in approximately an anterior to posterior sequence—in the hindlimb the sequence is digits 2, 1, 3, 4, 5 but in *Rana* and *Xenopus* the sequence is from posterior to anterior–digits 4, 3, 5, 2, 1.

In the chick, where the forelimb bud is routinelý used, the sequence is 4,3,2. Furthermore, the chick limb bud condensation pattern does not pass through a primitive amphibian stage as the common ancestor theory would have as believe. A more detailed examination of the size of elements when they first appear in the developing limb has also revealed differences between species. In chicks, the size of the elements in each of the four segments (humerus, radius and ulna, wrist digits) when they appear in proximo-distal sequence is approximately 300 μm. We might

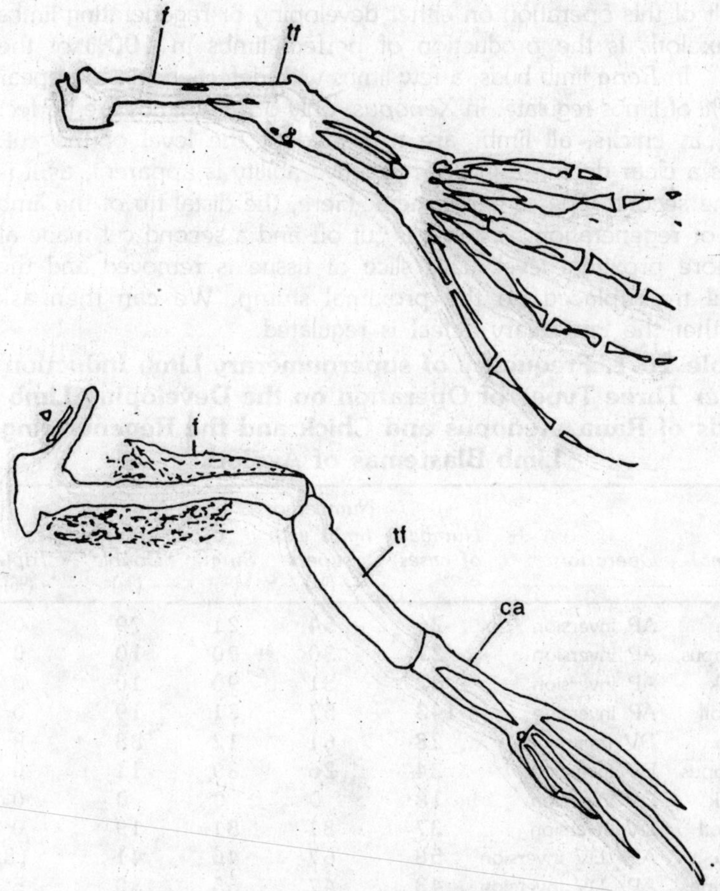

Fig. 10.1. Typical limbs resulting from the grafting of a distal hindlimb bud tip into the proximal stump. Victoria blue staining. f–femur, tf–tibia and fibula, ca–calcaneumand astragalus.

expect such a fundamental mechanism to be common to all vertebrates, but again this seems to be so–each species has its own characteristics element size. In axolotls, the figure is approximately 250 μm, in *Xenopus* 150 μm and in *Rana* 130 μm.

Work on the Proximo-distal Axis

For the present study we can perform two types of operation. In the first, the tip of the limb bud or regeneration blastema is simply cut off, leaving about 50% of the original amount of tissue. Can the remaining tissue regulate to form a complete limb? The result of this operation on either developing or regenerating limbs of axolotls is the production of perfect limbs in 100% of the cases. In *Rana* limb buds, a few limbs with defects begin to appear –95% of limbs regulate. In *Xenopus*, only 38% of limbs are perfect and, in chicks, all limbs are truncated at the level of the cut. Thus a clear decrease in the regulative ability is apparent, as it is in the second type of experiment. Here, the distal tip of the limb bud or regeneration blastema is cut off and a second cut made at a more proximal level. This slice of tissue is removed and the distal tip replaced on the proximal stump. We can then ask whether the intercalary defect is regulated.

Table 10.1. Frequency of supernumerary Limb Induction after Three Types of Operation on the Developing Limb Buds of Rana, Xenopus and Chick and the Regenerating Limb Blastemas of Axolotl.

Animal	*Operation*	*Number of cases*	*Number of limbs with supers* (%)	*Number of supernumeraries* *Single* (%)	*Double* (%)	*Triple* (%)
Rana	AP inversion	26	54	21	79	0
Xenopus	AP inversion	37	30	90	10	0
Chick	AP inversion	32	31	90	10	0
Axolotl	AP inversion	143	57	81	19	0
Rana	*DV inversion*	28	61	12	88	0
Xenopus	DV inversion	34	26	89	11	0
Chick	DV inversion	18	0	0	0	0
Axolotl	DV inversion	37	81	81	19	0
Rana	AP/ DV inversion	58	67	46	41	13
Xenopus	AP/ DV inversion	43	47	65	30	5
Chick	AP/ DV inversion	31	68	76	24	0
Axolotl	AP/ DV inversion	83	94	57	40	3

AP = antero-posterior; DV = dorso-ventral.

Again, in axolotl limb buds and regeneration blastemas, perfect regulation takes place and 100% of the limbs are normal. In *Rana* limb buds, this operation only results in 42% of perfect limbs, and in *Xenopus* and chicks all limbs have deficiencies of one kind or another, although the results are highly stage dependent. Thus again a decreasing regulative ability is apparent throughout these vertebrate species.

Work on the Transverse Axes

Investigations of the organization of the transverse axes (antero-posterior and dorso-ventral axes) have involved the grafting of limb buds or regeneration blastemas to invert either one or both axes. By grafting a left limb bud to right stump (or vice versa) each axis can be inverted independently, and by grafting a limb bud onto its own stump after 180° rotation both transverse axes are inverted at once.

The Antero-posterior axis

Inversion of the antero-posterior (AP) axis results in the production of *supernumerary limbs*, their frequency varying according to the species concerned. They arise at the points of maximum incongruity, i.e., at the anterior and/or posterior poles, and can be either single or double. The relatively low incidence

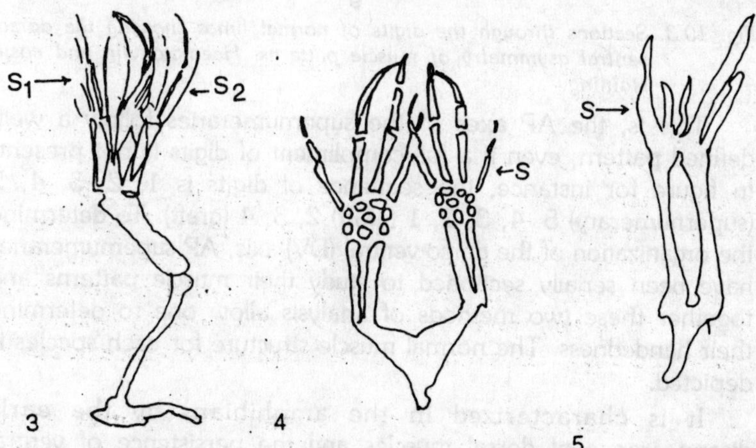

Fig. 10.2. Supernumerary limbs produced by contralateral grafting of limb buds of regeneration blastemas to invert the antero-posterior axis, Victoria blue or Alcian green staining. Rana hindlimb with two supernumeraries S_1 and S_2 at the anterior and Posterior poles of the limb.

of supernumerary limbs produced by AP inversion in chick limbs should be contrasted with the effect of grafting the organizing region from the posterior margin of the limb bud to the anterior margin. Using equivalent stage embryos, supernumeraries can be typically produced in more than 90% of the cases. Analysis of cleared whole mounts reveals entirely normal sequences of digits, carpals etc.

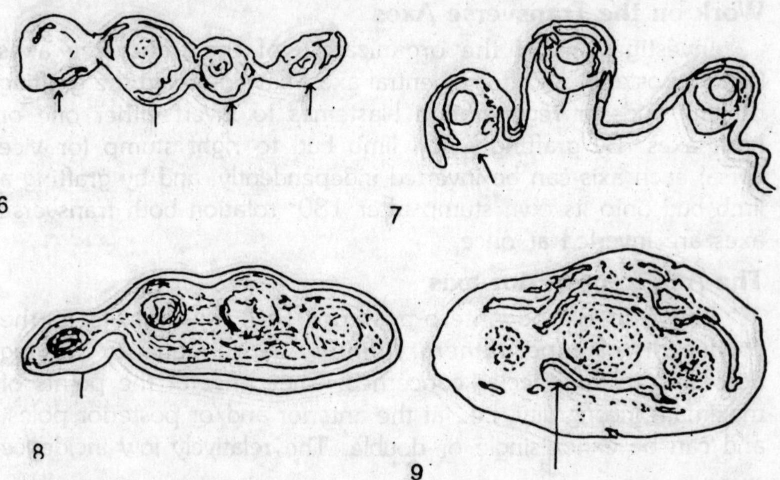

Fig. 10.3. Sections through the digits of normal limbs showing the dorso-ventral asymmetry of muscle patterns. Haematoxylin and eosin staining.

This is, the AP axes of the supernumeraries follow a well-defined pattern, even if a full compliment of digits is not present. In figure for instance, the sequence of digits is 1, 2, 3, 4, 5 (supernumerary) 5, 4, 3, 2, 1 (graft) 2, 3, 4 (graft). To determine the organization of the dorso-ventral (DV) axis, AP supernumeraries have been serially sectioned to study their muscle patterns and together these two methods of analysis allow one to determine their handedness. The normal muscle structure for each species is depicted.

It is characterized in the amphibians by the early disappearance of dorsal muscles and the persistence of ventral muscles to the tips of the digits. The muscles of the chick, too, have a characteristic dorsoventrality. Analysis of total of 38 supernumeraries revealed that in every case their muscle patterns were entirely normal. From this we can conclude that, after inversion of the antero-posterior axis in limb buds or regeneration blastemas, the supernumerary limbs which are produced are structurally normal and show *stump handedness*. Thus in this series there is considerable uniformity within the species studied.

Table 10.2. Cartilage Structure, Muscle Structure and Handedness of Those AP and DV Supernumeraries Which have been sectioned in the Four Species

Note that chick limbs do not form supernumeraries after DV inversion. Rana, Xenopus and chick results are from operations on developing limb buds, the axolotl results are from operations on regenerating blastemas.

Animal	*Type of Supernumerary*	*Number analysed*	*Structure of cartilage*	*Structure of muscle*	*Handedness of supernumeraries*
Rana	AP	11	Normal	Normal	Same as stump
Xenopus	AP	6	Normal	Normal	Same as stump
Chick	AP	10	Normal	Normal	Same as stump
Axolotl	AP	11	Normal	Normal	Same as stump
Rana	AP	11	Normal	Normal	Same as stump
Xenopus	AP	6	Normal	Normal	Same as stump
Axolotl	AP	10	Normal	Normal	Same as stump

AP = antero-posterior; DV – dorso-ventral.

The Dorso-ventral Axis

Inversion of the dorso-ventral axis results in the production of supernumeraries in the amphibian species, but not in the chick. In the former, one to two supernumerary limbs are produced at varying frequencies, usually at the points of maximum incongruity which, this time, are the dorsal and ventral poles of the host limb. Analysis of the muscle patterns in *Rana*, *Xenopus* and axolotl regenerating limbs revealed that, as in the previous case, all supernumeraries were normal in their dorso-ventral organization as well as in their AP sequence of digits and were of *stump handedness*.

Inversion of Both Axes

After 180° rotation of limb buds or regeneration blastemas, a high frequency of supernumerary limbs is produced in both amphibians and birds. It is again of interest to note that the incidence in chick limb buds is lower after this type of operation than after organizing region grafts. Several differences begin to emerge between the effect of uniaxial and biaxial inversion. Instead of just single or double supernumeraries, triples appear in a small, but consistent number of cases in *Rana*, *Xenopus* and *axolotl*. The position of origin of supernumeraries in *Rana*, *Xenopus* and *axolotl* is unpredictable rather than being at defined poles, although in the chick they only appear at the posterior and occasionally at the anterior pole as well. Analysis of their AP organization by cartilage staining reveals surprising differences between the species. In *Xenopus* and chick developing limbs, all supernumeraries had a normal sequence of digits, i.e., the AP axis was an expected. However, in *Rana* developing limbs, a new category of supernumeraries appeared for 28% (11 out of 39) were double posterior in structure, the remainder being normal. These double posterior supernumeraries were composed of either three four or five digits.

Table 10.3. Dorso-ventral Organization of Supernumeraries General by Inversion of Both Transverse Axes, Deducted from studying the Muscle patterns in Serial Sections

		Number of each type of muscle pattern in AP/DV supernumeraries			
Animal	*Number Of super-numeraries analysed*	*Normal*	*2D or 2V*	*Part normal/ Part mirror-imaged*	*Part normal/ part inverted*
Rana	16	5	3	4	6
Xenopus	12	9	1	1	1
Chick	17	17	0	0	0
Axolotl	100	22	22	32	24

Examples of each type are—normal; double ventral (2V); double dorsal (2D); part normal/part mirror-imaged; part normal/ part inverted. *Rana*, *Xenopus* and Chick results are from operations of developing limb buds, the axolotl results are from operations on regenerating blastemas. Analysis of their Dorso-ventral organization by muscle patterns revealed further surprises.

In axolotl regenerates and *Rana* limb buds, there were approximately equal numbers of four types of muscle patterns. First, there were normal supernumeraries as had been found before. Secondly, there were perfect mirror-imaged limbs, either double ventral or double dorsal. Thirdly, there were limbs of mixed symmetry and asymmetry, that is partly double dorsal or double ventral and partly normal. Fourthly, there were limbs which were partly normal and partly inverted. The double posterior *Rana* supernumeraries were all found to be of normal muscle structure. The only one transverse axis can be mirror-imaged at a time. In *Xenopus* the majority of supernumeraries (9 out of 12) were normal in their muscle patterns.

Only one of each of the other three classes of structure was found. Supernumeraries produced in chick limbs often had a reduced complement of muscles, although those that could be identified were always arranged normally. Some effect on the dorso-ventral axis may be involved here, since AP supernumeraries had well-formed muscles. Thus, after antero-posterior/dorso-ventral inversion, there seems to be a gradation in the structure of supernumerary limbs generated from 100% normal (chicks) to various frequencies of four (*Xenopus* and *axolotl*) and ultimately five different classes of structure in *Rana*, including double posteriors.

The Effect of Vitamin A and Retinoids

One further method of analyzing axial organization which has recently come to light is the effect of vitamin A and its derivatives, the retinoids, on developing and regenerating limbs. If the amputated limbs of axolotls are immersed in a solution of retinol palmitate, then, instead of replacing those elements removed by amputation, extra tissue in the proximo-distal sequence is regenerated. For example, after amputating through the mid-radius and ulna, an extra radius and ulna can be produced in tandem as well as the hand. If the concentration of vitamin A is increased, a whole limb can be regenerated from the cut stump. The same

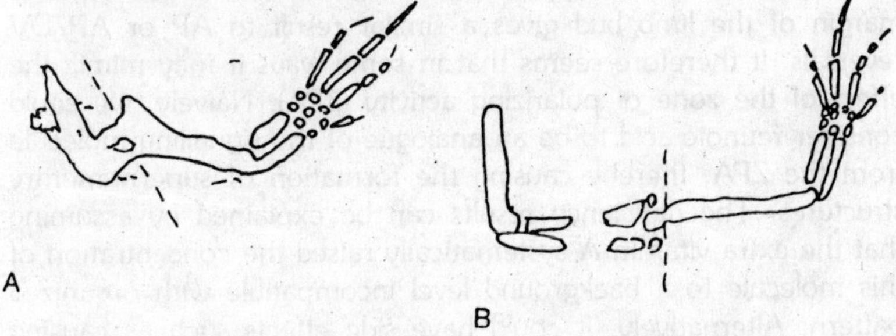

Fig. 10.4. Abnormal limbs resulting from Vitamin A treatment of regenerating axolotl limbs. Victoria blue staining.

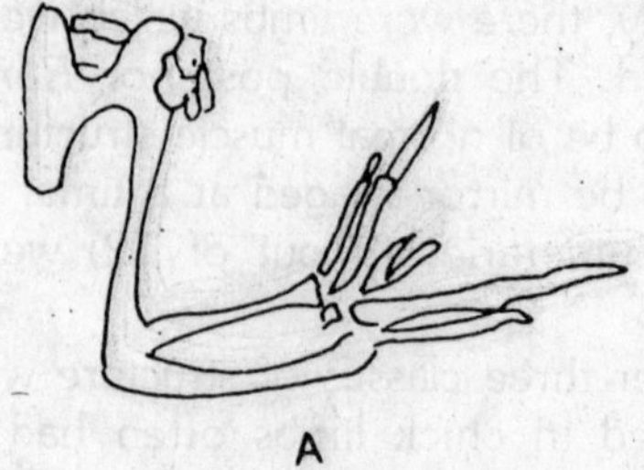

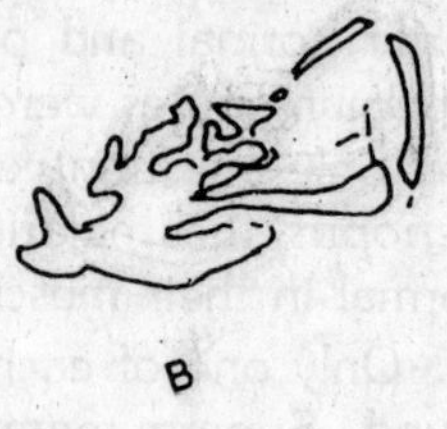

Fig. 10.5. Abnormal limbs resulting from Vitamin A treatment of chick limb buds. Alcian green staining.

phenomenon occurs more dramatically after amputation through the hand.

It is important to note that, in both these figures, the antero-postena sequence of digits is normal, as is the dorso-ventral organization of muscles when such limbs are sectioned. The effects of various retinoids have been rested and retinoic acid was found to be the most potent. In the developing chick limb bud, the effect of vitamin A analogues seems to be very different. We have attempted to produce homologous effects on the organization of the proximo-distal axis to those that we have shown in the Amphibia. However, ten different experiments extending over a dose range of six orders of magnitude and using over 1400 chick embryos have so far failed to produce any cases showing unequivocal additional skeletal elements along the proximo-distal axis. Instead, Vitamin A causes modifications of the AP axis including the formation of mirror-image reduplicate supernumerary limbs or deficiencies in which parts of the limb are deleted.

In particular, local application of retinoic acid to the anterior margin of the limb bud gives a similar result to AP or AP/DV reversals. It therefore seems that in some ways it may mimic the effect of the zone of polarizing activity (ZPA). Naively, we could consider retinoic acid to be an analogue of the signalling molecule from the ZPA, thereby causing the formation of supernumerary structures. The deficiency results can be explained by assuming that the extra vitamin A systematically raised the concentration of this molecule to a background level incompatible with organized pattern. Alternatively, it could have side effects such as causing cell death, the inhibition of proliferation or the Inhibition of cell differentiation.

Discussion

It is clear from the work described above that the four species of vertebrate limbs used here (three amphibians and one bird) behave differently in some aspects of development and similarity in others. The similarities, far fewer in number, seem to be only two. First, all limbs develop in a proximal to distal sequence of elements (stylopodium, zeugopodium, autopodium) rather than, for example, distal to proximal or various combinations of the two. Secondly, after inversion of the AP axis limbs all seem to develop supernumeraries of normal cartilage and muscle patterns with same handedness as the stump. On the other hand, the differences are many. During normal development the digits appear in an anterior to posterior sequence in the axolotl limb bud and in a posterior to anterior sequence in anurans and birds. The size of the elements when they first appear is characteristic of the species, as is the number and disposition of the elements.

Experiments revealed that there is an increasing mosaicism in the proximo-distal axis from the axolotl to the chick which results in the loss of the ability to replace missing tissue. After inversion of the DV axis, amphibians produce supernumerary limbs of stump handedness, but chicks do not. After inversion of both transverse axes there are fives types of supernumerary limb structure that appear in *Rana* limb buds, four in *Xenopus* and axolotl, and only one (normal) in chicks. Thus, the normal pattern seems to be most stable in the higher vertebrate class.

Other differences between these limbs have also been observed such as the presence of an apical ectodermal ridge (AER), which plays an essential role in distal outgrowth, and its lack of regeneration in chicks. In anurans an AER is present and it is presumably capable of regeneration, but in urodeles there is no AER at all. From this array of developmental behaviour in vertebrate limbs, we can draw one important conclusion. There seems to be more than one way of developing limb—it can involve highly regulative mechanisms such as those utilized by axolotl limb buds and regeneration blastemas, or it can involve mosaic mechanisms such as those in chick limb buds. The degree of mix of regulative and mosaic mechanisms seems to determine the developmental behaviour of the limbs of any one species.

The conclusion is clearly contrary to the evolutionary view of the development limbs, where such fundamental mechanisms as the organization of the axes should be common to all vertebrates. Neither does the·e seems to be a common ancestral stage in development through which all limbs pass. Therefore not only have evolutionary concepts been of no value in advancing our understanding of developmental mechanisms, they have been positively misleading. There is surely no reason a *priori* why developmental mechanisms should not evolve and we concur with Sander (this volume) that homology does not imply conservation of mechanisms. Finally, it is interesting to consider the relation between development and regeneration. Only urodeles are capable of limb regeneration in adults, and it is surely no coincidence that urodeles have the most highly regulative form of limb development. *Rana* can regenerate for a short period after limb development and have a less regulative form of limb development. *Xenopus* loses regenerative ability before limb development has terminated and has a relatively mosaic type of development. Chicks, which can never regenerate, are the most mosaic of all. It is possible therefore, that the reason why the limbs of higher vertebrates have not evolved regenerative ability is because their mosaic type of development precludes it.

11

DIFFERENTIATION

Differentiation is the formation of identical cell from the properly differentiated cells to make the regeneration of cell possible. On the basis of molecular approval it is extremely difficult to interpret these changes in nucleic acids because it is uncertain which species of RNA are functional. Since proteins are the important products of nucleic acid messages and since their synthesis or lack of synthesis can be definitively determined, many investigators study proteins as markers of development and differentiation. In this chapter the main emphasis has been given to proteins in early development, methods of analysis in the protein field, and some of the specific proteins that differentiate cells from one another.

Protein in Early Development

As mentioned earlier, the sea urchin embryo is ideal for studying development because a vast number of embryos can easily been obtained and because normal development occurs in sea water. We can add a radioactive amino acid (such as radioactive methionine) to the sea water and measure the amount of this amino acid that is incorporated into newly synthesized protein. This is done by incubating the embryos with the "hot" amino acid for a short time, then placing the embryos in seawater without the radioactive amino acid. If we do this at various development stages, we can determine how much protein was synthesized at each specific stage by measuring how much radioactive protein was formed after we administered each pulse of "hot" amino acid. The amount of new protein that is synthesized can be measured with reasonable accuracy if we make sure that the radioactive amino acid is being taken up by the cells at all stages. We must also make sure that the intracellular supply (pool) of available amino acid remains constant at all stages. An example of the results obtained by investigators examining protein synthesis in the sea urchin embryo.

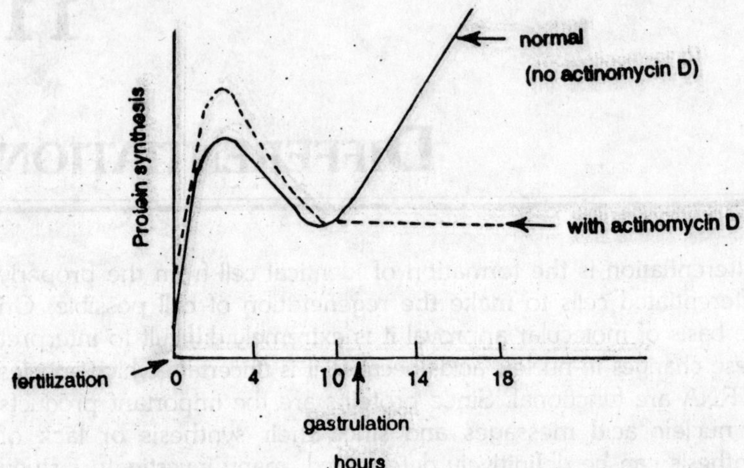

Fig. 11.1. Protein synthesis at various stages in sea urchin embryo development.

Protein synthesis rapidly increases for the first three or four hour after fertilization and then declines, only to increase again at about the gastrula stage. In this experiment radioactive methionine was used and was shown to be taken up by-unfertilized eggs and other stage. Amino acid uptake thus was not a problem in these experiments. Also, the supply (pool) of intracellular methionine was shown not to change appreciably in the different stage examined. Thus the pool of methionine was not a problem. These experiments suggest that different amounts of total protein are synthesized at various stages of embryonic development. If embryos are treated with the RNA synthesis inhibitor actinomycin D, protein synthesis increases after fertilization but not at gastrulation. These results suggest that the proteins synthesized during early development use messenger RNA formed in the egg before fertilization. At gastrulation, however, new species of mRNA may be transcribed, a process that would be blocked by actinomycin D.

More recent work has shown, however, that some specific proteins, such as histones, are synthesized in the early hours of

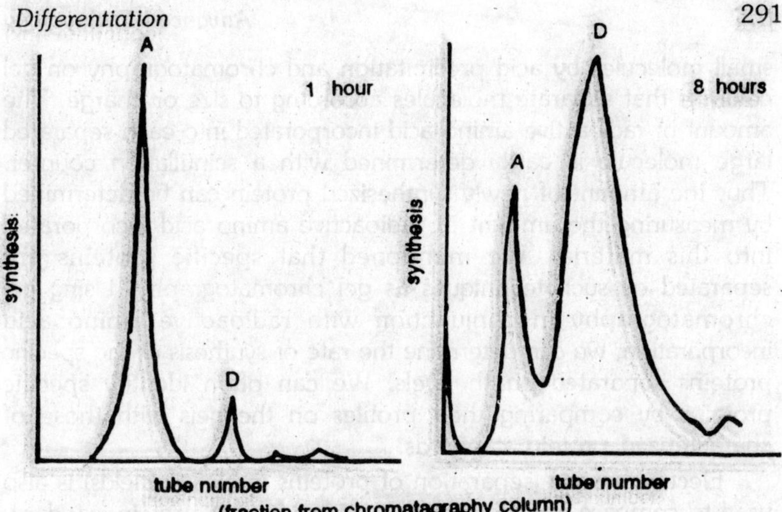

Fig. 11.2. Specific proteins synthesized at different stages of sea urchin embryo development.

sea urchin embryonic development using newly formed mRNA, as well as that stored in the egg before fertilization. We have described the synthesis of total protein at various developmental stages. What about specific proteins? Do all proteins in the cell follow the same pattern? Figure shows that if individual marked proteins are separated chromatographically at different developmental stages, different amounts of particular proteins are found to be synthesized at different developmental stages.

A lot of protein A, for instance, is synthesized one hour after fertilization, in the sea urchin, but only a little of protein D is synthesized at this time. Eight hours after fertilization, however, less of protein A is synthesized, and a lot of protein D is synthesized. The experiment does not identify proteins A and D. In fact, these proteins may be similar and may represent different numbers of identical subunits or an association of the protein with cofactors. Later in the chapter-we shall examine specific proteins and reaffirm that given proteins are often synthesized in different amounts at different developmental stages.

Measurement of Synthesis of Specific Protein

Protein synthesis is generally measured by incubating embryo cells in a solution containing a radioactive amino acid such as ^{14}C methionine or ^{14}C leucine. The cells are then homogenized, and large molecules are separated from the amino acids and other small molecules by acid precipitation and chromatography on gel columns that separate molecules according to size or charge. The amount of radioactive amino acid incorporated into each separated large molecule is easily determined with a scintillation counter. Thus the amount of newly synthesized protein can be determined by measuring the amount of radioactive amino acid incorporated into this material. We mentioned that specific proteins are separated by such techniques as gel chromatography. Using gel chromatography in conjunction with radioactive amino acid incorporation, we can determine the rate of synthesis of the specific proteins separated on the gels. We can often identify specific proteins by comparing their profiles on the gels with those of characterized protein standards.

Electrophoresis (separation of proteins in electric fields) is also used to compare a protein sample with a known protein standard. Immunological methods are also widely used to identify specific proteins. The specific protein is injected into rabbits or goats, which produce antibodies that can then be isolated from the blood of the animal. When the antibody is added to a specific protein in a mixture of many proteins, it complexes with the specific protein, forming a precipitate. Newly synthesized specific protein can be identified using the immunological technique coupled with the radioactive amino acid method. Cells are incubated with radioactive amino acid for a short time and then homogenized.

The homogenate (or an extract thereof) is then incubated with antibodies against a single specific protein, causing the protein to precipitate from solution. The amount of radioactivity in the precipitate is then measured. This tells how much radioactive amino acid was incorporated into specific protein precipitated by the antibody. In this way we can determine how much of the specific protein was synthesized while the cells were being incubated with radioactive amino acid. This technique is very useful in identifying the pattern of synthesis of specific proteins during differentiation. Remember that just measuring the activity of a given enzyme or the presence or absence of a specific protein at a given developmental stage does not determine the time or rate of synthesis per se. Synthesis must be measured by a method such as that described above, in which the amount of newly formed protein is determined.

The simple presence or absence of a specific protein or specific activity could result from the activation or inactivation of an enzyme by specific cofactors or from the differential degradation of the protein by degradative enzymes present in the cells. Before considering some specific proteins formed during differentiation, let us see how the immunological technique can be used to determine specific protein synthesis in one model system. These experiments are interesting because they will show not only how the technique is used, but also how protein synthesis and degradation are regulated.

The system involves the synthesis of the enzyme tryptophan pyrrolase by the rat liver. This enzyme opens the indole ring of tryptophan, a precursor of the vitamin nicotinamide. The activity of this enzyme rises rapidly at birth. Its activity is also increased by the hormone hydrocortisone and by the amino acid tryptophan. Purified tryptophan pyrrolase was injected into rabbits, which produced specific antibody against this enzyme. Rats were inoculated with hydrocortisone, tryptophan, or saline and, after a few hours, radioactive amino acid. Their livers were then removed and homogenized.

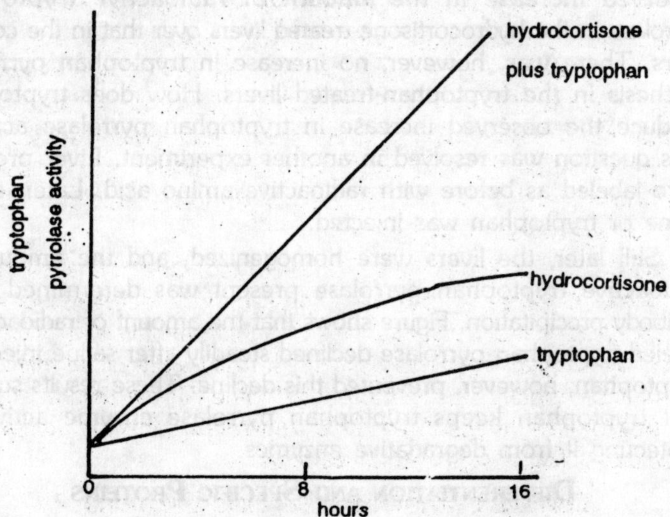

Fig. 11.3. Stimulation of tryptophan pyrrolase activity by hydrocortisone, tryptophan, or both.

Tryptophan pyrrolase antibody was added to the homogenate to precipitate out tryptophan pyrrolase, and the amount of newly synthesized enzyme was determined by measuring the radioactivity

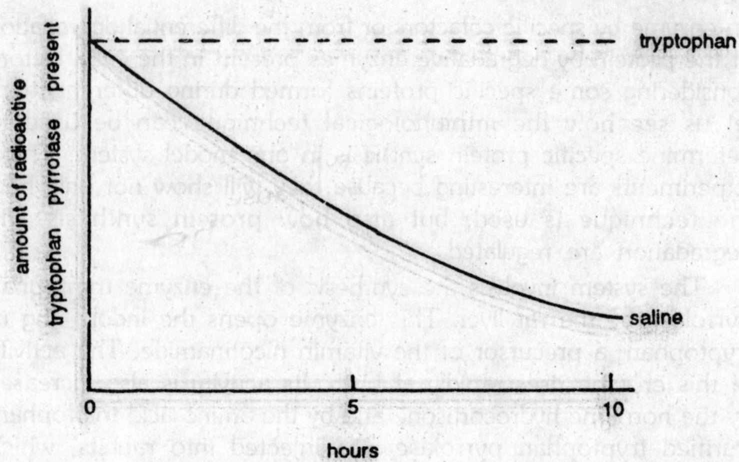

Fig. 11.4. Prevention of tryptophan pyrrolase degradation by tryptophan.

of the precipitate. Figure shows that hydrocortisone increases the synthesis of tryptophan pyrrolase. This conclusion is based on the observed increase in the amount of radioactive tryptophan pyrrolase in the hydrocortisone treated livers over that in the control livers. There was, however, no increase in tryptophan pyrrolase synthesis in the tryptophan-treated livers. How does tryptophan produce the observed increase in tryptophan pyrrolase activity. This question was resolved in another experiment. Liver proteins were labeled as before with radioactive amino acid. Later, either saline or tryptophan was injected.

Still later, the livers were homogenized, and the amount of radioactive tryptophan pyrrolase present was determined after antibody precipitation. Figure shows that the amount of radioactively labeled-tryptophan pyrrolase declined steadily after saline injection. Tryptophan, however, prevented this decline. These results suggest that tryptophan keeps tryptophan pyrrolase enzyme active by protecting it from degradative enzymes.

Differentiation and Specific Proteins

Haemoglobin

Haemoglobin is a *tetramer*, a molecule composed of four folded polypeptide chains. Each polypeptide chain carries a non-protein heme group. The heme groups contain iron and bind oxygen. The most important function of hemoglobin is to carry oxygen to different parts of the body. Haemoglobin is the major protein contained in red blood cells and thus is distinct differentiation marker in these highly specialized cell types. There are three different types of haemoglobin: embryonic haemoglobin, foetal haemoglobin, and adult haemoglobin. Stage-specific differences in haemoglobin probably evolved as a result of the different respiratory needs of the embryo foetus, and adult. Mammalian foetuses obtain oxygen by diffusion across the placenta from the maternal blood. Thus the haemoglobin of the foetal red blood cells must have a higher oxygen affinity than that of the mother in order for such an exchange to occur. What is the nature of these stage-specific differences?

As mentioned previously, adult mammalian haemoglobin consists of four polypeptide chains. Two of these chains are the identical α-chains; they are encoded by one gene. The other two chains are the identical β chains; they are encoded by an entirely different gene that is not even on the same chromosome as the α-chains gene. Adult haemoglobin can thus be designated as αδ ββ, indicating that two polypeptide chains are coded for by gene A and two by gene B. In the early mouse embryo, other gene code for haemoglobin chains. The A gene is active in the embryo, as are the X, Yβ and Z genes.

Embryonic haemoglobin does not contain β chains because the B gene is not active until late in foetal development. Thus important changes in haemoglobin take place when the B gene is activated, making β polypeptide chains part of the new haemoglobin molecules. Abnormal types of adult haemoglobin include haemoglobin S and haemoglobin C. Haemoglobin S is found in individuals who have sickle-cell anemia. In this disease, the red blood cells are sickly shaped. Haemoglobin S differs from normal haemoglobin only in the β subunits; the β subunits are exactly like those in normal adult haemoglobin. The normal β subunit consists of 146 amino acids. The only difference between the B subunits of normal and sickle-cell haemoglobin is that in the latter a single valine residue replaces the number 6 amino acid, glutamic acid. This single amino acid change can be caused by a single base change in the mRNA coding for the β subunit.

Glutamic acid is coded for by the base sequence GAA, whereas valine is coded for by GUA. Thus a difference of a single base in a large protein can produce a disease that can be fatal. This change probable arises as the result of a single point mutation in the DNA of the β subunit gene Haemoglobin C, another abnormal adult haemoglobin, also result from a single amino acid change. The same β chain amino acid replaced in haemoglobin S is also replaced in haemoglobin C. Instead of a valine, however, a lysine substitutes for the glutamic acid residue at position 6. The codon AAA codes for lysine, whereas GAA codes for glutamic acid. Again a single base change, an A for a G, leads to an amino acid change in the β subunit. Haemoglobin C, however, does not cause severe anemia in people. Thus, if one specific amino acid out of 146 amino acids in the β subunit is changed to valine, major disease results, if it is changed to lysine, on the other hand, only a minor problem arises.

Now that we know a little about haemoglobin, let us look at the differentiation of red blood cells. Red blood cells in the embryos of humans and mice originate first in the yolk sac, then (later in development) in the liver. After, red blood cells are produced by the bone marrow. Several important questions now arise. When do the cells that give rise to red blood cells begin to differentiate? What events mark the beginning of differentiation? Recall that embryonic haemoglobin differs from adult haemoglobin. Do the same cells, or cells derived from the same parent cell, give rise to the different types of haemoglobin? Or do different cell lines give rise to the different haemoglobins?

In the mouse, haemoglobin begins to appear by the eighth day of gestation, in red blood cell precursors located in the blood islands of the yolk sac. On the ninth day, these precursor cells enter the circulation and begin to synthesize embryonic haemoglobin. After they divide four times, their nuclei condense and they become inactive. On the Twelfth Day of gestation, the liver begins to produce red blood cells. This new cell population synthesizes adult haemoglobin. The yolk sac continues to produce red blood cells through the fourteenth day, so both types of red blood cells (yolk sac and liver) are found in the foetal circulation.

Spleen and bone marrow take over the production of red blood cells between day 15 and birth and liver production declines. Thus, there are at least two populations of red blood cells, one that synthesizes embryonic haemoglobin and another than synthesizes adult haemoglobin. No one knows whether a common ancestor cell originally seeded the yolk sac and the liver, spleen, and bone marrow. In the yolk sac haemoglobin (globin) mRNA is synthesized before day ten of gestation. After day ten, haemoglobin synthesis is not inhibited by actinomycin D. This suggest that the mRNA made before day ten is stable and used later on for synthesizing the protein chains of haemoglobin. Thus, the main event that triggers the formation of red blood cells is probably the activation of the genes that code for the polypeptide chains of haemoglobin (globin).

Messenger RNA that codes for the polypeptide chains of haemoglobin has been isolated. Such mRNA is identified by its ability to direct the synthesis of globin in a cell-free system and by hybridization experiments such as those described in the last chapter. Early foetal mouse liver cells do not contain globin messenger RNA. The glycoprotein hormone, *erythropoietin*, cause these cells to synthesize globin mRNA. Haemoglobin synthesis begins shortly thereafter. Erythropoietin is produced by the kidney in response to decreases in tissue oxygen content. It is a key factor in inducing prospective erythrocytes to differentiate into mature red blood cells. It probably acts by turning on globin genes. Although the exact mechanism is unknown, it may actin in a manner similar to that proposed for estrogen (estradiol). Its final action may be on specific chromosomal proteins that function to regulate gene action. Exactly how certain cells commit themselves to forming red blood cells is unknown. However, the events that immediately lead to differentiation-globin message transcription followed by haemoglobin synthesis-are becoming better understood.

Myosin

The protein myosin is probably present in many cell types. Muscle cells, however, contain a large amount of this protein. Myosin makes up 10-12% of the fresh weight and nearly 50% of the dry weight of adult muscle. Myosin is thus a major marker for muscle cell differentiation. Myosin is a part of the contractile networks present in all eukaryote cells. Since the main function of muscle cells is to contract, these cell types should be enriched in myosin.

Myosin is a fibrous protein composed of two identical subunits wound around each other. In muscle, myosin molecules aggregate into filaments. This assembly phenomenon will be examined in the next chapter. Myosin combines with other proteins, such as actin. Units composed of myosin, actin, and certain other proteins are contractile and serve as a cellular contractile apparatus.

Proteins associated with actin filaments include α-actin, β-actinin, tropomyosin and troponin. These proteins serve a variety of functions that keep the contractile apparatus intact and allow it to work smoothly. Let us look first at developing muscle cells and then focus on the differentiation of these cell types. In this chapter we shall stress myosin synthesis. In the next, we shall focus on how muscle proteins assemble into functional contractile units. Experiments using fluorescent-labeled myosin antibody have shown that muscle cell contractile proteins are present in many prospective muscle cells before any organized contractile apparatus is visible there.

Prospective muscle cells (myoblasts) are derived from the myotome region of the somite and hypomere mesoderm. The prospective muscle cells divide, forming many additional myoblasts. When they stop dividing, they fuse together, forming long muscle tubes (myotubes) containing many nuclei in a common cytoplasm. In two-day-old chick embryos, somite myotome cells are actively dividing. Some of these cells form spindle shapes containing detectable myosin filaments. By the fourth day most of the myoblasts contain muscle contractile fibrils (myofibrils). All this occurs before the cells fuse into the multinucleated myotubes. In limb muscle, however, myofibrils appear only in the multinucleated myotubes and not in the prefusion myoblasts.

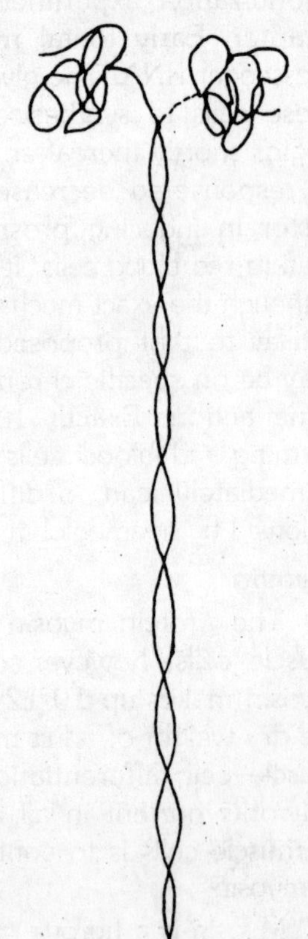

Fig. 11.5. Myosin consists of two subunits wound around units wound around each other.

We have said that elongated myotubes arise as a result of fusion of single cells rather than continued nuclear divisions without cell division. This was elegantly shown by an experiment in which cells from two different mouse embryos were aggregated. Each embryo strain contained a different form of the enzyme isocitrate dehydrogenase, which is made up of subunits. The mosaic embryos were implanted in the uterus of a foster mother mouse where they developed into normal mice consisting of cells from the two different strains of mice. The skeletal muscle of these mice contained some enzyme like that of each parent, as well as some hybrid enzyme. This hybrid enzyme consisted of some subunits from one parent's enzyme and some from the other parent's enzyme. This type of hybrid enzyme could have been produced only if the myotubes contained nuclei from cells of both parents in a common cytoplasm.

Each type of nucleus produced a message that coded for its own enzyme subunits. These subunits, when synthesized then

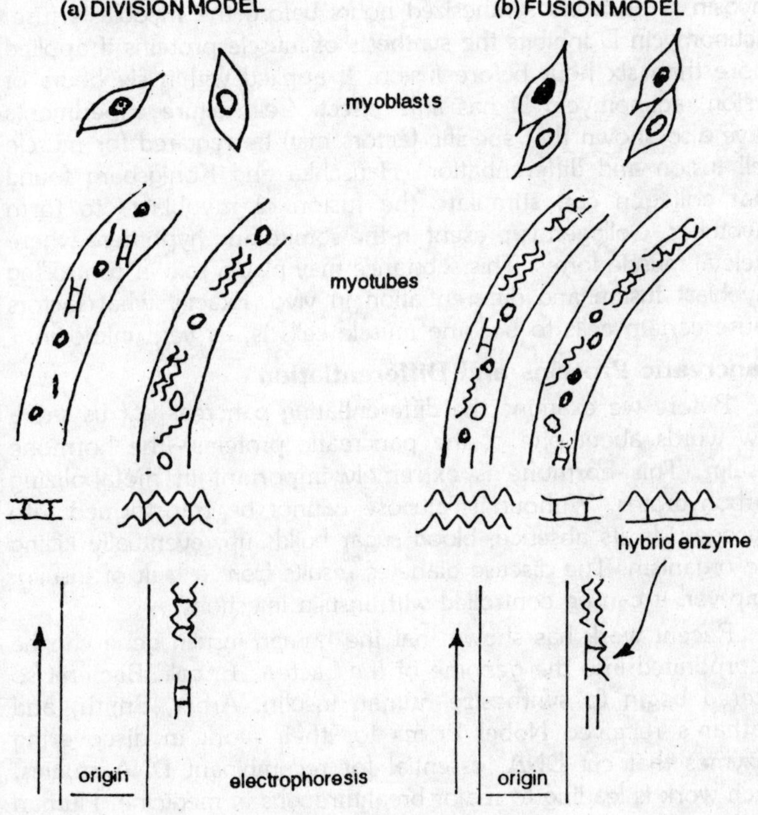

Fig. 11.6. Isocitrate dehydrogenease in myotubes.

polymerized into the hybrid enzyme observed. Such enzymes are easily detected using electrophoresis, which separates proteins on the basis of size and charge. Thus skeletal muscle forms when single cells fuse into elongated myotubes. Myoblast fusion has also been studied *in vitro* to learn about the specific factors that influence this phenomenon. Myoblasts in culture do not fuse until six to eight hours after cell division has stopped. Presumably these cells use this time to acquire specific cell-surface properties that facilitate the fusion process.

Limb myoblasts in culture synthesize myosin and assemble myofibrils even if fusion is prevented by chelating agents that bind divalent cations. Thus even in limb muscle, fusion is not absolutely necessary for muscle differentiation to occur. Yaffe and others have shown that in culture much of the mRNA that codes for myosin is probably synthesized hours before the myoblasts fuse. Actinomycin D inhibits the synthesis of muscle proteins if applied more than six hour before fusion. If applied within six hours of fusion, actinomycin D has little effect. Cell culture experiments have also shown that specific factors may be required for muscle cell fusion and differentiation. Hauschka and Konigsberg found that collagen can stimulate the fusion of myoblasts to form myotubes. Collagen is present in the somite and hypomere where skeletal muscle forms. This substance may play a role in controlling myoblast fusion and differentiation in vivo. Exactly what factors cause certain cells to become muscle cells is, as yet, unknown.

Pancreatic Proteins and Differentiation

Before we examine the differentiating pancreas, let us say a few words about one of the pancreatic proteins—the hormone insulin. This hormone is extremely important in metabolizing carbohydrates. Without it, glucose cannot be transformed into glycogen. In its absence, blood sugar builds up, eventually killing the organism. The disease diabetes results from a lack of insulin: however, it can be controlled with insulin injections.

Recent work has shown that the human insulin gene can be incorporated into the genome of the bacteria E. coli. Bacteria so altered begin to synthesize human insulin. Arber, Smith, and Nathan's received Nobel Prizes for their work in discovering enzymes that cut DNA, essential for recombinant DNA studies. Such work is leading to major breakthroughs in medicine. Human insulin, which can now be produced by the insulin gene in *E. coli*,

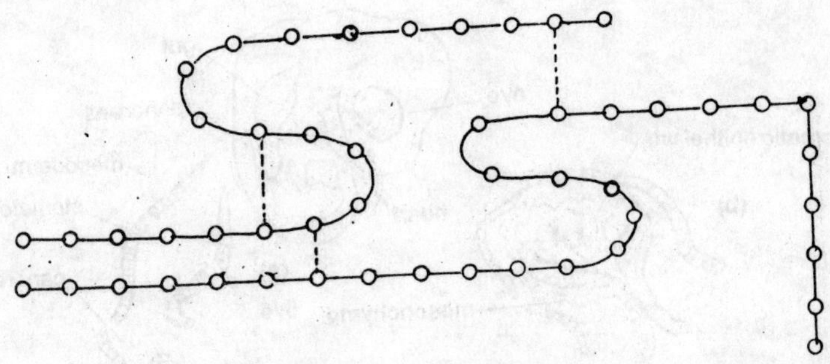

Fig. 11.7. The hormone insulin is composed of two polypeptide chains. Disulfide bridges form between the sulfhydryl groups of specific amino acids (cystein residues).

is very useful in treating many difficult cases of diabetes that do not respond well to animal insulin. Virtually unlimited supplies of this hormone are now available. Insulin is a small protein hormone with a molecular weight of only 6000 daltons. It consists of two polypeptide chains.

Insulin is one of many important proteins produced by the pancreas. In the mouse embryo, insulin is first detected in the pancreas rudiment nine or ten days after fertilization. The cells that produce insulin pinch themselves off from other pancreas tissue, forming separate masses called Islets of Langerhans. The islet cells differentiate into two distinct cell populations: A-cells, which synthesize the hormone glucagon, and B-cells, which produce insulin. Both hormones are involved in carbohydrate metabolism. B-cells are easily distinguished from A-cells by their dense cytoplasm and beta granules. The cells that produce digestive enzymes are located at the ends of a branched duct system in pouches called acini.

The digestive enzymes produced in the acini include lipase, amylase, trypsin, and chymotrypsin. When the mouse pancreas rudiment appears as an evagination from the gut tube, pancreatic enzymes and hormones are detectable, but in very small amounts. A period of rapid cell division follows during which the enzyme and hormone levels remain low. By about day 15, cell division stop and the synthesis of specific enzymes and hormones rapidly increases. During this time the cells of the pancreas also differentiate cytologically. Endoplasmic reticulum, Golgi apparatus, and zymogen granules become increasingly abundant in the non-

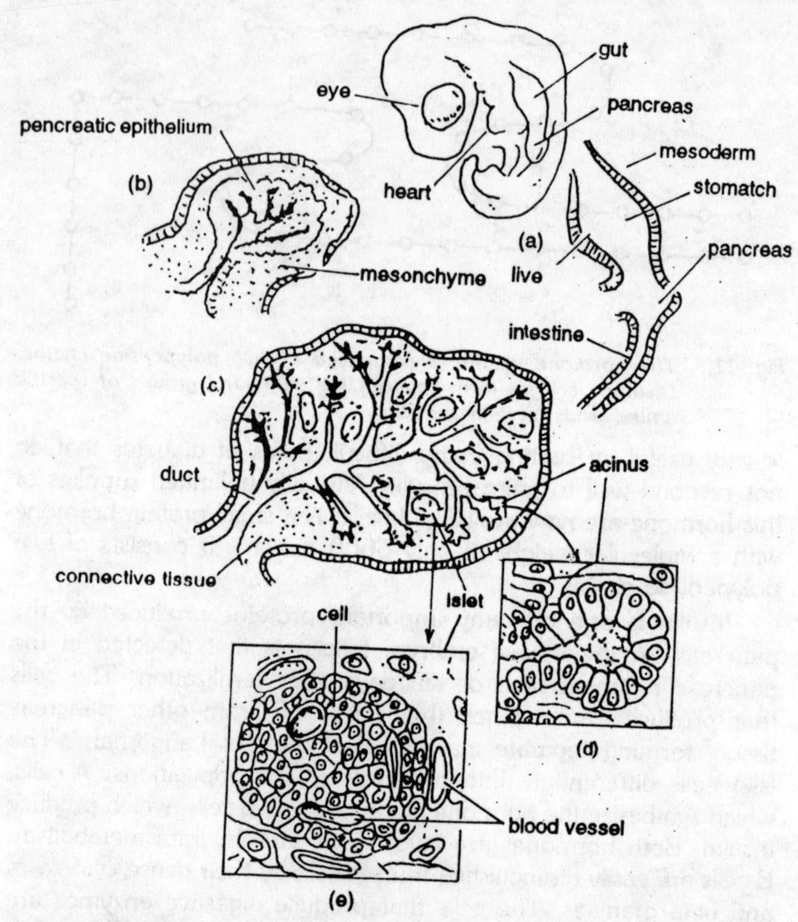

Fig. 11.8. Development of mouse pancreas. (a) Initial elevation of pancreas epithelium, which would have low levels of specific proteins (about 12-day embryo). (b) Acini and islets are now present. The acini composed of exocrine cells with zymogen granules. The islets are endocrine in function. (c) Secretory acini begin to form (about 12-day embryo). (d)-(e) Enlargement of an acinus, containing secretory (zymogen) granules and an islet, respectively.

dividing acinar cells. The differentiation of the endodermal epithelium of the mouse pancreas is apparently induced by the mesenchyme that overlies the gut.

The mesenchyme merges with the endodermal pancreatic rudiment as the rudiment evaginates from the gut tube at day 9 or 10. If an 11-day pancreas rudiment is separated from its mesenchyme and cultured on a porous filter, the endodermal epithelium does not proliferate and differentiate. If the mesenchyme is placed on the other side of the porous filter, however, cell division and differentiation occur. Specific proteins are synthesized, and characteristic tubules develop. Mesenchyme from other tissues—and even from other species—also induces differentiation. The mesenchyme apparently produces a factor that can pass through the filter pores to stimulate cell division in the epithelium. Adding a cell-free extract of whole chick embryos or of mesenchymal tissues to the medium in which pancreas epithelium is cultured causes the epithelium to divide and differentiate in the absence of mesenchyme.

Cell division is needed for differentiation, because if DNA synthesis is inhibited in predividing cell by FUdR (5-fluorodeoxyuridine) differentiation does not occur. BUdR (5-bromodeoxyuridine) is a thymidine analog that replaces thymidine in the DNA of dividing cells but does not prevent DNA synthesis or cell division. The DNA synthesized, however, is abnormal. BUdR also prevents the differentiation of pancreas cells treated during the division period. In the presence of BUdR, mitosis continues and some common cell proteins are synthesized, but levels of pancreas-specific proteins and differentiation do not increase. BUdR may prevent the production of mRNA that codes for cell-specific proteins.

The RNA synthesis inhibitor actinomycin D also inhibits the increase in protein-synthesis if it is applied during the cell division period or before the increase in specific protein synthesis begins. It has no effect one protein synthesis accelerates. This implies that stable messages are produced before the increase in specific protein synthesis. To sum up, period of normal DNA synthesis must occur before a high rate of specific protein synthesis can begin. The mesenchymal factor (MF) that stimulates cell division and differentiation in pancreas epithelium, was isolated by Rutter and colleagues. This factor was covalently bound to insert beads, and the beads were cultured with pancreas epithelium cells. The cells attached to the beads and continued DNA synthesis and cell division. Zymogen granules formed in the cells, indicating that differentiation had occurred. Since it is unlikely that MF attached to the beads entered the cells, these results suggest that MF stimulate mitosis and differentiation by some action at the surface

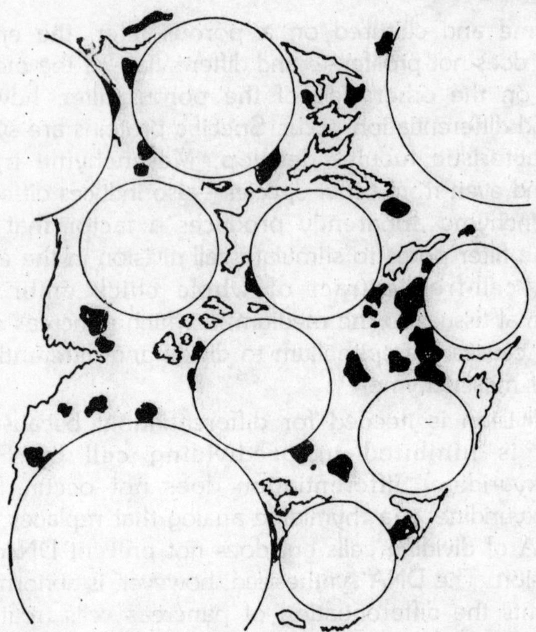

Fig. 11.9. Pancreas epithelial cells attached to sepharose beads coated with mesenchymal factor (MF).

of the pancreas epithelial cells. In summary, the pancreas develops as an evagination of the gut tube.

The pancreatic rudiment joins with the mesenchyme surrounding the gut tube. In the presence of mesenchyme, morphogenesis begins and specific proteins are synthesized at a low rate. Cell division and morphogenesis continue. Finally, specific protein synthesis increases rapidly and cytological differentiation is completed. Stable mRNA coding for the specific proteins is apparently synthesized before the synthesis of specific proteins accelerates. Mesenchyme apparently induces cell division in the pancreas epithelium. Wessels and others have proposed that increased cell mass in the developing pancreas (and in other systems) may be a key factor in the increase in the levels of specific protein synthesis and cytodifferentiation. Large masses of cells could alter their immediate environment, creating conditions that accelerate specific protein synthesis. This idea may suggest additional experiments that will uncover some of the control mechanisms of differentiating systems such as the pancreas.

Tissue-and Stage-Specific Isozyme Patterns

We have seen that proteins often consist of subunits coded for by separate genes. Haemoglobin subunits, for instance, are coded for by several genes. The relative activities of these genes vary at different developmental stages, so foetal haemoglobin and adult haemoglobin clearly have different subunit compositions. These different haemoglobins are presumably those best adapted to the different environments of the embryo and the adult. *Isozymes* are enzyme molecules that are functionally similar but that behave differently during electrophoresis because of differences in charge: some forms are more negatively charged than others. These charge differences occur because the enzymes are composed of subunits and, as in haemoglobin, different subunits can combine to form the isozymes.

Each different subunit is coded for by a different gene. Thus, if particular genes are more active than others at a given stage or in a given tissue, the resulting proteins will also be different. *Lactate dehydrogenase* (LDH) is an enzyme that exists in multiple isozymes. Different isozymes of LDH predominate in different species, in different tissues, and at different developmental stages. LDH catalyzes the interconversion of pyruvate and lactate. Pyruvate is a key intermediate when glucose is oxidized to carbon dioxide and water, releasing energy that is essential for various metabolic activities. LDH makes it possible for cells to function effectively when oxygen is temporarily unavailable. As LDH converts pyruvate to lactate, it generates NAD from NADH. NAD is essential for completion of an early step in a set of metabolic reactions called glycolysis.

Like different haemoglobins, different LDH isozymes presumably function best under different metabolic conditions. Thus one isozyme might predominate at a specific stage of in a specific tissue because it is best adapted to function under those cytoplasmic conditions. Before we briefly consider LDH isozymes during development, let us see how LDH was found to be composed of several isozymes in the first place. When crude homogenates of tissues such as mouse skeletal muscle are subjected to electrophoresis, the various proteins present in the homogenate separate on the basis of their net electrical charge.

After a time, the gel block is immersed in a staining solution that contains lactic acid (an LDH substrate), the required cofactor

NAD, and a tetrazolium salt. If LDH is present in the gel, the lactic acid is oxidized and the tetrazolium salt is reduced to coloured formazan that precipitates at the site of LDH activity. Five of these coloured zones appear after a mouse muscle homogenate is subjected to electrophoresis. Each coloured zone identifies an LDH isozyme. Five isozymes of LDH have been found in most of the mammal and bird tissues examined. Why do five LDH isozymes form? Markert and Ursprung purified LDH using several methods.

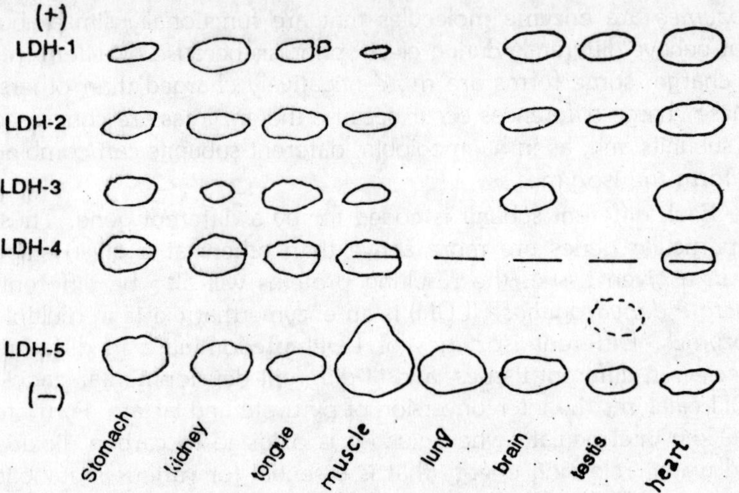

Fig. 11.10. Changing LDH isozyme pattern in heart muscle at different stages of mouse development. LDH-1 migrates toward the anode (+) and LDH-5 migrates to the cathode (–) during electrophoresis.

The pure LDH also formed five isozyme bands after electrophoresis. By analytical ultracentrifugation in sucrose density gradients, they determined that the molecular weight of each LDH isozyme is about 140,000 daltons. When the pure LDH isozymes were subjected to agents or conditions that separated proteins into subunits, only two, not five, electrophoretic bands appeared that stained for protein. Each band represented a subunit of LDH. Each subunit had a molecular weight of about 35,000 daltons. The two bands were designated LDH subunit A and LDH subunits B. The separated LDH subunits recombined to form the five LDH isozymes as follows:

LDH-1 BBBB
LDH-2 ABBB

LDH-3 AABB
LDH-4 AAAB
LDH-5 AAAA

Such combinations of subunits probably occur in cells, forming the five LDH isozymes. Different genes code for LDH subunits A and B. Thus, the amount of each isozyme that appears in cells of different tissues or at different developmental stages reflects, in part, the ratio of A subunits to B subunits produced at that time in that tissue. For example, mouse eggs produce mostly LDH-1 (the B tetramer); thus the B gene is the most active subunit during oogenesis. As development proceeds, the A gene becomes activated and the B gene is suppressed. LDH-5 (the A tetramer) begins to predominate. By birth, the B gene has been activated in many tissues of the body. Then, the isozyme pattern shifts again toward LDH-1. Thus shifts in LDDH patterns reflect the shifting relative activities of the A and B gene. Adult rat tissues also differ with respect to LDH pattern. For example, skeletal muscle shows

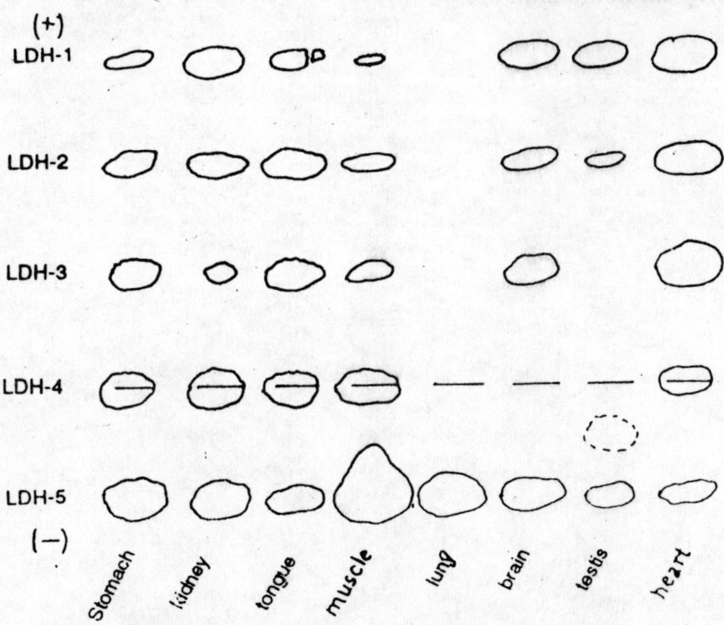

Fig. 11.11. LDH isozyme patterns in eight adult rat tissues. Electrophoretic patterns show tissue specific differences in LDH isozymes. Dotted circle in testis pattern represents an isozyme that may be testis-specific.

a lot of LDH-5, whereas heart muscle has more LDH-1, LDH-2 and LDH-3. Here, too, shifts occur relative to activities of the A and B genes.

Specific cytoplasmic factors such as pH and ionic species probably control the rates of subunit synthesis. Such conditions may also affect subunit synthesis. Such conditions may also affect subunit accumulation and assembly. Remember that the amount of a given protein present is a function not only of the rate of synthesis of the protein, but also of its rate of degradation (recall the work on tryptophan pyrrolase). Thus the presence of specific LDH isozymes represents a balance between the synthesis and degradation of those isozymes. To sum up, both stage-specific and tissue-specific protein patterns appear during the differentiation process. Such protein patterns result from differential synthesis, degradation, and activation of specific proteins. Now that we have considered differentiation at the nucleic acid and protein levels, we can look at how larger components of cells are assembled during differentiation.

12

Morphogenesis

Cell contact and adhesion take place at the cell surface. Cells in embryos move over one another and stick to one another at their surfaces. Thus in order to examine embryonic cell rearrangements and associations, we must look at the cell surface—the place where these associations begin. It has already described the cell surface as a lipid bilayer in which proteins and glycoproteins are imbedded. The cell surface has fluid properties that allow the proteins and glycoproteins to move laterally in the lipid. This movement is sometimes restricted by rod-shaped proteins, such as microtubules and microfilaments, that anchor some of the surface proteins from the cytoplasmic side of the membrane. Singer and Nicolson's fluid-mosaic model of the cell surface, currently the most popular and widely accepted model, describes the cell surface as having fluid properties and being mosaic in nature—that is, it consists of various proteins interspersed in the lipid.

Fluid-Mosaic Model of the Cell Surface

An elegant experiment by Frye and Edidin provided the basis for determining that the cell surface has fluid properties. Using immunological techniques, these workers developed complex stains that specifically bind to certain cell-surface proteins. Thus stains specific to the surface proteins of given cell types could be produced and coupled to fluorescent dyes. Two different cell types (a human cell and a mouse cell) were fused together, with Sendai virus used to promote cell fusion. The surface proteins specific for each cell type were then traced using the complex stains and a fluorescence microscope.

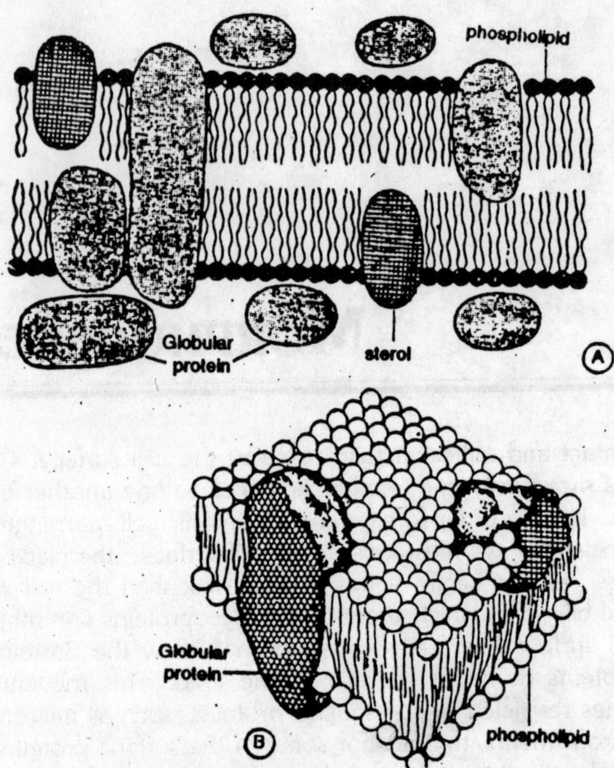

Fig. 12.1. A—Schematic representation of a section through a eukaryote fluid-mosaic membrane. B—The fluid-mosaic model of Singer and Nicolson (1972): bacterial membrane.

By watching the movement of each fluorescent stain bound on the surface of the hybrid cell, Frye and Edidin showed that the surface proteins of the human cell and mouse cell intermixed. The rate at which these proteins intermixed suggested that the proteins diffused laterally (moved sideways) in the plane of the cell surface). Numerous subsequent experiments supported these results. Thus investigators concluded that the cell surface is fluid in nature. We must stress, however, that under certain conditions (for example, low temperatures that solidity the lipid phase of the membrane) the movement of membrane proteins becomes restricted.

In addition, there is increasing evidence that certain cell-surface proteins are anchored by cytoskeletal elements (microtubules and microfilaments) that restrict their movement. Some of the proteins associated with the cell membrane are loosely associated and easily removed; others are imbedded deeply in the lipid bilayer, sometimes spanning its entire thickness. The former are called peripheral membrane proteins, and the latter are called integral membrane proteins.

Integral membrane proteins can often be viewed using the freeze-fracture technique. In this technique the membrane is split down the middle and one of the lipid bilayers is peeled away. The surface of the cleaved membrane is coated with a heavy metal that forms a replica of the fractured surface. The replica is examined with the electron microscope. Integral membrane proteins are observed as particles on the replica surface. To sum up, the cell surface has fluid properties and is a mosaic consisting of lipid interspersed with proteins, many of which can move laterally in the lipid.

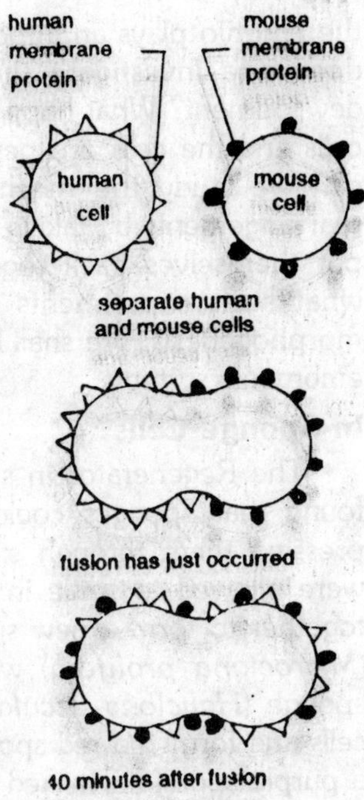

Fig. 12.2. Lateral mobility of proteins in the membrane lipid.

Surface Sugars

It has been observed that some of the proteins and lipids of the cell surface carry attached sugar chains. These sugar chains—and especially the terminal sugar residues of the sugar chains—reach farthest away from the cell surface and thus offer a first line of contact with approaching cells and molecules. Some of the molecular models proposed to explain the adhesive recognition of cells in embryonic systems emphasize the role of the surface sugars.

Selective Cell Adhesion

We have seen examples of how specific cell adhesion plays a role in developmental events. For example, sperm-egg recognition appears to depend on specific adhesion between the gametes. Similarly, the preferential adhesion of the filopodia of the secondary mesenchyme cells in the sea urchin gastrula to the animal end of the embryo plays an important role in forming the gut tube. How does one investigate the role of cell adhesion in embryonic development? What happens if embryos are dissociated into single cells and the cells are permitted to reassociate ? In other words, can we study the mechanisms controlling embryonic cellular rearrangements by taking embryos apart to see whether they can put themselves back together? The answer is yes. Let us see what such experiments can tell us about the mechanisms of morphogenesis. We shall begin with sponges and then move onto embryonic systems.

In Sponge Cells

The Regenerator in sponges was studied by H.V Wilson. He found that sponges could be disaggregated into single cells by pressing them through silk cloth. When the single sponge cells were allowed to settle in a dish of sea water, they moved back together to form a new sponge. When cells from the red sponge (*Microciona prolifera*) were mixed with cells from the purple sponge (*Haliclona occulata*), the red cells sought out other red cells and formed a red sponge and the purple cells likewise formed a purple sponge. The red cells seldom stuck to purple cells. Thus, sponge cells can specifically recognize cells of their own species. This specificity has served as a model for studying adhesive recognition in developing systems. Moscona and Humphreys continued to investigate specific reaggregation in the sponge system. Using a rotary shaker method developed by Moscona, these workers found that when mixtures of red and purple sponge cells were rotated together, they still formed aggregates consisting mainly of only one type of cell. Red stuck to red, purple to purple. These results suggested that the major reason sponge cells form species-specific aggregates is that each type of cell preferentially sticks to its own kind.

The shaker method allowed cell aggregates to form as a result of adhesive stability only without the directed movement (migration) that could occur in a stationary (nonrotated) system. Moscona and Humphreys tried to determine the molecular nature of this species-specific adhesion. They found that sponges could also be disaggregated into single cells by immersing them in seawater free of calcium and magnesium. These cells, however, did not reaggregate if returned to sea water and rotated at low temperature (5°C). When calcium was added to the disaggregation supernatant (the solution left over after the disaggregated cells were removed) and the supernatant was added back to the cells, they did reaggregation of only red sponge cells.

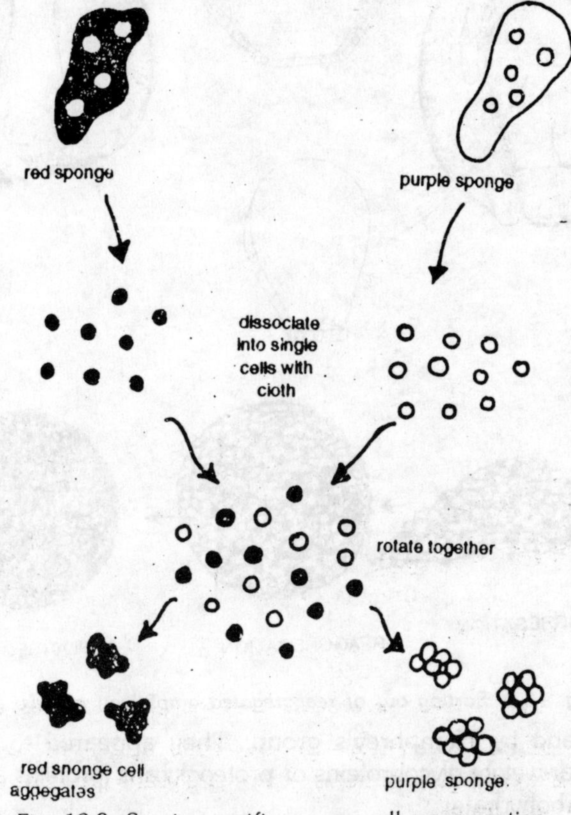

Fig. 12.3. Species-specific sponge cell reaggregation.

Similarly, the supernatant from the purple sponges promoted reaggregation of only purple sponge cells. Moscona and Humphreys concluded that a molecule responsible for species-specific sponge cell adhesion was released when the sponge disaggregated in sea water free of calcium and magnesium. If these molecules were added back to the cells from which they were removed, the cells would stick back together. Apparently, cells can also resynthesize this molecule, because they reaggregated to some extent at higher temperatures (24°C) but not at low temperatures (5°C). At higher temperatures, cellular synthetic reaction could occur. The molecules promoting adhesion in sponge cells were purified both by Moscona's

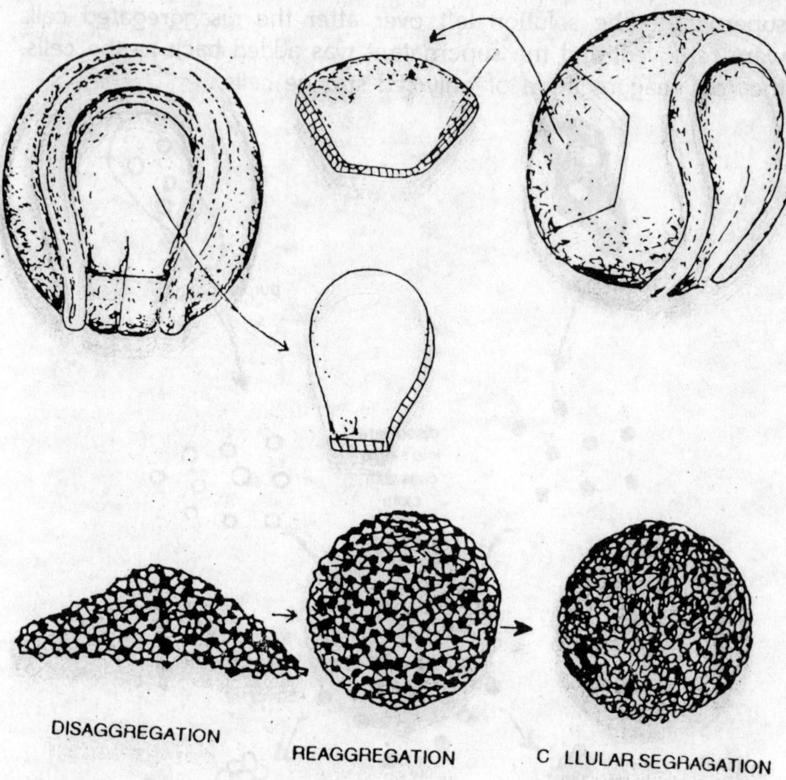

Fig. 12.4. Sorting out of reaggregated amphibian embryo cells.

group and by Humphrey's group. They appeared to be high-molecular-weight glycoproteins or proteoglycans (proteins associated with carbohydrate).

Turner and Burger found that the sugar D-glucuronic acid was on the adhesion-promoting molecule of *Microciona prolifera*. This sugar may be responsible for the binding with the receptor site on the sponge cell surface. The nature of the chemical bonds between readhering sponge cells is not well understood, however. In any case, molecules that appear to be responsible for allowing sponge cells to adhere to their own kind have been isolated, purified, and characterized.

In Embryonic Systems

The methods used to study sponge cell adhesion have also been used to study embryonic cell adhesion. Townes and Holtfreter,

Moscona, and Steinberg performed some pioneering adhesion experiments with embryonic cells. Embryos of all types can be disaggregated into single cells using alkaline solutions, solutions free of calcium and magnesium, or solutions containing proteolytic (protein-degrading) enzymes such as trypsin. Embryonic cells disaggregated by such means can be rotated in the medium using a gyratory shaker. The cells from a whole embryo or from a variety of embryonic tissues recombine into aggregates of many cell types. After a day or two, however, cells tend to "sort out" into groups of only one specific cell type.

The mixed-cell aggregate is gradually transformed into an aggregate containing several areas, each composed of only one cell type. For example, ectoderm cells stick to ectoderm cells, retina cells stick to retina cells, and preheart cells stick to preheart cells. The nature of this "sorting out" phenomenon is unclear, but it appears to involve the relative adhesiveness of the different cell types. The most tightly adhesive cells squeeze together toward the center of the aggregate, pushing the less adhesive cells to the periphery. McGuire and co-workers found that if embryonic cells were disaggregated from the embryos very gently, they behaved like the sponge cells. That is, embryonic liver cells mainly adhered immediately to other liver cells and not to cells from other embryonic tissue cell types. Thus tissue-specific recognition occurs between like embryonic cells. This adhesive recognition may govern embryonic cell rearrangements.

Let us now turn to some experimental evidence suggesting that specific adhesive recognition plays an important role in morphogenesis. Roth's group used the chick embryo retina-tectum system in their study. Nerves grow from the retina of the eye to specific regions of the optic tectum, the brain center that interprets visual impulses. Nerves from the dorsal portion of the retina always grow to the ventral portion of the optic tectum, not to the dorsal portion. Roth and co-workers sought to determine whether adhesive recognition plays a major role in these specific nerve connections. They first cut retinas and tecta into dorsal and ventral halves. They then labeled retina cells from each half of the retina with radioactive precursors (molecules that become incorporated into cells) or with retinal pigment. Single labeled dorsal or ventral retina cell suspensions were agitated with either dorsal tectal halves or ventral tectal halves. This technique of shaking chunks of tissue

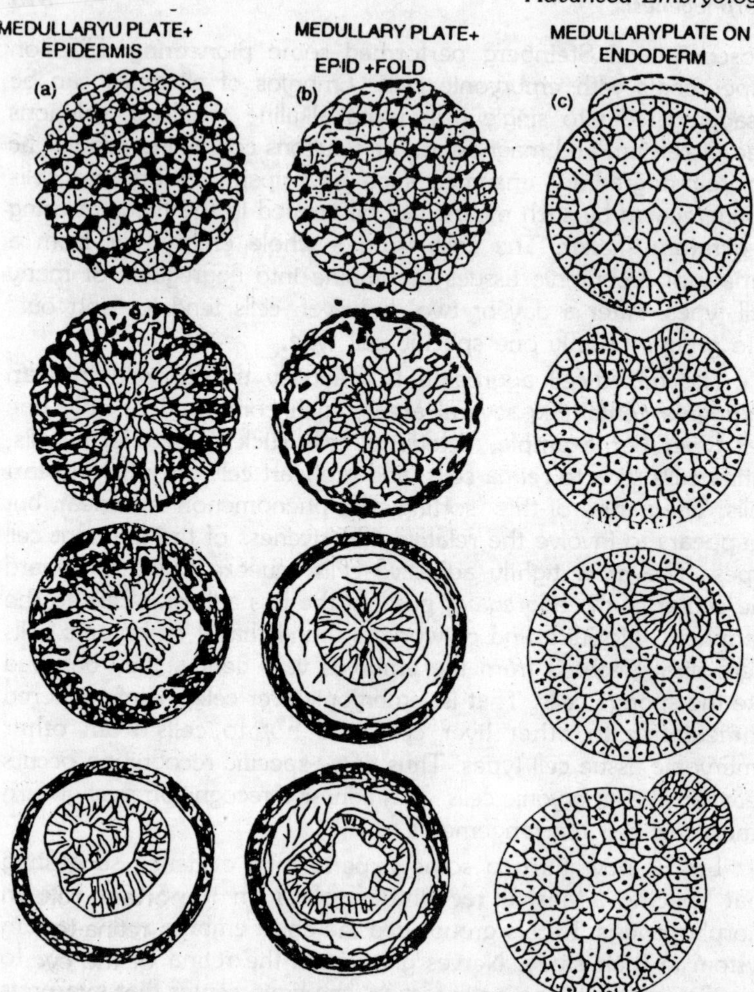

Fig. 12.5. Sorting out of reaggregated combination of amphibian embryonic cells. Sections through successive stages of composite reaggregates. (a) Randomly arranged cells of epidermis (black) and medullary plate (white) move in opposite directions and reestablish homogeneous tissues. (b) When cells from the neural fold are added, they move to occupy the space between the neural tissue and the epidermal covering. (c) A piece of medullary plate or of larval forebrain moves first into and then out of a matrix of endoderm.

with more than one cell type was developed by Roth and Weston to measure the adhesion of like to like and unlike to unlike cell types.

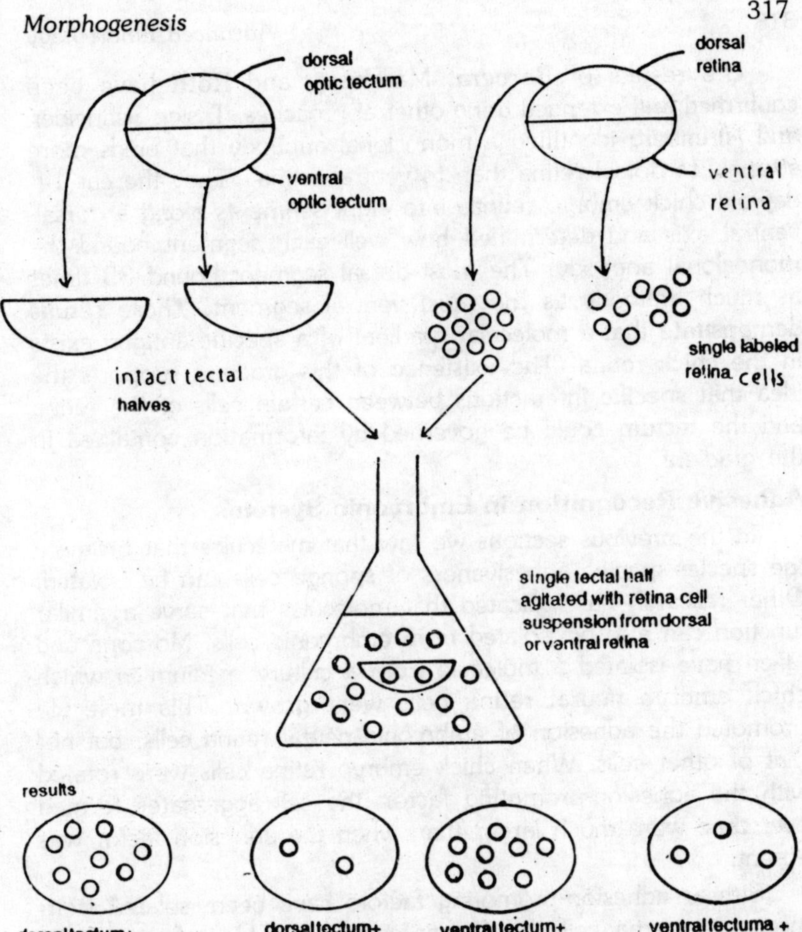

Fig. 12.6. Retinotectal adhesive recognition. Barbera, Marchase, and Roth found that dorsal retina cells adhere best to ventral tectum and ventral retina cells to dorsal tectum. The time needed for adhesion varied in these experiments.

In general, the dorsal retina cells were picked up more often by ventral tectal halves than by dorsal tectal halves. Ventral retina cells, on the other hand, were picked up more often by dorsal tectal halves than by ventral halves. Roth and co-workers concluded that there is specific adhesive recognition between dorsal retina and ventral tectum and between ventral retina and dorsal tectum. This adhesive recognition may help determine where nerves from the retina become located on the tectum.

The results of Barbera, Marchase, and Roth have been confirmed and extended using other approaches. Trisler, Schneider, and Nirenberg identified a monoclonal antibody that binds more strongly to dorsal retina than to ventral retina.. They the cut 14-day-old chick embryo retinas into eight segments along a dorsal-ventral axis and determined how well each segment bound the monoclonal antibody. The most dorsal segment bound 35 times as much antibody as the most ventral segment. These results demonstrate that a molecular gradient of a specific antigen exists in the chick retina. The existence of this gradient supports the idea that specific interactions between certain cells of the retina and the tectum could be governed by information contained in the gradient.

Adhesive Recognition in Embryonic Systems

In the previous sections we saw that molecules that promote the species-specific adhesiveness of sponge cells can be isolated. Other research has indicated that molecules that serve a similar function can also be isolated from embryonic cells. Moscona and Lilien have isolated a molecule from a culture medium in which chick embryo neural retina cells were grown. This molecule promoted the adhesion of embryonic neural retina cells, but not that of other cells. When chick embryo retina cells were rotated with the adhesion-promoting factor, the cell aggregates formed over time were much larger than when the adhesion factor was absent.

Similar adhesion-promoting factors have been isolated from mouse teratoma cells be Oppenheimer and Humphreys, from cellular slime molds by Rosen and Barondes, and from sea urchin sperm by Vacquier and colleagues. Edelman and colleagues have identified a number of *cell-adhesion molecules* (CAMs) in specific tissues of various vertebrates. One of these molecules controls neuron-neuron and- neuron-muscle adhesion. During embryonic development, this neuron cell-adhesion molecule (N-CAM) loses much of it sialic acid content. Such a molecular change could account for the changing adhesive properties of developing cells and tissues.

In the cerebellums of mice with the staggerer mutation, a mutation in which nerve-cell interactions are defective, the N-CAMs do not change. This observation is consistent with the hypothesis that changing CAMs help to control cellular interactions during embryonic development. Embryonic nerve cells actually produce N-CAMs. These cells grow along the outer surface of the brain, guided by a trail of N-CAMs contained on projections of glial cells. N-CAMs also appear to guide formation of synaptic connections between nerve and muscle cells. Antibodies to N-CAMs alter the route taken by the optic nerve and disrupt synapse formation. Glial cells make N-CAMs just when the nerve reaches them, apparently to guide the nerve in the right direction. Other molecules called *cadherins* also appear to play an important role in morphogenesis. These are calcium-dependent cell-cell adhesion molecules, at least three cadherins are known, E-cadherin (epithelial cadherin), N-cadherin (neural cadherin), and P-cadherin (placental cadherin).

These are cell-surface proteins with molecular weights ranging from 118,000 to 127,000 daltons. Cadherins help maintain associations between cells of the same type and foster separation of cells of different types during tissue and organ formation. When the ectoderm separates from the mesoderm, the cells that form mesoderm stop producing E-cadherin and make N-cadherin. Also, when nervous tissue forms from ectoderm, the ectoderm switches from producing E-cadherin to producing N-cadherin. This switch suggests that cadherins may play a role in breaking old cellular connections so that a new layer or organ can be formed.

A number of other cell adhesion molecules have also been discovered, including very large particles such as *adherons* and *toposomes*. These cell-surface particles have been found in a wide variety of organisms ranging from sea urchins and fruit flies to chick embryos, and they may play a role in embryonic cellular interactions and morphogenesis. Oppenheimer and Meyer isolated an aggregation-enhancing protein from blastulas of the sea urchin *Strongylocentrotus purpuratus* by disaggregating them in sea water free of calcium and magnesium.

The isolated protein dramatically promoted the reaggregation of *Strongylocentrotus purpuratus* blastula cells, but not the corresponding gastrula cells. It had no effect what so ever on blastula cells from *Lytechinus pictus*, a different species of sea urchin. Evidence suggests that this protein binds cells together at their surfaces, possibly by attaching to carbohydrates. This molecule, like the sponge aggregation factor, is clearly species specific. It is also stage specific in its action. The fact that this

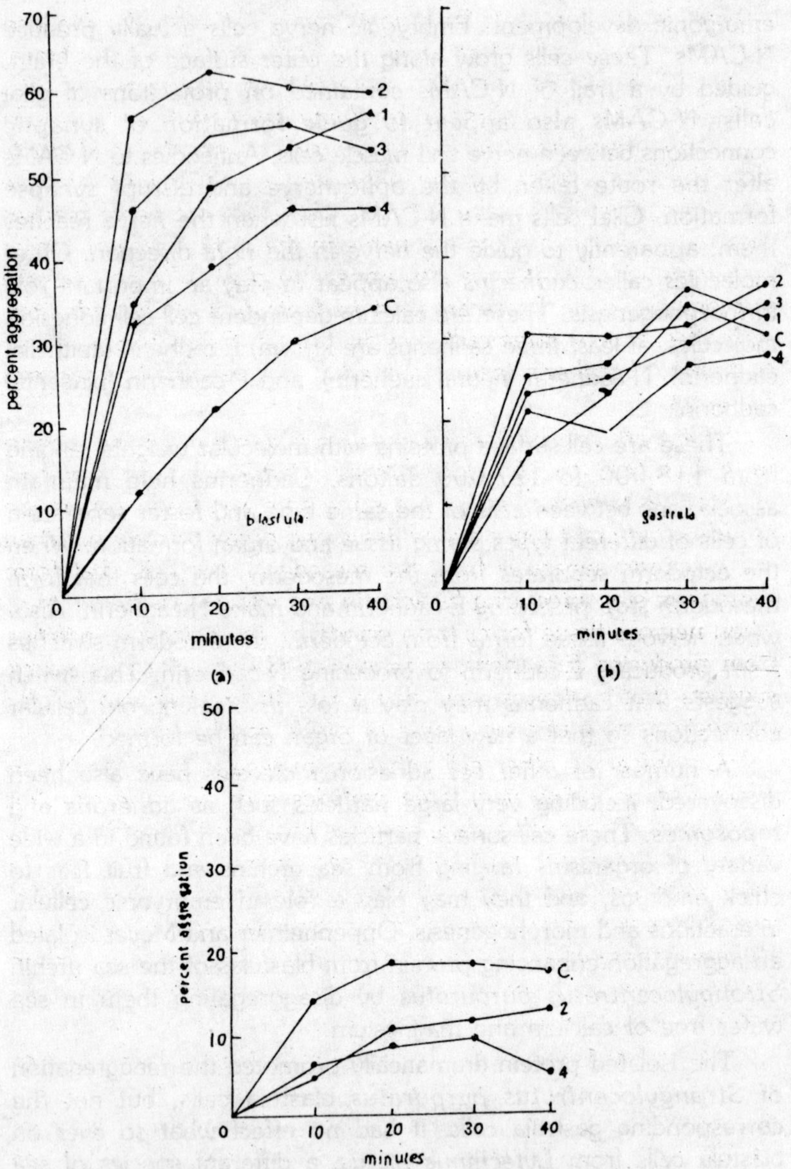

Fig. 12.7. (a-b) Developmental stage-specificity of Strongylocentrotus purpuratus blastula aggregation protein. (c) Species specificity of Strongylocentrotus purpuratus blastula aggregation protein.

material, isolated from intact embryo cells, promotes the adhesion of specific embryo cells at specific stages in their natural medium, sea water, supports the contention that the molecule helps mediate cell adhesion in the embryo. Continued work with this material and with other factors isolated in different systems should illuminate the nature of specific cellular adhesiveness during embryonic development. Isolating, purifying, characterizing, and determining the functional groups of these adhesion factors should help elucidate the nature of specific cell adhesion at the molecular level.

The mechanisms controlling selective cell association are probably not the same in all systems. A couple of models have been proposed to explain cell adhesion in dynamic systems such as developing embryos and tumors. One model, the cell-surface glycosyl transferase—carbohydrate acceptor model, was proposed by Roseman and developed further by Roth and others. This model suggests that embryonic cell adhesion and "de-adhesion" result from sell-surface enzymes called *glycosyl transferases*, which attach to or detach from the carbohydrate chains of glycoproteins and glycolipids at the cell surface. Glycosyl transferases are enzymes that catalyze the transfer of single sugar residues from nucleoside

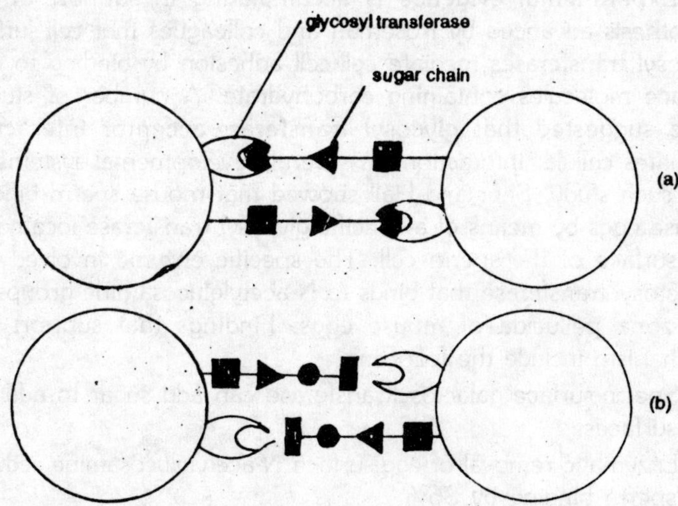

Fig. 12.8. Hypothetical model of embryonic cell adhesion. (a) Adhesion takes place when the appropriate glycosyl transferases and sugar chains are present. (b) Adhesion fails when the glycosyl transferases and sugar chains do not match.

diphosphate sugars to the ends of the carbohydrate chains of glycoproteins and glycolipids. Thus, these enzymes are responsible for synthesizing sugar chains. Are these enzymes present of the cell surface? Most experiments suggest that they are. Are the sugar chains present on the cell surface? Definitely yes. So it makes sense that a glycosyl transferase on one cell would grab a sugar chain on an adjacent cell.

Many such enzyme-substrate complexes would cause the cells to adhere. Release of the sugar chains would allow the cells to separate. The beauty of this hypothesis is that it can explain the complexity of specific cell associations and disassociations in embryos. Each transferase is specific for a given receptor chain and for the sugar it transfers. Thus certain cells could stick together if the right transferases and the right sugar chains were present. It should be emphasized that this model explains only initial cell contacts. It does not deal with the more stable associations that result from the secretion of cementing substances in differentiated tissues.

Glycosyl Transferase

Experimental evidence is accumulating in support of the hypothesis advanced by Roseman and colleagues that cell-surface glycosyl transferases mediate cell-cell adhesion by binding to cell-surface molecules containing carbohydrate. A number of studies have suggested that glycosyl transferase-acceptor interaction mediates cellular interactions in several developmental systems. In one such study, Shur and Hall showed that mouse sperm bind to mouse eggs by means of a specific glycosyl transferase located on the surface of the sperm cell. The specific enzyme involved is a galactosyl-transferase that binds to N-acetylglucosamine groups on the zona pellucida of mouse eggs. Findings that support this mechanism include the following:

1. Sperm-surface galactosyltransferase can add sugar to add cell surfaces.
2. Enzymatic removal of egg-surface *N*-acetylglucosamine reduces sperm binding by 86%.
3. Two reagents that interfere with surface galactosyltransferase activity inhibit sperm-egg binding. The milk protein alpha-lactalbumin changes the ability of the transferase to bind to *N*-acetylglucosamine and also inhibits sperm-egg binding: UDP-

dialdehyde inhibits sperm-egg binding and sperm-surface galactosyltransferase activity to identical degrees.

The molecular basis of sperm-egg binding is indeed becoming clearer!

Although glycosyl transferases had been found on the cell surface, until lately their role in mediating adhesion had been suggested only indirectly (as in the previously mentioned experiment). Recently a pure glycosyl transferase (a fucosyl transferase) become available in quantities sufficient to test its role in cellular adhesion. A variety of cell types, such as embryonic human skin cells (fibroblasts), were found to attach and spread on surfaces on which the pure transferase was attached. The adhesion of cells to the transferase was specific in that the cells would not attach to other proteins and specific alterations of the transferase prevented adhesion.

Adhesion of the cells to the enzyme-covered surface was inhibited by compounds that specifically bind to the transferase. Thus evidence suggests that protein-sugar interactions, such as glycosyl transferase-sugar or lectin-sugar binding play a role in mediating cellular adhesion. Substantial evidence points to a role for cell-surface carbohydrates in cell associations; experimenters whose work supports this view include Roseman, Roth, Oppenheimer, Steinberg, Moscona, Burger, Rosen, and Barondes. When embryonic cells are treated with purified glycosidases (enzymes that catalyze the removal of sugar residues from carbohydrates), their adhesion and aggregation rates change. Molecules that promote aggregation rates change.

Molecules that promote aggregation are often inactivated when treated with glycosidases. In addition, specific carbohydrate-binding proteins have been isolated from slime mold and embryo cells. These proteins promote cell association when the cells are rotated with these molecules. The proteins apparently bind specific sugar residues on the cell surface. Thus protein-carbohydrate interaction may be of key importance in controlling cell-cell associations. In summary, the molecular nature of cell associations in embryos is not well understood. Cell adhesion, however, plays an important role in the cellular rearrangements that occur in developing embryos. Carbohydrate-protein interaction may control embryonic cell associations in some systems. Because the carbohydrate chains of cell-surface glycoproteins and glycolipids reach farthest away from the cell, these chains must play a role at least the initial contact with approaching cells and substrates.

Other Mechanisms

Specific adhesive recognition among cells is clearly not the only mechanism that controls the development of form in embryos. We have treated cell adhesion at length because this area has been intensively investigated in recent years, and because specific testable models of the molecular nature of adhesive recognition among cells have been developed. Now let us briefly examine some of the other specific mechanisms important in controlling embryonic morphogenesis.

Cytoskeletal Elements in Morphogenesis

We have mentioned that cytoskeletal elements such as rod-shaped proteins, microtubules, and microfilaments—the "microskeleton" of the cell—are involved in a variety of embryonic cell activities. For instance, cytoplasmic microtubules are associated with elongating cells. Cell elongation often occurs during embryonic cellular rearrangements and during morphogenesis of embryonic organs. When cell elongation takes place, microtubules become visible, lined up in the long axis of the cell. Microtubules are observed in elongating cells around the amphibian blastopore, in the neural plate cells, in cells moving into the primitive groove in bird embryos, and in outgrowing nerve cells. Cell elongation is prevented by colchicine, which specifically disrupts microtubules.

Colchicine prevents nerve cell outgrowth and inhibits cell elongation in the chick primitive streak and in the neural plate. Colchicine is not known to affect anything other than microtubules. Microtubules thus play a role in embryonic cell elongation. They may directly force the cells to elongate, or they may serve as tracks for cytoplasmic flow in the direction of elongation. Cytoplasmic microfilaments cause cell narrowing. Microfilaments serve as a simple cellular contractile system.

The morphogenesis of tubular glands, for example, may be aided by bundles of microfilaments that contract and exert force on the sides of cells, narrowing the cells at one end. If a sheet of cells narrows at one side, the sheet will bend. Other hypotheses, however, have been advanced to explain how certain cell layers (epithelia) bend or fold. For example, suppose that the outer surface cells of the layer are tightly attached to each other so

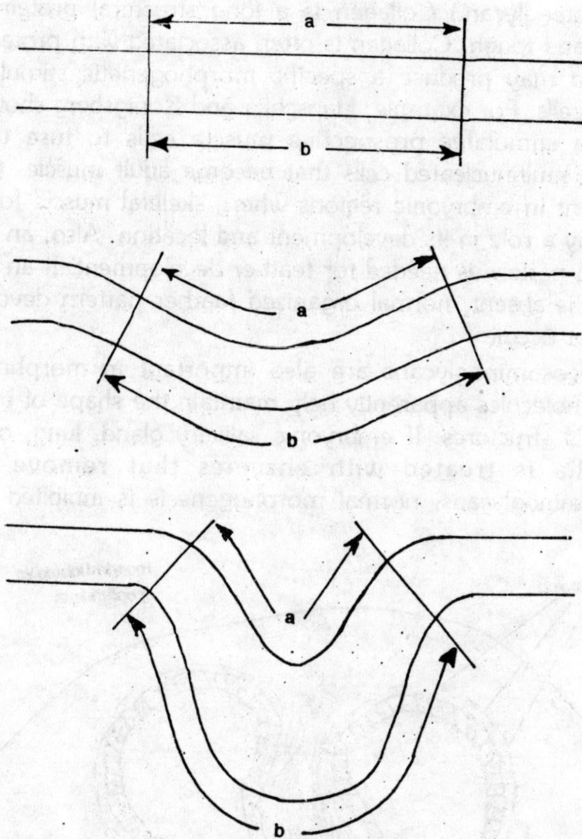

Fig. 12.9. Epithelial folding: a hypothetical mechanism. Surface a (outer surface) is tightly adherent. Cell division takes place within the layer, and the sheet buckles.

that they cannot more laterally at all. If cell division takes place in the layer, the cell sheet-will buckle inward. Microfilaments may also act here by causing the outer cells to become less elastic. Bundles of microfilaments have been observed in the surface regions of embryonic pancreas, lens, thyroid, and lung.

Extracellular Materials and Morphogenesis

Two groups of extracellulaı molecules play an important role in morphogenesis: glycosaminoglycans and collagen. Glycosaminoglycans are sugar polymers consisting of uronic acids and amino sugars. They are often linked to proteins, forming proteoglycans. (Recall that one of the sponge cell adhesion factors is a proteoglycan.) Collagen is a long structural protein that is fibrous and tough. Collagen is often associated with proteoglycan. Collagen may produce a specific morphogenetic stimulation of certain cells. For example, Hauschka and Konigsberg showed that collagen stimulates prospective muscle cells to fuse together, forming multinucleated cells that become adult muscle. Collagen is present in embryonic regions where skeletal muscle forms and may play a role in its development and location. Also, an oriented collagen pattern is needed for feather development. If an oriented pattern is absent, normal organized feather pattern development does not occur.

Glycosaminoglycans are also important in morphogenesis. These molecules apparently help maintain the shape of branching epithelial structures. If embryonic salivary gland, lung, or kidney epithelia is treated with enzymes that remove surface glycosaminoglycans, normal morphogenesis is inhibited and the

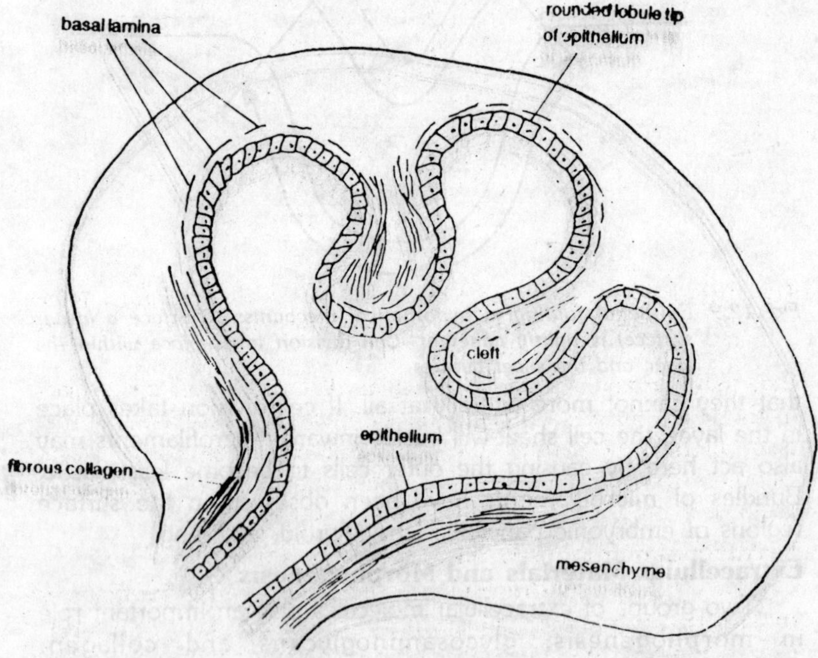

Fig. 12.10. Branched gland showing relationships among the epithelium, the mesenchyme, the basal lamina, and the fibrous collagen.

structures tend to round up and form a ball. The branching or lobulation of epithelial layers in structures such as the salivary gland, mammary gland, lung, kidney, and thymus is a morphogenetic event that has been carefully studied. Although the mechanisms involved are not completely understood, a well-documented picture is emerging.

Branching glands consist of an epithelium associated with mesenchyme. The entire organ rudiment can be removed from the organism and studied in culture. In this way the roles of the different parts have been studied, helping us understand the nature of branching morphogenesis. Branching begins when the epithelial layer folds inward, forming clefts in the epithelium. One of the mechanisms that control this infolding is the contraction of cytoplasmic microfilaments at one end of the cells in the folding layer. Also, the outer surface cells of the epithelial layer may be tightly attached to each other, so that when cell division occurs within the layer the cell sheet buckles inward. Cleft formation in epithelial layers is only one step in the formation of a lobulated gland. Another component involved in epithelial lobulation is the mesenchyme that envelops the epithelium. The epithelium will not branch in the absence of mesenchyme.

The influence of mesenchyme on epithelial branching is not well understood. A great deal of attention, however, is being paid to extracellular materials between the epithelium and the mesenchyme. Some of these materials are synthesized by the epithelium, others by the mesenchyme. In the well-studied mouse salivary gland, the extracellular material between the mesenchyme and the epithelium is composed of two layers. One consists of collagen fibres together with glycoprotein and proteoglycan. The other layer, the *basal lamina*, is rich in collagens, glycoproteins, and polysaccharides such as glycosaminoglycans. It adheres tightly to the epithelium adjacent to the plasma membrane of the epithelial cells and is important in maintaining the branched gland morphology.

Bernfield and colleagues showed that if the basal lamina is removed, the epithelium making up the gland rounds up. In one experiment the epithelium of the gland was isolated and its basal lamina was removed enzymatically. When the epithelium was recombined with mesenchyme, the epithelium lost its lobular shape and rounded up. When the epithelium resynthesized basal lamina, the branching began again. Thus the epithelium produces the basal lamina, and the basal lamina maintains the branched morphology of the epithelium. The basal lamina acts like a tight-fitting glove, preventing the epithelium from losing its form. The basal lamina is thickest in the clefts and thinnest at the rounded lobule tips. This arrangement should keep the clefts stable while allowing new clefts to form at the tips, Collagen fibres are deposited outside the basal lamina; these are also most densely packed in the clefts. Collagen fibres apparently help stabilize the indentations.

A major role of the mesenchyme is apparently to degrade the basal lamina near the rounded lobule tips so that clefts can form in these regions. Bernfield and colleagues labeled the glycosaminoglycans of the basal lamina with ^{3} H glucosamine. When mesenchyme was present, the label was lost from the basal lamina outside the rounded lobule tips. Very little label was lost from the basal lamina in the clefts. When mesenchyme was absent , no significant label was lost in any region of the basal lamina. These experiments suggest that the mesenchyme selectively degrades the basal lamina in the morphogenetically active regions of the branching epithelium (that is, the rounded tips). Once the basal lamina has been degraded, some sort of signal might stimulate epithelial cleft formation in that region. For instance, Ca^{2+} ions released during glycosaminoglycan degradation can trigger microfilament contraction in the adjoining epithelium.

Alternatively, basal lamina degradation might stimulate cell division in the epithelium, which could also lead to cleft formation. The exact mechanism by which mesenchyme influences epithelial branching is not yet clear. Because of the active investigations in this area, however, more of the answers to these question should soon be in hand. In summary, some of the factors involved in morphogenesis are cytoskeletal elements (microtubules and microfilaments), localized cell division, and extracellular substances such as collagen and glycosaminoglycans.

Mechanisms of Neural Crest Migration

We noted earlier that neural crest cells differentiate into many structures some far from the original neural crest ectoderm. These cells must travel long distances to their final locations. Work from many laboratories, indicates that neural crest cells migrate along well-defined pathways. Neural crest cells of the trunk migrate in two streams. Crest cells in one stream move ventrally into the mesenchyme located between the somites and the nerve tube. Some of the cells migrate into the somites, forming dorsal root ganglia; others from sympathetic ganglia and the adrenal medulla. Crest cells in the other stream move into the surface layers of the embryo, forming pigment cell precursors.

The environment in which neural crest cells find themselves seems to play a major role in controlling their differentiation. This conclusion is based on experiments in which neural crest cells transplanted from one trunk region could substitute for those of other trunk regions, and neural crest cells transplanted from one cranial region could substitute for those of other cranial regions. In one experiment, neural crest cells from the mesencephalon areas, which normally give rise to the rostral parts of the skull, where transplanted to the more posterior metacephalon region. The resulting cells differentiated into the lower jaw, which is normally formed from crest cells in the metacephalon region. Thus the final environment of the crest cells can change the fate of the cells to produce the structure that normally develop at the location. These transplantation experiments used chick and quail embryos, so that the donor cells in specific structures could be distinguished from the recipient cells. Although trunk crest can be interchanged with other levels of trunk crest and cranial crest can be interchanged with other levels of cranial crest, trunk crest cannot be completely interchanged with cranial crest, and vice versa.

In other words, cranial crest cells are somewhat restricted or predetermined to form cartilage, whereas trunk crest cells cannot form cartilage no matter what their final environment is. Finally we come to one of the most intriguing questions: what are the mechanisms that control the migration of neural crest cells to distant sites in the body? The work of Marianne Bronner-Fraser has led to a better understanding of this problem. We noted above that neural crest cells migrate along well-defined pathways. Bronner-Fraser injected neural crest cells, retinal pigment cells (non-neural crest cells), or latex polystyrene beads (uncoated or coated with bovine serum albumin) onto the ventral neural crest pathway of chick embryos. All of these, including the beads, translocated ventrally along the pathway in the same way as normal neural crest cells. When the latex beads were coated with the glycoprotein fibronectin, however, the beads remained near the site of implantation and did not move ventrally. These results suggest that neural crest cell migration is influenced by some driving force imparted by the embryonic environment.

Also, molecules such as fibronectin on the cell surface may serve as a recognition mechanism that prevents non-neural crest cells from entering the ventral neural crest migratory pathway. Cells of the somite, fibroblasts, and beads coated with fibronectin—all of which have fibronectin on their surfaces—cannot move onto the ventral pathway. Neural crest cells, retinal pigment epithelial cells, and latex beads without fibronectin—all of which lack surface fibronectin—can move along the ventral pathway. Exactly what forces control the translocation of cells along neural crest pathways is unclear. These experiments show, however, that the entrance to the ventral pathway is controlled by the presence of specific surface substances. Once cells enter the pathway, cellular forces (not necessarily originating in the translocating cells) may move them to their final location. These forces may involve movements in cells surrounding the translocating particles that allow the particles to move in response to, for example, adhesiveness toward substances such as fibronectin.

Fibronectin on the cells apparently prevents cells from entering the ventral pathway, whereas fibronectin concentrated at specific points along the pathway stops translocating cells at these points. This concepts is supported by the observation that both beads and cells locate in areas where fibronectin is present. If fibronectin disappears from the substratum, cells adhering to the substratum may come loose and form clusters with one another. This could account for the aggregation of some neural crest cells into ganglia. This is only speculation, but such speculations form good working hypotheses that surely will lead to additional discoveries in this exciting area.

Cell Death in Morphogenesis

Before concluding our discussion of the mechanisms that influence the development of form in embryos, let us briefly mention another important factor that shapes certain structures. Some structures are molded by cell death that occurs in specific regions of the structures. For example, early in embryonic development our hands and feet resemble paddles or flippers. Our fingers and toes form as the result of cell death in the regions between the prospective digits. If cell death fails to take place, certain abnormalities result, such as webbed hands or feet. Cell death

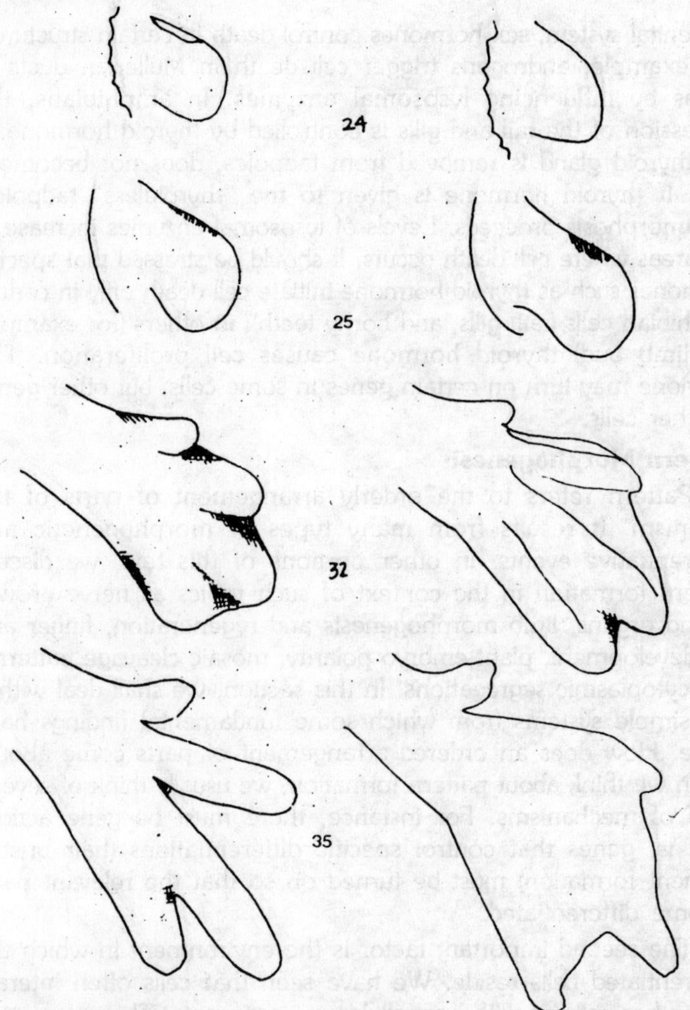

Fig. 12.11. Cell death during digit development in chick and duck embryo hindlimb. Shaded areas die in forming limb.

occurs in many developing embryonic structures. We saw, for example, that certain structures in each sex degenerate after the indifferent stage of sexual development.

The tadpole tail regresses during metamorphosis as the result of cell death. Increased levels of lysosomal enzymes (for example, *acid phosphatase*, *ribonuclease*, and *cathepsins*) are observed in many embryonic regions that are "programmed to die." In the urogenital system, sex hormones control death in certain structures. For example, androgens trigger cell death in Mullerian ducts in males by influencing lysosomal enzymes. In amphibians, the regression of the tail and gills is controlled by thyroid hormone. If the thyroid gland is removed from tadpoles, does not become a frog. It thyroid hormone is given to the "thyroidless" tadpoles, metamorphosis proceeds. Levels of lysosomal enzymes increase in the areas where cell death occurs. It should be stressed that specific hormones such as thyroid hormone initiate cell death only in certain amphibian cells (tail, gills, and horny teeth); in others (for example, the limb bud) thyroid hormone causes cell proliferation. The hormone may turn on certain genes in some cells, but other genes in other cells.

Pattern Morphogenesis

Pattern refers to the orderly arrangement of parts of the organism. It results from many types of morphogenetic and differentiative events. In other portions of this text, we discuss pattern formation in the context of such topics as nerve growth to end organs, limb morphogenesis and regeneration, finger and toe development, plant embryo polarity, mosaic cleavage patterns, and cytoplasmic segregations. In this section, we shall deal with a few simple systems from which some fundamental findings have come. How does an ordered arrangement of parts come about? When we think about pattern formation, we usually think of several sorts of mechanisms. For instance, there must be gene action. That is, genes that control specific differentiations (hair bristle, pigment formation) must be turned on so that the relevant parts become differentiated.

The second important factor is the environment in which the differentiated cells reside. We have seen that cells often interact with other cells or with noncellular environments. This interaction initiates selective gene activation or aids functioning in responding cells. A third set of events is cell migration and selective adhesion. In many systems these events get the relevant cells to the future sites of pattern development. In the fruit fly *Drosophila*, the first mechanism mentioned above comes into play in forming bristle patterns in the epidermis.

The presence or absence of a bristle at a specific site in the epidermis depends on the genotype of that portion of the epidermis. If the tissue is mutant at a given locus, no bristle forms even if this section of tissue is surrounded by nonmutant, bristle-forming epidermis. If the tissue is wide type at that locus, a bristle forms even if the tissue is surrounded by mutant tissues. Such results indicate that specific genes control bristle formation in tissue. In other cases, however, surrounding tissue also plays a role. *Melanocytes* are pigment-forming cells, most of which are derived from the neural crest. Only the melanocytes in the pigmented retina of the eye are derived from the neural tube itself.

The neural crest melanocyte precursors migrate from the neural crest to a variety of sites in the body, such as the skin and hair follicles. Pigment pattern in mammals can be studied in depth because the controlling genes in such mammals as mice are known. One can transplant a piece of skin from a black mouse onto the back of a yellow mouse, or vice versa. When yellow skin is placed on the back of a black mouse, the melanocytes from this skin move out into adjacent hair follicles in the host tissue and begin to produce black pigment despite their own genotype.

In the reciprocal experiment, the melanocytes from the black transplant migrate into the hair follicles of the yellow host and produce yellow pigment. Thus the formation of yellow or black pigment appears to be controlled by the hair follicle cells, not by the melanocytes. Sulfhydryl compounds, such as glutathione, may be the hair follicle factors that stimulate the formation of yellow pigment in the melanocytes. That is, the glutathione present in yellow hair follicles may inhibit black pigment synthesis and stimulate yellow pigment synthesis. This suggestion is supported by the finding that isolated, *in vitro* cultured yellow melanocytes form black pigment. If sulfhydryl compounds are added to the culture medium, however, these melanocytes revert to synthesizing yellow pigment.

The most dramatic examples of pigment patterns are the spotted coats of many mammals and birds. What about an animal with white spots on a dark background? How does such a pattern form? Cytological examination of hair follicles in the white spots of such an animal indicates that differentiated melanocytes are absent in these areas. Can the hair follicles in the white spots permit melanocytes to form pigment? Yes. Melanocytes form a black skin transplant can migrate into a white spot region, enter the hair follicles, and make pigment. Thus the white-spotted areas can sustain the differentiation of melanocytes that are already mature. The white-spotted pattern is probably caused by some factors present in prewhite spots that prevent melanocyte differentiation. Even if melanocyte precursors enter the white spot areas, they do not differentiate. Pigment pattern formation , therefore, involves all three of the mechanisms mentioned earlier:

1. Migration of premelanocytes from the neural crest to the skin and hair follicles;
2. Interaction of the melanocytes with the environment (that is, with products of the hair follicle cells); and
3. Specific gene activation in the melanocytes that results in the synthesis of pigment.

Pattern formation, therefore, is not unique. It results from mechanisms that operate in many embryonic processes. We shall completely understand the formation of organized arrangements in organisms—that is, patterns—only when we understand the molecular nature of cell-cell interactions and differentiation.

Morphogenesis In a Simple System: Cellular Slime Molds

Morphogenesis in vertebrates is complex and difficult to study. As yet, little is known about the mechanisms that control morphogenesis in vertebrate embryos. Progress is being made slowly; the major problem is that such studies must usually be performed with isolated embryo cells *in vitro*. Thus, for example, studies of cell-cell interactions are often done in culture, an environment that is not the same as that in the embryo. *In vivo* studies of a few simple systems, however, have increased our understanding of the forces that shape the organism. One of these systems is the cellular slime mold. What are cellular slime molds?

At one stage in their life cycle, these organisms are amoebae. At another stage, the amoebae aggregate to form a multicellular fruiting body composed of a stalk and a mass of spores at the tip of the stalk. The spores give rise to amoebae. This fairly simple system goes through morphogenetic processes similar to some of those occurring in higher organisms. There is an aggregation phase in which the free amoebae congregate together to form in integrated migratory cellular mass (slug). This process resembles some of the cellular rearrangements that occur in vertebrate embryos. The cells in the slug differentiate, and morphogenesis occurs in which the slug is transformed into the fruiting body (sporocarp). What is known about the mechanisms of slime mold morphogenesis? The aggregation phase in several slime molds, such as *Dictyostelium discoideum*, has been examined by several groups.

The free amoebae continue to divide and remain single as long as there is an adequate supply of the bacteria they feed on. When the bacteria are exhausted, the amoebae in dense cultures begin to aggregate, forming the multicellular slug. What causes this aggregation to occur? Certain cells secrete the nucleotide cyclic AMP (cyclic adenosine monophosphate). The amoebae move in the direction of increasing concentration of cyclic AMP. Such movement of cells is called chemotaxis. Thus specific chemotaxis controls one phase in slime mold development, the aggregation phase. Cyclic AMP caused amoebae to move together. Other work by Rosen's and Barondes's groups suggests that the amoebae stick together by protein-carbohydrate interaction. These workers isolated carbohydrate-binding proteins (lectins) from slime mold cells that were competent to aggregate.

The carbohydrate-binding proteins apparently occur on the cell surfaces of amoebae. These molecules bind specific sugar residues on carbohydrate chains that apparently occur also on the amoebae cell surfaces. Thus cyclic AMP gets the amoebae together, and the carbohydrate-binding protein-sugar chain interactions stick the amoebae together. What causes the mass of amoebae, the slug to differentiate into the fruiting body? Cells at the leading end of the migrating slug become prestalk cells; the remainder become prespore cells. The position of the cells in the slug therefore plays a role in determining the fate of the cells.

The fate of cells is reversible, because if the slug is divided into prestalk cells and prespore cells, each part can produce a complete fruiting body. Thus some cells shift their fates to produce the missing cell types. Position with respect to other cells and the external environment probably plays a major role in the fates of cells in a cut slug, just as in an intact slug. When previously internal cells come to be at the tip of the cut slug, their new position

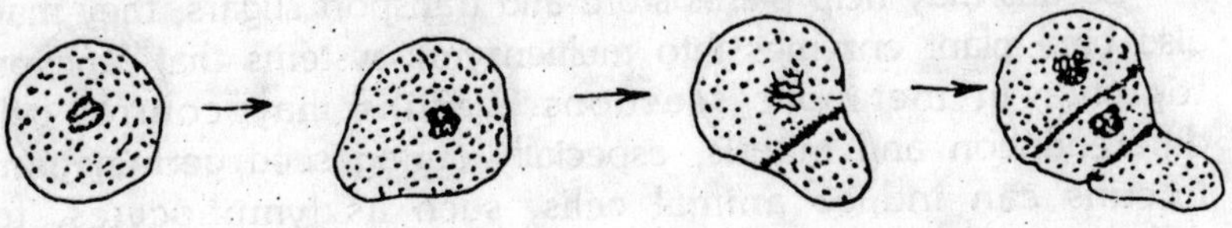

Fig. 12.12. Early development of Focus embryo.

appears to determine a new fate. In summary, scientists have investigated specific factors that control development in the cellular slime molds, such as cyclic AMP. Other factors such as carbohydrate-binding proteins also play roles, but these roles are somewhat less well understood. Cell position also governs differentiation in slime molds, but no one yet knows how position "*works*" at the molecular level. Still, simple systems such as this one help as understand the mechanisms that operate in the morphogenesis of higher organisms.

Tools for Studying the Role of the Cell Surface in Morphogenesis and Malignancy

We mentioned that lectins, carbohydrate-binding proteins or glycoproteins, have helped us understand certain cell-surface phenomena. These molecules have provided major tools for cell biologists in their study of the cell surface. We shall now discuss these widely used molecules in more detail. *Lectins* have been isolated from the seeds of plants and from extracts of sponges, snails, crabs, fish, slime molds, and vertebrate cells. What is the function of lectins *in vivo*? We already mentioned that lectins are important in sperm-egg recognition. The sea urchin sperm binding isolated by Vacquier is a lectin that recognition. The sea urchin sperm binding isolated by Vacquier is a lectin that recognizes carbohydrate groups on the cell surface of eggs.

Lectins isolated from slime molds by Rosen's and Barondes' groups may play a role in the cell recognition and adhesiveness of slime mold amoebae as they aggregate to form a multicellular slug. Lectins, therefore, are important mediators of the cell recognition aspects of morphogenesis. What about the function of lectins in plants, which are a major source of lectins that have been studied to date? Lectins may function as plant antibodies, protecting the plant against microbial attack. Lectins can agglutinate harmful microbes and can inhibit the hydrolyzing enzymes used by fungi to attack the plant cell wall.

Lectins may help plants store and transport sugars; they may also bind plant enzymes into multienzyme systems that function together in metabolic reactions. Lectins may control cell differentiation and mitosis, especially during seed germination. Lectins can induce animal cells, such as lymphocytes, to differentiate. The lectins bind and cross-link specific cell-surface receptors that contain carbohydrate. Such cross-linking sets into motion a series of reactions that lead to differentiation and cell division. An important function of lectins in legumes such as clover and beans is to bind nitrogen-fixing bacteria to the roots of the plant.

Legumes possess nodules of the nitrogen-fixing bacterium *Rhizobium* on their roots. These bacteria process gaseous nitrogen into a form usable by the plants and thus are of major importance to the plants. Lectins are secreted on the root surfaces of such plants. Specific lectins have been shown to bind only the strains of *Rhizobium* found on the roots of the plant from which they were extracted. For example, soybean agglutinin, a lectin isolated from soybean plants, binds only strains of *Rhizobium* found on the roots of soybeans. It does not bind other *Rhizobium* strains that form nodules on the roots of other plants. Dazzo and his

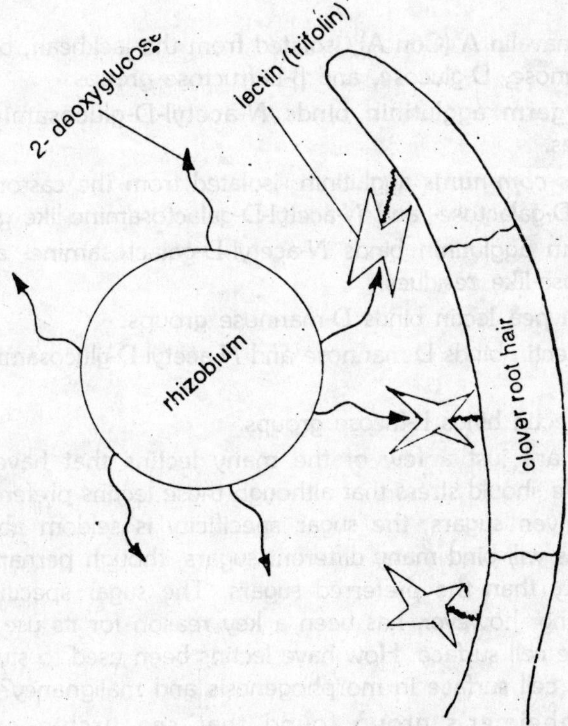

Fig. 12.13. Lectin-mediated binding of nitrogen-fixing bacterium (Rhizobium) to clover root hair.

colleagues have done some elegant work on the function of lectins in plants. Dazzo's group isolated a lectin from clover roots. They named the lectin *trifoliin* because it specifically binds to sugars on the nitrogen-fixing bacterium *Rhizobium trifolli*, which adheres to clover root hairs. We can speculate that evolution produced a pair of organisms, clover and *Rhizobium trifolli*, that live together symbiotically, each deriving benefit from the association.

The means by which they bind together probably evolved as a lectin-carbohydrate interaction at the surfaces of clover root hair and *Rhizobium* cells. Before we move on to discuss how lectins are used to study the role of the cell surface in morphogenesis and malignancy, a brief description of the different lectins is in order. Lectins bind terminal sugars of the cell-surface oligosaccharide chains or groups of sugars in the chains. The following is a list of lectins and the sugar(s) they preferentially bind.

Concanavalin A (Con A), isolated from the jackbean, binds β-D-mannose, D-glucose, and β-D-fructose groups.

What germ agglutinin binds *N*-acetyl-D-glucosamine-like residues.

Ricinus communis agglutinin, isolated from the castor bean, binds D-galactose- and *N*-acetyl-D-galactosamine-like groups.

Soybean agglutinin binds *N*-acetyl-D-galactosamine- and D-galactose-like residues.

Garden pea lectin binds D-mannose groups.

Lentil lectin binds D-mannose and *N*-acetyl-D-glucosamine-like groups.

Lotus lectin binds L-fucose groups.

These are just a few of the many lectins that have been isolated. We should stress that although these lectins preferentially bind the given sugars, the sugar specificity is seldom absolute. Most lectins will bind many different sugars, though perhaps with less tenacity than the preferred sugars. The sugar specificity of lectin binding, however, has been a key reason for its use in the study of the cell surface. How have lectins been used to study the role of the cell surface in morphogenesis and malignancy?

Oppenheimer's group found that sea urchin embryo micromeres and mesomere-macromeres had different reactions to the lectin concanavalin A. Although the micromeres, mesomeres, and macromeres all bound concanavalin A, only the micromeres showed a lectin-induced capping of concanavalin A cell-surface receptor sites. In other words, the cell surfaces of the micromeres allowed the lectin to move the lectin receptor sites into a cap at one end of the cell. This suggested that the concanavalin A receptor sites are more mobile in the cell surfaces of the micromeres than in those of the macromeres and mesomeres. What does this have to do with morphogenesis? We know that the micromeres form the migratory primary mesenchyme cells that play a role in the gastrulation of the sea urchin embryo.

The above-mentioned study suggests that the surfaces of these premigratory cells are indeed different from the surfaces of the other types of cells. The difference in lectin receptor site mobility in the cell surface may directly aid cell migration during morphogenesis. In any case, the use of lacin in this experiment showed that the cell surfaces of known populations of embryonic cells are very different in terms of the mobility of specific receptor sites. We shall briefly mention some of the many other studies that have used lectins in connection with morphogenesis and malignancy.

In 1888, Stillmark found that extracts from castor bean seeds can agglutinate (clump) human erythrocytes (red blood cells). Since then lectins have been extracted from the seeds of over 800 species of flowering plants, especially the legumes. Lectins have been found not only in seeds but in the leaves, roots, and bark of plants and also in extracts from many animal cell types. Numerous studies have shown that, in general, malignant tumor cells, transformed cells, and early embryo cells are agglutinated at rather low concentrations of many lectins, whereas normal adult cells are seldom agglutinated at these concentrations. Why are embryonic cells and tumor cells agglutinated when normal cells are not? Studies using lectins labeled with the fluorescent dye ferritin or with a radioactive isotope have shown that, in general, most cells bind lectin. That is, normal adult cells, tumor cells, and embryonic cells have lectin receptor sites in their cell surfaces.

It has also been shown, however, that cells that are agglutinated with lectins (tumor and embryo cells, for example) tend to have mobile lectin receptor sites in their cell surfaces. Thus capping and clustering of these sites will occur in the presence of lectins, because lectins are multivalent molecules that can cross-link several surface receptor sites. This clustering or capping of the lectin receptor sites may explain how lectins agglutinate cells. A buildup of lectin-bound receptor sites in specific regions of the cell surface may facilitate lectin-bound cell agglutination. It has also been proposed that early embryo cells and malignant cells have altered cytoskeletal elements (microtubules and microfilaments) that fail to restrict the movement of cell-surface lectin receptor sites. In normal adult cells, these cytoskeletal elements are attached to the inner membrane surface, restricting the movement of surface receptor sites. These suggestions are supported by the finding that drugs that disrupt cytoskeletal elements, such as colchicine, local anesthetics, cytochalasin B, dramatically alter the mobility and distribution of cell-surface lectin receptor sites in a variety of cell types.

Lectins, therefore, have been useful in identifying specific differences in the surfaces of tumor cells, normal cells, and embryonic cells. Other extensive studies with lectins have led to additional conclusions. Moscona's group showed that young chick embryo cells are agglutinated by concanavalin A, whereas older cells are not. Such studies were extended to other systems by several investigators. The findings in general suggest that young, motile embryonic cells are agglutinable with lectins, whereas older embryonic cells and those that do not move much are less so. When follow-up studies with labeled lectin were carried out, the results showed that agglutinable embryo cells display mobile surface lectin receptor sites like those in the sea urchin micromeres. In summary, we can say that lectins have helped us identify cell-surface molecules containing sugar and have led to the conclusion that active cells in embryos and tumors possess mobile lectin receptor sites on their surfaces. It remains to be seen whether this characteristic plays a key role in facilitating the movements and interactions that occur during morphogenesis and the spread of cancer cells.

Flies and Worms

The Puzzle of Differentiation: Differential Gene Activation

Every cell of an embryo has exactly the same gene content, and genes control, directly or indirectly, all of the processes of development. How then can cells differentiate into the different body structures? One possibility is that as development proceeds, particular genes are lost from particular embryonic cells during mitosis. At the same time, other genes are lost from other cells. In that case, if gene loss could be controlled so that the same set of genes were always lost from the same cells at the same stage of development, an explanation for cellular differentiation would be possible. Gene loss, however, does not seem to lie at the bottom of the problem of how differentiation occurs, as the following two experiments illustrate. If a small piece of carrot is placed in a culture dish and then dissociated in to individual cells, one such cell, when properly stimulated, is fully capable of developing into a complete carrot plant.

Similarly, if a nucleus is taken from an intestinal cell of a white tadpole and injected into an enucleated ovum of a green frog, this artificial, uniparental zygote can develop into a normal white frog. These experiments show clearly that, in two entirely different organisms, no genes required to direct the full course of development are lost during the process of cell differentiation; all genes are still present, ready to be activated in the right cells at the right times to assure normal development.

Polytene Chromosome Puffs

Directly observable evidence of gene activation and deactivation is provided by the giant polytene chromosomes of *Drosophila* and other Dipteran insects such as *Chironomus*. At regular times during larval development, "puffs"—local, fuzzy enlargements of the chromosome-appear at and later disappear from specific places along the chromosomes. A puff, which forms as a result of the loosening of the tightly coiled DNA around a gene, provides visible evidence that a gene at the location of the puff is in the process of synthesizing mRNA (that is, is turned on).

When the puff disappears, we know that the DNA has returned to its tightly coiled condition and the gene is turned off. The reason that the DNA must uncoil (producing the visible puff) before the gene can be turned on is that RNA polymerase molecules needed for mRNA synthesis cannot bind to the gene and initiate synthesis when the DNA is tightly coiled. When mRNA synthesis is finished, the DNA becomes tightly coiled again and the puff disappears. The location of the puff and the timing of their appearance vary from tissue to tissue, because the particular genes that must be activated at a given time in one tissue are often different from those whose products are needed at that same time in some other tissue.

The Control of Puffing

Intact salivary glands can be removed from *Drosophila* larva and put in insect Ringer's solution, where they can survive in tissue culture for several hours. When chromosome preparations are made from these cultured glands at successive intervals, the specific puffs that are produced can be correlated with the age of the larva at the time of its dissection. In *Drosophila*, for example series of puffs arise at specific chromosome locations just before the onset of puparium formation. If a larva is dissected before that time, relatively few puffs will be seen, and the locations of most of these puffs will be different from those of the puffs that appear just prior to pupation. Some years ago attempts were made to find out whether any specific chemicals could artificially induce the set of pupation-related puffs in glands taken from larva too young to have begun to express them.

In due course success was achieved: the insect hormone *ecdysone* was indeed able to induce the formation of those puffs. This finding was not altogether surprising, because ecdysone was known from previous studies to be needed for developing insects to undergo molts, including pupation. Further study showed that

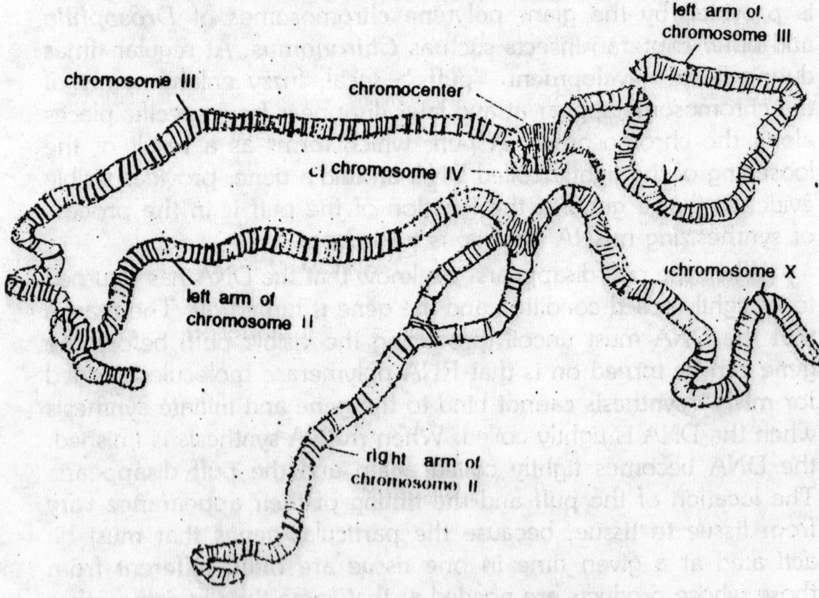

Fig. 12.14. Polytene chromosome.

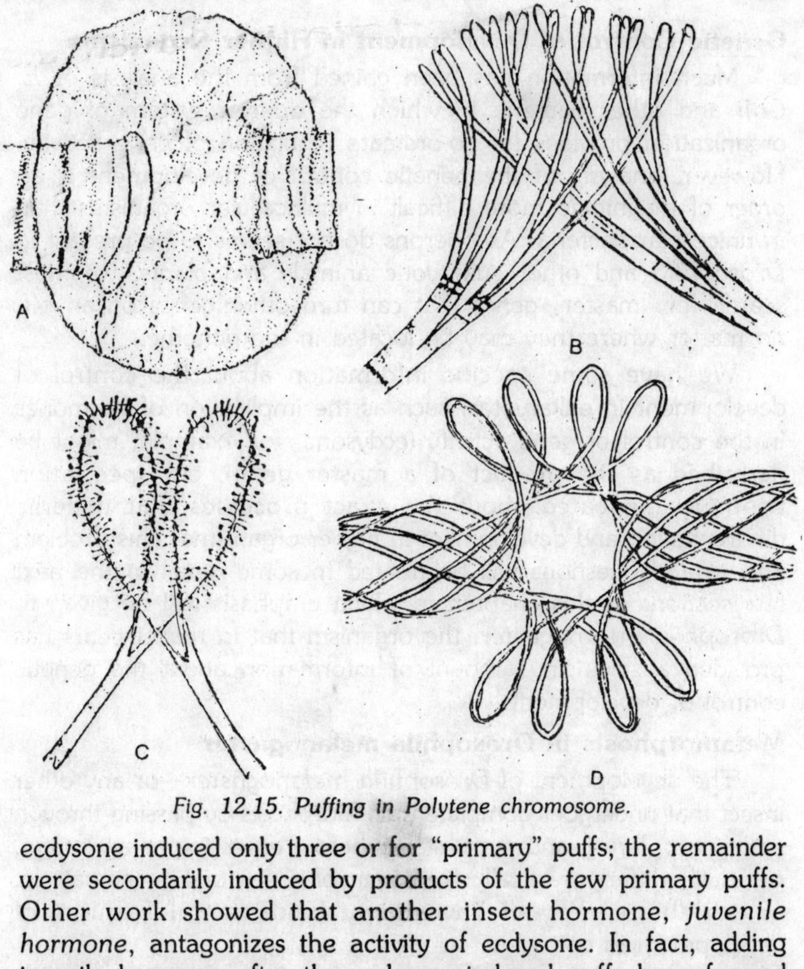

Fig. 12.15. Puffing in Polytene chromosome.

ecdysone induced only three or for "primary" puffs; the remainder were secondarily induced by products of the few primary puffs. Other work showed that another insect hormone, *juvenile hormone*, antagonizes the activity of ecdysone. In fact, adding juvenile hormone after the ecdysone-induced puffs have formed results in a recession of the puffs. If the appearance of a puff marks the turning on of a gene and its recession indicates that the gene has been turned off, as we believe to be the case, then we must conclude that these hormones, ecdysone and juvenile hormone, are intimately involved in the regulation of gene activity in Dipteran insects. Unfortunately, very few other organisms have polytene chromosomes in which the control of puffing can be studied so directly. Thus more indirect kinds of evidence relating to gene-control mechanisms must be sought in the case of most plants and animals.

Genetic Control of Development in Higher Organisms

Much information has been gained from the analysis of *E. Coli* and other bacteria in which the operon system of gene organization provides for co-ordinate regulation of gene activity. However, analysis of the genetic control of development is an order of magnitude more difficult in multicellular organisms than in unicellular bacteria. As operons do not seems to be present in *Drosophila* and other eukaryotic animals and plants, we must search for "master" genes that can turn other genes on an off, no matter where they may be located in the genome.

We have some specific information about the control of development in eukaryotes, such as the implication of hormones in the control of gene activity (ecdysone, for example, might be described as the product of a master gene), but speculation continues unabated about the exact processes that underlie differentiation and development in higher organisms. This problem and related questions will be treated in some detail in the next few sections of this chapter, in which emphasis will be given to *Drosophila melanogaster*, the organism that in recent years has provided a surprising amount of information about the genetic control of development.

Metamorphosis in Drosophila melanogaster

The development of *Drosophila melanogaster*—or any other insect that undergoes complete metamorphosis by passing through embryonic, larval, and pupal stages before emerging as an *imago*, or adult—differs drastically from that of a sea urchin, nematode, or vertebrate. Yet all have many fundamental features of development in common.

The Egg

Like most insects, *Drosophila* possesses *centrolecithal* eggs. That is, the yolk is concentrated centrally in the cigar-shaped egg; it is completely surrounded by nonyolky cytoplasm. The egg nucleus is confined to the cortical region between the yolk and the cell membrane at one end of the egg. The egg is covered by a tough, impermeable (but transparent) *cuticle*, which cannot be penetrated by sperm. However, at the posterior end of the egg is a small conical *micropyle* containing a hole that allows sperm to pass through the cuticle and penetrate the egg.

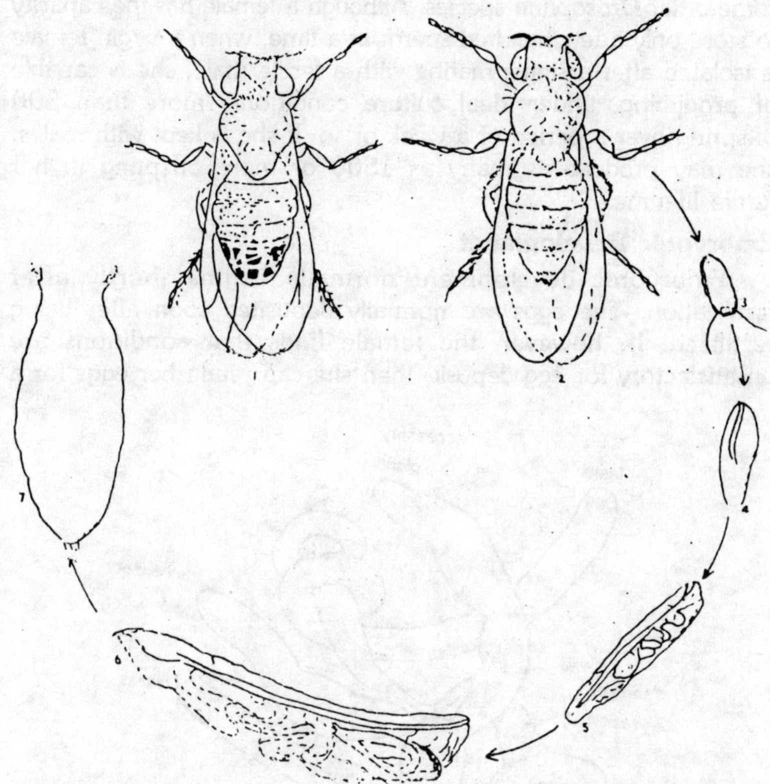

Fig. 12.16. Drosophila life cycle. (1) Adult male. (2) Adult female. (3) Egg. (4) Larva, first instar. (5) Larva, second instar. (6) Larva, third instar. (7) Pupa.

Fertilization

Fertilization is internal. Following mating, sperm from the male are deposited in one of the female's three sperm-storage organs: the larger *ventral receptacle* and two smaller *spermathecae*. Sperm can survive in these female organs for ten days or more, during which time a stored sperm is capable of entering an egg through its micropyle when the egg passes down the oviduct and conveniently pauses with the micropyle positioned precisely opposite the opening from the ventral receptacle into the oviduct. Thus fertilization itself is highly efficient, not requiring astronomically large numbers of sperm to fertilize an egg.

On the other hand, *Drosophila* sperm are immensely large, being 2 mm long in *D. melanogaster* and up to 16 mm long in some other *Drosophila* species. Although a female has the capacity to store only a few hundred sperm at a time, when a virgin female is isolated after a single mating with a fertile male, she is capable of producing, under ideal culture conditions, more than 500 offspring over a period of a week or so. If she is kept with males, she may produce as many as 1500 or more offspring in her fertile lifetime.

Embryonic Development

Embryonic development normally begins shortly after fertilization. The eggs are normally deposited soon after being fertilized. If, however, the female finds that conditions are unsatisfactory for egg deposit, then she can retain her eggs for a

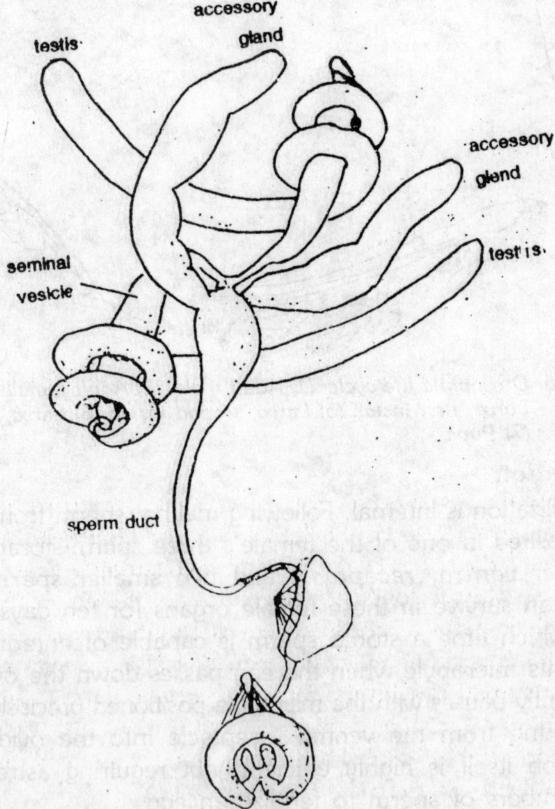

Fig. 12.17. Male Drosophila reproductive system.

few days, before finally depositing them out of necessity. Some of the retained eggs may have been fertilized, in which case some eggs that were retained before being deposited may be seen to hatch as crawling larvae even as they are being laid. The earliest period of embryonic development within the egg cuticle begins after the sperm and egg pronuclei fuse in the clear cytoplasm of the cortical region of the embryo.

An extremely rapid series of 12 or 13 mitoses occurs, giving rise within about two hours to a *synkaryon* of over 4000 nuclei, which migrate peripherally all around the central yolk mass to form a monolayer of nuclei just under the egg surface. Only then do cell membranes appear that separate individual cells from one another. This signals the blastula stage. A cluster of *polar granules* is present at the posterior end of the egg cytoplasm. At the time

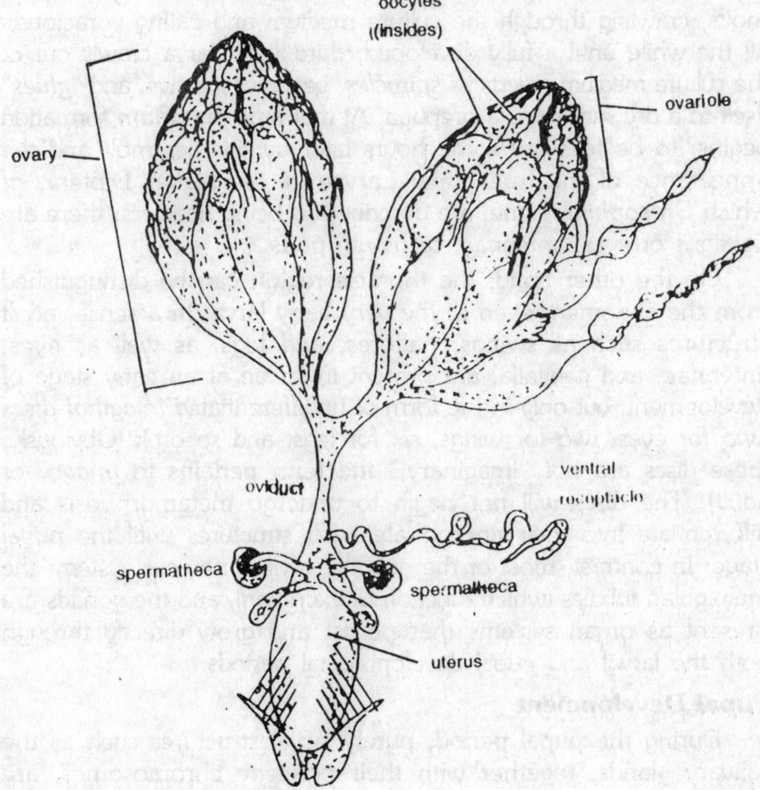

Fig. 12.18. Female Drosophila reproductive system.

of cellularization, the cells that come to enclose these granules, the *polar cells*, are predetermined to become gametes if they are later included in the gonads, which most are. Mutant females that produce defective polar granules, or none at all, are sterile. After about three hours of embryonic development at 25°C, a series of cellular movements takes place, leading to the appearance of a ventral furrow, which is the first externally visible evidence of gastrulation. Not long thereafter organ formation begins in the embryo. Organ formation can be directly observed microscopically through the transparent cuticle of a living embryo.

Larval Development

Embryonic development continues until about 24 hours after the eggs are deposited, at which time the embryo hatches from the eggs as a crawling, segmented larva. During the next 96 hours or so, the larva develops through three successive *instars*, or larval molts, crawling through the culture medium and eating voraciously all the while until a fully developed, third-instar larva crawls out of the culture medium, everts its *spiracles*, becomes inactive, and "*glues*" itself to a dry surface as a prepupa. At that time *puparium* formation begins, to be followed some hours later by another molt and the appearance of the true pupa. Larvae of the higher Diptera, of which *Drosophila* is one, are described as being headless; there are no signs of eyes, antennae, or mouth parts.

On the other hand, the thoracic region can be distinguished from the abdomen, even in the very early larva. In a sense, adult structures such as swings, halteres, and legs, as well as eyes, antennae, and genitalia, are present in larvae at an early stage of development, but only in the form of undifferentiated *imaginal discs* (two for eyes, two for wings, six for legs, and so on.). (Obviously, these discs are not "imaginary;" the term pertains to *imago*, or adult). The discs will not begin to undergo metamorphosis and differentiate into their appropriate adult structures until the pupal stage. In contrast, most of the gut, the central nervous system, the Malpighian tubules (which function in excretion), and the gonads are present as organ systems that persist and grow directly through both the larval and pupal developmental periods.

Pupal Development

During the pupal period, purely larval structures such as the salivary glands, together with their polytene chromosomes, are broken down and disappear. At the same time, all of the adult structures arise, as cells of the various imaginal discs differentiate and grow. After a four-day pupal period, the imago emerges from its pupal case as a result of gene activity initiated by the hormone ecdysone (which also is required for the transformation of egg into larvae and larvae into pupae). Mutants that cannot synthesize ecdysone will not develop very far.

Some mutants with reduced ecdysone activity become larvae, but they eventually die as "*giant*" larvae, unable to pupate. It is interesting to note that because of its ability to antagonize the activity of ecdysone, juvenile hormone, when overproduced, also prevents development (or "*growing older*") from proceeding beyond an early stage. Many plants produce chemical substances that process juvenile hormone-like activity, thereby protecting themselves against the ravages of parasitic insects that develop from eggs deposited in or on them.

Development of the Segmented Body

In many kinds of organisms, ranging from tapeworms to mammals, a segmented body plan underlies the greater part of the anatomy. In the most primitive organisms, a typical segment includes a full complement of virtually all of the structures required for living, including respiration, locomotion, sensation, excretion, and even reproduction. Except for a mouth in the first segment and a gut running all the way through, the organism, for all practical purposes, is formed by a series of virtually identical segments, giving an impression of development by Xerox. This condition is closely approached by the tapeworm.

Most higher organisms, however, through the course of evolution, have diverged from that plan. Some, such as round worms and echinoderms, show no segmentation; others have varied segmental expressions, with some segments fused together or retaining only one or two of the different organ systems. Instill higher forms, segmentation is seen not in every system, but only in a few, such as the nervous and circulatory systems. In fact, often the different segments no longer much resemble each other, so the underlying segmental pattern is not readily apparent. In the following material we shall examine segmentation in *Drosophila melanogaster*, which is commonly used in genetic studies. (*D. melanogaster*, though called a "*fruit fly*," is a harmless insect, not the highly destructive Mediterranean fruit fly.)"

The Adult

Although the segmental body plan is first discernible about halfway through the period of embryonic development, not until the adult stage does the segmental body plan become obvious. The adult *Drosophila* has an abdomen with seven easily counted, similar segments (A final posterior segments obscured and distorted by the genitalia, and it probably is a product of the fusion of three terminal abdominal segments.) Three pairs of legs reflect the much less obvious fact that the thorax contains three segments; the *prothorax*, the *mesothorax*, and the *metathorax*, each of which bears a pair of legs, or ventral appendages. Although

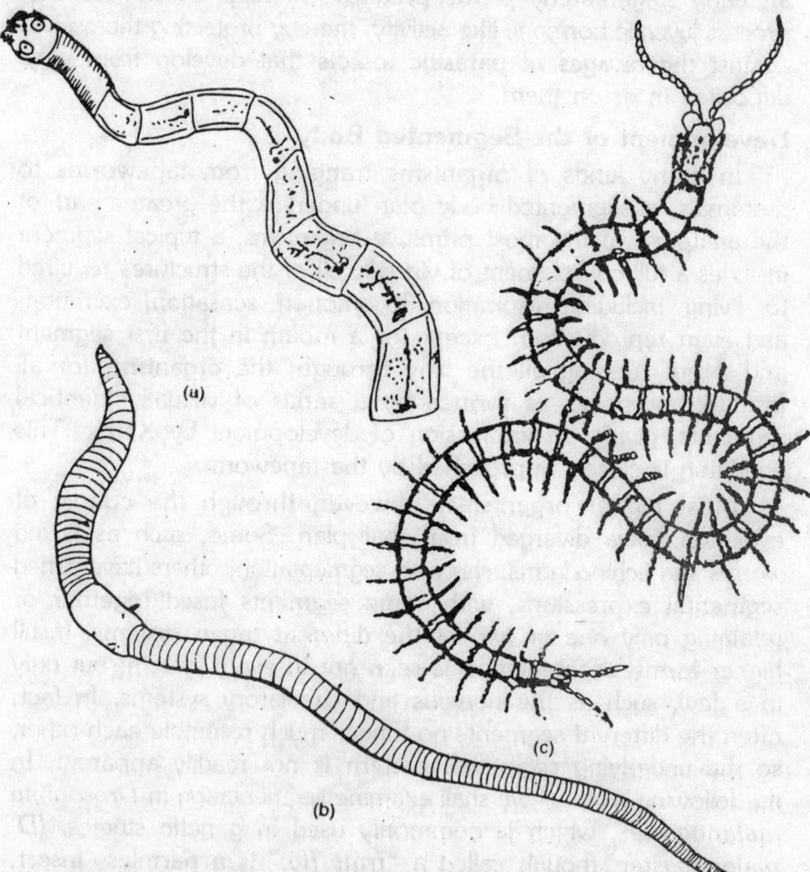

Fig. 12.19. Some animals that illustrate the segmented body: (a) tapeworm; (b) earthworm; (c) centipede.

primitive Diptera also had a pair of dorsal appendages, or wings, on each thoracic segment, no modern insect develops prothoracic wings. Almost all insects have both mesothoracic and metathoracic wings, but the Diptera (true flies) have lost their metathoracic wings, which leaves only the two mesothoracic wings for flying.

Dipteran metathoracic wings have been reduced to small knob-like *halteres*, the so-called balancers. The head shows no obvious segmentation, but the antennae, the eyes, and the mouthparts are, developmentally speaking, appendages. The suggests that the head is a completely fused product of what, in evolution, was once at least three, but more likely five, or even more, head segments.

Genetic Control of Segmentation in Drosophila

The genetic control of the segmental body plan of *D. melanogaster* has been most thoroughly analyzed by E.B. Lewis and his colleagues at Caltech. They have identified a cluster of ten or more genes lined up near the center of the left arm of chromosome 3 at salivary chromosome position 89E. This group of genes, now known as the *bithorax complex* (BX-C for short), controls the differentiation of the thoracic and abdominal segments. Lewis has shown that a gene at one end of the BX-C becomes activated first; it is involved in the control of development of the prothoracic segment. The next gene in line is then activated to control development of the mesothoracic segment; the third, the metathorax; the next, the first abdominal segment; and so on down to the last gene and the last abdominal segment.

Through observations of the phenotypic effects of mutations of these genes and molecular analyses of most of them, Lewis and his collaborators, particularly W. Bender, have provided remarkable insights into the genetic control of segmentation. Historically, the first mutation in a BX-C gene was called *bithorax* (bx), because at first glance the thorax seemed to be partly duplicated and the halteres were enlarged into rudimentary wing-like structures. In time, other BX-C mutations were discovered that produced similar, but different, modifications of the thoracic and abdominal phenotypes. For example, *post-bithorax* (*pbx*) is much like bx in a appearance, but it can be separated from *bx* by crossing over. The homozygous double mutation, *pbx bx/pbx bx,* has a strong, four-winged phenotype. At least 12, and by now surely more, different mutations of BX-C genes have been found and analyzed by Lewis and others.

As we said above, BX-C genes control the development and differentiation of the thoracic and abdominal segments. At first it appeared that the most common effect of a mutation in a BX-C gene was the transformation of a structure characteristic of one segment into an inappropriate structure characteristic of a different segment, such as that of a haltere (a metathoracic structure) into a wing (a mesothoracic structure). Closer study showed, however, that the *bx* transformation is not simply a conversion of haltere to wing, but rather a partial or complete transformation of about half the metathorax into mesothorax, after which the transformed mesothorax, like the normal mesothorax, develops a wing. Changed of this sort are thought to result from a mutational inactivation of the *bx* locus.

A different mutation is able to convert the first abdominal segment, which has no appendages, into a metathorax, which has halteres. Inactivation (or loss) of one of the other BX-C genes generally transforms the segment it controls into the next anterior one. There is some evidence to suggest that duplication of a BX-C gene transforms the segment it controls into the next posterior one. If the entire BX-C is deleted, however, all of the segments develop as a mesothorax, which appears to be the basic segment from which the others were derived. Mutations like these, which replace a given structure with an inappropriate one, such as a leg with an antenna, are now called *homeotic mutations*.

Transmission of Developmental Signals

Gerald Rubin and his colleagues successfully cloned a gene needed for the proper development of one of eight specific photoreceptor cells present in each of the 750 to 800 *ommatidia* (facets) of the fly compound eye. Normally each of these eight cells forms a structure called a *rhabdomere*, which contains the visual pigments. Six of the eight rhabdomeres (R1-R6) form a trapezoidal arrangement around the other two (R7-R8), which are shorter and are stacked on top of each other. A cross section of an ommatidium cannot cut through both R7 and R8 simultaneously. The cell bodies of the photoreceptor cells surround the group of rhabdomeres. In the mutant condition, no R7 is formed; hence the mutation is called *sevenless* (*sev*). Developmental studies have shown that in *sev* R7 is replaced with a cone cell.

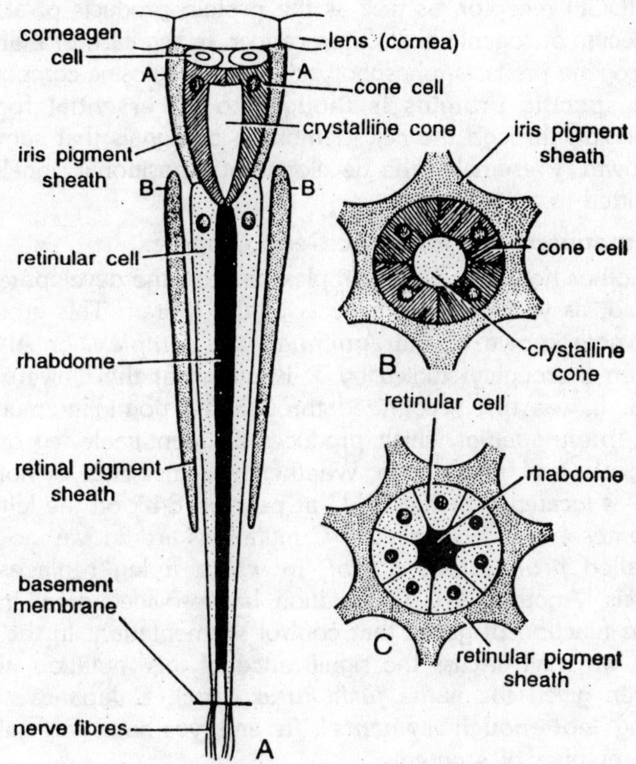

Fig. 12.20. Drosophila ommatidium.

Cone cells normally lie just above the rhabdomeres and have no photoreceptive properties at all. For that reason, *sev* qualifies as a homeotic mutation, even though its effect is at the cell level, not at the organ level. The interpretation of *sev* is as follows. The normal allele produces a protein membrane receptor that responds to positional signals from surrounding cells. These signals provide the information for proper development. However, the mutant allele produces an abnormal (or perhaps no) protein membrane receptor; thus the cell differentiates improperly. The normal *sev* receptor has a molecular weight of about 220,000 daltons and exhibits the properties of protein receptors that span cell membranes. The part of the *sev* receptor that lies inside the cell resembles the tyrosine kinase portion of the epidermal growth factor (DGF) receptor, as well as the protein products of at least two specific oncogenes that cause cancer. In the case of EGF and the oncogene products, phosphorylation of the tyrosine components of the specific proteins is thought to be essential for the transmission through the cell membrane of signals that stimulate cell growth. Presumably the developmental positional signals are transmitted in a similar way.

The Antennapedia Homeotic Gene Complex

Another homeotic gene complex controls the development of the head, as well as some aspects of the thorax. This group of genes, now known as the *Antennapedia complex*, or ANT-C, has been thoroughly studied by T. Kaufman at the University of Indiana. It was first recognized through the dominant mutation called *Antennapedia*, which produces an identifiable leg on the head in place of the antenna. Whether by coincidence or not, the ANT-C is located near the BX-C at position 84A on the left arm of chromosome 3. Several ANT-C mutations are known including one called *proboscipedia* (*Pb*), in which a leg replaces the proboscis. Another ANT-C mutation has provided great insight into the function of genes that control segmentation. In the next section we shall discuss the significance of this mutation, which has been given the name *fushi tarazu* (*ftz*), a Japanese word meaning "*not enough segments*"; *ftz* embryos have only half the normal number of segments.

The Homeobox and its Potential Significance

Among ANT-C mutations, *fushi tarazu* is most interesting because it does not give rise to segmental transformation; rather, it appears, to control the segmentation process itself. In addition, a thorough molecular analysis of *ftz* led to the identification of a special nucleotide sequence, long enough to encode 60 amino acids, that is exactly the same as (or almost identical to) sequences found not only in many of the homeotic BX-C and ANT-C genes of *Drosophila* but also in genes of several other organisms ranging from yeast to humans. Not all of these genes give rise to homeotic mutants; several, like *ftz*, appear to play a regulatory role during development. Because the significance of this DNA sequence was first recognized in a gene of the ANT-C homeotic group and because a majority of the genes known to have this particular nucleotide sequence can produce homeotic mutants, the sequence is now called the homeobox. The sequence of the 60 amino acids encoded by the homeobox is now known. The question is, what does this small protein do? Work with yeast has provided some insight.

To a significant degree, the protein produced by the homeobox sequence is homologous to the yeast mating-type proteins MAT $\alpha 1$ and MAT $\alpha 2$. Because the mating-type genes control fundamental aspects of cell differentiation in yeast, they are considered to be master control genes. The protein is believed to be able to bind specifically to DNA so as to turn certain genes on or off. Thus in all likelihood the homeobox protein in organisms ranging from fungi to primates plays a major role in controlling gene activity during development. If so, the homeobox sequence of nucleotides must have been highly conserved throughout a long period of organic evolution. As provocative as this concept may seem, it is too early to be sure of it; much more research is needed to reach a clear-cut decision about the fundamental significance of the homeobox. For the moment, however, it is quite inviting to speculate that the homeobox protein can regulate the activity of the appropriate genes to ensure not only that the proper number of segments will develop in their proper sequence in *Drosophila*, but also that the various segments will become equipped with the appendages normally associated with that segment. When a homeobox-containing gene is in the mutant condition, the genes it controls do not turn on or off at the correct times; as a result, some structure such as a segment does not differentiate correctly.

Other Genes Regulating Early Development in Drosophila

Several genes that are not known to contain the homeobox sequence have an effect on early development in *Drosophila*. One, called *dorsal*, is involved in establishing dorsal-ventral polarity. Females homozygous for an extreme allele of *dorsal* produce dorsalized embryos that lack all lateral and ventral structures; however, dorsal structures are found throughout the body. Various other nonallelic genes besides *dorsal* have also been found to affect dorsal-ventral polarity. One of these other genes is *toll*.

In the homozygous recessive condition, *toll* dorsalizes the embryo, but if cytoplasm from an embryo with a normal genotype is carefully injected into a *toll* embryo, the embryo can be "rescued" from the dorsalizing effects of the *toll* gene. The ability to rescue organisms from the effects of these genes appears to reside in a fraction of the poly-A mRNA that is stored in the egg during oogenesis. The *dorsal* gene has been cloned; it codes for a 2.8-kilobase length of mRNA that can be found both in the ovaries and in embryos up to 2.5 hours old. A remarkable number of genes are now known to control one or another specific aspect of *Drosophila* development, although the protein products that they encode and the mechanisms by which these products produce their effects remain largely unknown. As we shall describe next, a nematode worm has provided the material needed for progress in this exciting area of study.

Caenorhabditis Elegans: The Embryologist's Choice

Let us begin consideration of the small, free-living nematode worm with the long name *Caenorhabditis elegans* (*C. elegans* for short). Some years ago S. Brenner, a well-known British molecular geneticist, recognized the potential of *C. elegans* for the analysis of the genetic control of development. Overnight he became an embryologist and set up the first laboratory for the study of the developmental genetics of this free-living worm, so small that the adult is only one millimeter in length and has less than 1000 cells altogether. Because *C. elegans* has so few cells, it is possible to map and follow all of the cell lineages throughout the embryonic period of development.

In other words, cells making up a give structure can be traced back to the single cell from which they all originated in the early embryo. This kind of information is vital to gaining insights into the mechanisms involved in development. Moreover, a very large number of mutations are known to affect specific facets of

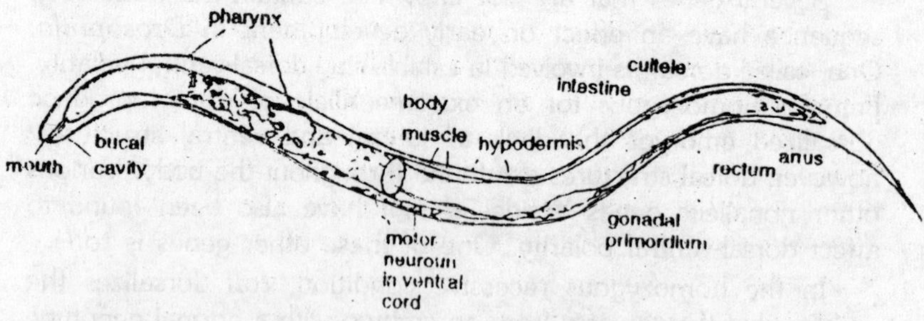

Fig. 12.21. Larva (L1) of C. elegans. The nematode worm at this stage is little more than a tube from mouth to anus. Every cell is in exactly the same position in every normal individual.

development in *C. elegans*. These features have persuaded many embryologists to join Brenner in the study of the development of *C. elegans*. A huge literature devoted to the genetics and development of *C. elegans* has now accumulated, far out of proportion to the size of the worm.

The Life Cycle

C. elegans passes through its short life in the fast lane. After only six hours of embryonic development, its full schedule of cleavage divisions is finished. Eight hours later, the embryo hatches as an active larva. In other day and a half, the larva becomes an adult hermaphroditic worm that can produce about 250 progeny in the next four days or so. When it first emerges from the egg, the larva is composed of precisely 558 cells. After passing through four larval molts, the larva becomes an adult worm with only 959 cells, each of which can be traced back to the zygote. This-simple structure allows for the construction of a complete fate map showing the developmental fate of each embryonic cell. *C. elegans* also has produced numerous mutations in which the developmental fate of particular cells or tissues has been modified. These characteristics have made *C. elegans* a favourite organism for the study of the genetic control of developmental processes.

Experimental Studies on Development

Many of the adult structures of *C. elegans* have been traced back to one or another single cell of the early embryo. By using a microlaser beam, an experimental embryologist can kill one such progenitor cell. Then, if the related adult structure fails to develop, the embryologist knows that the developmental potentialities of the destroyed cell cannot be assumed by other cells of the embryo (that is, the structure is completely determined early in development). On the other hand, if the structure does appear as it would normally, the embryologist can conclude that determination of the structure is liable (that is, other cells can assume the role of the destroyed cell). Many experiments of this sort have made it clear that the majority of structures in *C. elegans* are fully determined very early in development, though some are not.

Specific Genes That Control Development in C. elegans

The search for specific genes that control important developmental mechanisms has led to some interesting discoveries. In certain mutants, some cells continue dividing rather than differentiating into functional cells when they normally should. This result suggests that particular genes control the precise number of cell divisions that a given cell must undergo before a specific adult structure will differentiate. Other genes have been identified that seem to have a major role in controlling development. For example, *lineage* (*lin*) genes play a key role in determining whether a cell will have one developmental fate or another.

A specific example of how a *lin* gene works is provided by *lin*-14, which is active in determining the cuticle. R. Horwitz and his colleagues at MIT have shown that an active *lin*-14 allele assures that so-called seam cells secrete larval cuticle. When *lin*-14 is turned off, the seam cells produce adult cuticle. Thus development of a specific structure is dependent on whether a specific gene is turned on or off. Think of a train coming to a switch in the track. Which way the train will go thereafter depends entirely on whether the switch is turned to the right or to the left. The train does not control its own destiny (or destination); the switch does. What happens when *lin*-14 undergo mutation? If the mutant allele cannot turn off in the proper cells at the proper time, those cells retain larval cuticle, even through the remainder of the worm has adult cuticle.

In the case of other *lin*-14 mutants, the gene is licked off and cannot turn on. Cells in this situation never express larval cuticle. Another *lin* gene, *lin*-5, provides a different example of the genetic control of fundamental developmental mechanisms. When *lin*-5 in the mutant condition, no postembryonic cell division occurs, yet chromosome replication continues much as in larval tissues of *Drosophila*. (Unfortunately, the replicated chromatids do not stay together so as to produce analyzable polytene chromosomes.) Note that *lin*-5 affects only larval cell division, not embryonic cell division. Further, the consequences of mutation are restricted to cytokinesis, not mitosis. This demonstrates how finely tuned genetic control is, even in simple organisms. *C. elegans* may well turn out to be the organism of choice in the continuing search for an understanding of development. Many genes, not only in *C. elegans*, are known to play major roles in controlling the structure of organisms. Such genes may also play a fundamental role in evolution of the species because their mutants fail to survive the change in timing of various basic developmental processes. Often the consequences are lethal, but one some occasions such mutational changes might lead to a novel adaptive improvement that could start a new line of descent.

INDEX

C

D